Student Study Guide and Solutions Manual

to accompany

Denise Guinn's
Essentials of General, Organic, and Biochemistry
2nd edition

Rachel C. Lum, Ph.D.

W.H. Freeman and Company
New York
A Macmillan Higher Education Company

ISBN-13: 978-1-4641-2506-5
ISBN-10: 1-4641-2506-6

Printed in the United States of America

First Printing

W.H. Freeman and Company
41 Madison Ave.
New York, NY 10010

www.whfreeman.com/guinn2e

For Mateja and Alex

About the Author

Rachel Lum received her S.B. in chemistry from the Massachusetts Institute of Technology and her Ph.D. in organic chemistry from Harvard University. As a graduate student, she received the American Chemical Society Division of Organic Chemistry Fellowship sponsored by Procter & Gamble. She was postdoctoral fellow at the Georgia Institute of Technology. She was an adjunct professor at Regis University, where she taught courses in analytical chemistry, general chemistry, and environmental chemistry. She was an instructor at the University of Colorado at Boulder, where she taught courses in organic chemistry. She currently resides in Boulder, Colorado with her husband, Al, and her two children, Alex and Mateja, and works as a consultant.

Table of Contents

Preface

One way to learn chemistry is to practice many problems. This study guide has been designed towards that goal. The second edition has been edited so that it reflects the changes within the textbook. Each chapter starts with a brief summary of the material covered in the chapter. For each section, there is a summary of the key points, followed by a Worked Example. You can either follow along as the problems are solved in the Worked Example or attempt to answer the question yourself and then compare your answer to the Worked Example. The Worked Example is complemented by a Try It Yourself question, which is a similar question to the Worked Example broken down into steps for you answer. At the end of each section there are practice problems to perfect your skills. At the end of each chapter there is a quiz to test your knowledge. So pick up your pencils and start practicing!

Acknowledgments

I would like to thank Denise Guinn for giving me the opportunity to continue working with her on this project. I would like to thank Tue Tran at W.H. Freeman for his editorial help. Finally, I would like to thank my husband, Al for his unwavering support.

Chapter 1

Measuring Matter and Energy

Chapter Summary

In this chapter, you have learned about how important measurements are in medicine and science. You learned about the limitations of measurements and how they are expressed using significant figures, and the common units for measurements and using the metric system. You learned how to calculate physical properties such as density and specific gravity. You also learned about energy and the difference between kinetic and potential energy. As a healthcare professional, you will need to take many measurements and to perform critical calculations, such as the dosage of medicine needed to administer to a patient.

1.1 Matter and Energy

All matter exists in one of three physical states or phases: solid, liquid, or gas. A solid has a definite shape and volume, which is independent of its container. A liquid has a definite volume but does not have a definite shape; it conforms to the shape of the container. A gas has neither a definite volume nor a definite shape. A volume of gas expands to fill its entire container.

In a Nutshell: Kinetic and Potential Energy

Energy is the capacity to do work, where work is the act of moving an object. There are two fundamental forms of energy: kinetic energy and potential energy. Kinetic energy is the energy of motion, the energy a substance possesses as a result of the motion of its molecules or atoms. Faster-moving molecules have greater kinetic energy than slower ones. Potential energy is stored energy, the energy a substance possesses as a result of position, composition, and condition of its atoms. Heat is kinetic energy that is transferred from one object to another due to a difference in temperature. Heat energy always flows from a hot object to a cold object. While heat is a form of energy, temperature is a measure of the average kinetic energy of the molecules, ions, or atoms that make up a substance.

In a Nutshell: Kinetic-Molecular View of the States of Matter

In the gas phase, atoms or molecules have high kinetic energy. Atoms and molecules are moving faster than such particles in the solid or liquid phases, and they are much farther apart. Since the particles are so far apart, they only interact when they collide. In the liquid phase, the particles are much closer together; they are in contact with other molecules moving randomly and tumbling over one another. In the solid phase, atoms or molecules exist in a regular ordered pattern. Molecules in the solid phase have less kinetic energy than liquids, so they remain in a fixed position with mainly vibrational motions.

Worked Example #1

Match the following descriptions to the state that describes it: solid, liquid, or gas.

a. The particles in this state are far apart from one another.
b. The particles in this state have the least amount of kinetic energy.
c. This state has a definite volume, but it conforms to the shape of the container.

a. Particles in the gas state are spaced far apart from one another.
b. In the solid state, the particles are in a fixed position and do not move, except for vibrational motions.
c. Liquids have a definite volume and conform to the shape of the container.

Try It Yourself #1

Match the following descriptions to the state that it describes: solid, liquid, or gas.

a. This state has neither a definite volume nor a definite shape.
b. The particles in this state are very close to one another and have vibrational motion.
c. In this state, particles can be poured from one container to another.

a. State described: gas
b. State described: Solid
c. State described: Liquid

Worked Example #2

Indicate whether each of the following examples is a demonstration of potential energy or kinetic energy:

a. molecules colliding with the walls of their container
b. food particles

a. The molecules are in motion, so it is a demonstration of kinetic energy.
b. Food particles have potential energy; the energy is transferred to your body after you eat and digest it.

Try It Yourself #2

Indicate whether each of the following examples is a demonstration of potential energy or kinetic energy:

a. water in the lake behind a dam
b. water falling over a dam

a. Is there stored energy or is there motion?
b. Is there stored energy or is there motion?

In a Nutshell: Physical and Chemical Changes

A physical change is a process that does not affect the composition of a substance. A change in state is a physical change. A chemical change, also known as a chemical reaction, involves a change in the composition of a substance.

Worked Example #3

Indicate which of the following changes are physical changes or chemical changes.

a. roasting marshmallows
b. cutting hair
c. an ice cube melting

a. *A chemical change because the appearance of the marshmallow changes and it cannot return to the unroasted state*
b. *A physical change because the composition of the hair has not changed and it can return to the longer length*
c. *A physical change because water is undergoing a phase change from solid to liquid, but it is water*

Try It Yourself #3

Indicate which of the following changes are physical changes or chemical changes.

a. bleaching your hair
b. formation of ice cubes in the freezer
c. burning your toast

a. *Has the composition of the hair changed?*
b. *Has the composition of the water changed?*
c. *Has the composition of the toast changed?*

Practice Problems for Matter and Energy

1. Which of the following would conform to shape of a glass bottle: liquid water, steam, or ice?

2. Indicate whether the following examples are a demonstration of potential energy or kinetic energy.
   a. a bouncing ball

   b. a sightseer standing in the crown of the Statue of Liberty

   c. an airplane flying

3. Indicate whether the following changes are physical changes or chemical changes.
   a. snow forming from water vapor in clouds

   b. steam coming from your hot cup of coffee

   c. a match burning

1.2 Measurement in Science and Medicine

In a Nutshell: English and Metric Units

Every measurement consists of both a number and a unit. Every measurement has a margin of error associated with it, which is conveyed by the number of digits reported in the measurement. The English system is used only by the United States and a few other countries. The metric system is most often used in scientific measurements and consists of several base units such as the meter (for length), the gram (for mass), the second (for time), and the calorie (for energy).

In a Nutshell: Prefixes in the Metric System

Metric prefixes are used when a measurement is much larger or smaller than the base unit to avoid including many zeros in the numerical value. When prefixes are placed in front of the base unit, they represent a multiplier or a divider that makes the prefixed unit larger or smaller by some multiple of 10 or 1/10. The prefixes used are *giga-*, *mega-*, *kilo-*, *deci-*, *centi-*, *milli-*, *micro-*, *nano-*, *pico-*, and *femto-*. The meter is the base unit of length. Mass is a measure of the amount of matter, and the gram is the base unit of mass in the metric system. Volume is a measure of how much three-dimensional space a substance occupies, and the most common unit of volume in the metric system is the liter. The volume of a solid is often measured by determining how much water the solid displaces: $V_{object} = V_{final} - V_{initial}$. The common units of energy used in science are the calorie and the joule. A calorie is the amount of heat required to raise the temperature of 1 gram of water by 1 °C. A joule is defined as the amount of energy required to lift a 1-kilogram weight a height of 10 centimeters. The Calorie listings on nutritional labels, spelled with a capital "C," are actually kilocalories.

Worked Example #4

Which length is bigger: 5 pm or 5 fm?

Using Table 1-1, you see that there are 10^{12} pm in 1 m, while there are 10^{15} fm in 1 m. Therefore, a picometer is bigger than a femtometer, so 5 pm is bigger than 5 fm.

Try It Yourself #4

Which mass is bigger: 10 kg or 10 mg?

Tools: Use Table 1-1.

There are __________ kg in 1 g, while there are 1000 *mg in 1 g.*
A Kg *is bigger than a* mg.

Worked Example #5

When a rock is placed in graduated cylinder containing 50.0 mL of water, the water level rises to 54.5 mL. What is the volume of the rock in cm^3?

$V_{initial}$ = *50.0 mL* V_{final} = *54.5 mL*

The volume of the rock is equal to the difference in the volume before and after the rock is submerged: $V_{rock} = V_{final} - V_{initial}$.

Substituting the known variables into the equation:

V_{rcok} = *54.5 mL – 50.0 mL*

V_{rock} = *4.5 mL, which corresponds to 4.5* cm^3.

Try It Yourself #5

When a gold nugget is placed in a graduated cylinder containing 80.0 mL of water, the water level rises to 97.2 mL. What is the volume of the rock in cm^3?

V_{inital} = ________________ V_{final} = ________________

Equation for the volume of the gold nugget:

$V_{gold\ nugget}$ =

Substituting the known variables into the equation:

$V_{gold\ nugget}$ =

__________ *mL* = __________ *cm*^{3}

Practice Problems for Measurement in Science and Medicine

1. How many of the following are there in 1 gram?
   a. picogram
   b. gigagram
   c. nanogram
2. For each pair, indicate which represents the greater length. If they are the same, state so.
   a. 1 mm or 1 dm
   b. 1 m or 1 km
   c. 1 nm or 1 fm
3. What is the volume of an object, in cubic centimeters, if it is placed in 15 mL of water and causes the level of the water to rise to 19 mL?

1.3 Significant Figures and Measurement

Every measurement contains some degree of uncertainty. The uncertainty in a measurement depends on the precision and accuracy of the measuring device. The number of digits reported conveys information about the uncertainty of the measurement.

In a Nutshell: Precision and Accuracy

Precision is a measure of how close together repeated measurements are to each other; accuracy is a measure of how close the measurement is to the "true" value.

Worked Example #6

A person weighed himself on two different scales three times. The man weighs 77.3 kilograms. Which scale is precise and accurate?

Scale A	Scale B
77.6 kg	80.5 kg
77.0 kg	80.6 kg
77.3 kg	80.7 kg

Scale A *is precise and accurate. The average weight on scale A is 77.3 kg, and all the measurements on the scale are within ± 0.3 kg. The measurements on scale B are precise, but not accurate.*

Try It Yourself #6

A premature baby was weighed three times on two different scales. Which set of measurements has more precision?

Scale A	Scale B
1,619 g	1,619 g
1,622 g	1,659 g
1,616 g	1,579 g

The average weight for scale A: ________

The average weight for scale B: ________

In a Nutshell: Significant Figures in Measured Values

When reporting measurements, the convention is to record all of the certain digits plus one uncertain digit to a best approximation. Significant figures are all the certain digits as well as one uncertain digit in a value. The following digits are significant: nonzero digits, zeros between nonzero digits, zeros at the end of a number containing a decimal point, and all digits expressed in scientific notation are significant. If there is no decimal point, zeros following a nonzero digit are not significant, they are merely place holders. Zeros that appear before nonzero digits, whether or not there is a decimal point, are not significant figures. The zeros merely serve as place holders. Exact numbers contain an infinite number of significant figures because they carry no uncertainty.

Worked Example #7

How many significant figures are there in the following measured values?

a. 0.002030 L 4
b. 3.20×10^4 nm 3
c. 2,400 g 4

a. There are four significant figures in 0.002030. The zeros before the non-zero digits are place holders and are not significant. The zero after the 3 is significant.

b. There are three significant figures in 3.20×10^4 nm. The zero after the 2 is significant.

c. There are four significant figures in 2,400. The decimal point after the zeros indicates that the two zeros are significant.

Try It Yourself #7

How many significant figures are there in the following measured values?

a. 0.0004108 m 4

b. 7.10×10^{-3} g 3

c. 500. Seconds 3

In a Nutshell: Significant Figures in Calculations

When multiplying or dividing measured values, the final calculated answer cannot have more significant figures than the measurement with the fewest number of significant figures. When adding or subtracting measured values, the final answer can have no more decimal places than the measured value with the fewest decimal places.

Worked Example #8

Perform the following calculations and round the final answer to the correct number of significant figures.

a. 21.34 ÷ 0.105 =

Type of calculation: division

Number of significant figures in 21.34: four

Number of significant figures in 0.105: three

The calculation involves division, so the final answer cannot have more significant figures than the measured value with the fewest significant figures.

Measured value with the fewest significant figure: 0.0105 with three significant figures

Calculator result: 21.34 ÷ 0.105 = 203.2380952

Rounded calculator result to three significant figures: ***203***

b. 9.23 + 3.314 − 21 =

Type of calculation: addition and subtraction

Last decimal place in 9.23: hundredths place

Last decimal place in 3.314: thousandths place

Last decimal place in 21: ones place

The calculation is an addition and subtraction, so the final answer should have no more decimal places than the measured value with the fewest decimal places.

Measured value with the fewest decimal places: 21 with ones place

Calculator result: 9.23 + 3.314 − 21 = −8.456

Rounded calculator result to the ones place: ***−8***

Try It Yourself # 8

Perform the following calculations and round the final answer to the correct number of significant figures.

a. 34.56 × 0.0042 =

Type of calculation: ______________

Number of significant figures in 34.56: ________

Number of significant figures in 0.0042: ________

The calculation involves ________________, *so the final answer cannot have more significant figures than the measured value with the fewest significant figures.*

Measured value with the fewest significant figures: __________

Calculator result: 34.56 × 0.0042 = _____________

Rounded calculator result: ___________________

b. 31.5678 + 25.34589 =

Type of calculation: __________________

Last decimal place in 31.5678: _______________

Last decimal place in 25.34589: ________________

The calculation is ____________________________, *so the final answer should have no more decimal places than the measured value with the fewest decimal places.*

Measured value with the fewest decimal places: ______________

Calculator result: 31.5678 + 25.34589 = ___________________

Rounded calculator result: _________________

Practice Problems for Significant Figures and Measurement

1. A child is weighed at the doctor's office on two different scales three times. The man weighs 20.4 kilograms. Which scale is precise and accurate?

Scale A	Scale B
20.6 kg	25.5 kg
20.2 kg	25.6 kg
20.4 kg	25.7 kg

2. How many significant figures are in the following numbers?
   a. 0.00462

   b. 1,200

   c. 1.02003

3. Perform the following calculations and round the final answer to the correct number of significant figures.
   a. 23.6 × 0.05 =

   b. 145.876 + 96.36 =

   c. 476.23 – 145.9 =

1.4 Using Dimensional Analysis

A unit conversion is a type of calculation in which a measurement in one unit is converted into the equivalent value in another unit. In the sciences, conversions are performed using a method known dimensional analysis. There are three steps to dimensional analysis. Step 1: Identify the conversion(s) needed. Step 2: Express each conversion as two possible conversion factors. Step 3: Set up the calculation by writing the supplied unit multiplied by the correct conversion factor. As a final step in any calculation, you should double check that you have shown the correct number of significant figures.

Worked Example #9

The human ovum (egg cell) is about 150 μm in diameter. What is this diameter in mm?

Identify the conversion(s):

$10^6\ \mu m = 1\ m$ *and* $10^3\ mm = 1\ m$

Express each conversion as two possible conversion factors:

1st *conversion factor:* $\frac{10^6\ \mu m}{1\ m}$ *or* $\frac{1 m}{10^6\ \mu m}$

2nd conversion factor: $\frac{10^3\ mm}{1\ m}$ *or* $\frac{1\ m}{10^3\ mm}$

Set up the calculation by writing the supplied unit and then multiplying by the correct conversion factors. The correct conversion factor always has the unit that needs to be canceled in the denominator:

$$150\ \cancel{\mu m} \times \frac{1\ \cancel{m}}{10^6\ \cancel{\mu m}} \times \frac{10^3\ mm}{1\ \cancel{m}} = 0.15\ mm$$

Check that the answer has the correct number of significant figures.

Try It Yourself #9

How many kilograms are there in 2.34 pg?

Identify the conversion(s):

10^{12} pg = _______________

Express each conversion as two possible conversion factors:

Set up the calculation by writing the supplied unit and then multiplying by the correct conversion factors. The correct conversion factor always has the unit that needs to be canceled in the denominator:

2.34 pg × _______________ = _______________

In a Nutshell: Conversion Between Metric and English Units

Dimensional analysis is also used to convert between metric and English units. Use the same steps as described above.

Worked Example #10

The Eiffel tower in Paris, France, is 300.65 meters tall. What is the height of the Eiffel Tower in feet?

Identify the conversion(s):

1 m = 39.37 inches and 1 ft = 12 inches

Express each conversion as two possible conversion factors:

1st conversion factor: $\frac{1\text{ m}}{39.37\text{ inches}}$ *or* $\frac{39.37\text{ inches}}{1\text{ m}}$

and

2nd conversion factor: $\frac{1\text{ ft}}{12\text{ inches}}$ *or* $\frac{12\text{ inches}}{1\text{ ft}}$

Set up the calculation by writing the supplied unit and then multiplying by the correct conversion factors. The correct conversion factor always has the unit that needs to be canceled in the denominator:

$$300.65\ \cancel{\text{m}} \times \frac{39.37\ \cancel{\text{inches}}}{1\ \cancel{\text{m}}} \times \frac{1\text{ ft}}{12\ \cancel{\text{inches}}} = 986\text{ ft, rounded to 990 ft}$$

Try It Yourself #10

In the United States, a can of soda has 12 oz. Is this can of soda larger or smaller than the 333-mL cans of soda sold in Europe?

Identify the conversion(s):

Express each conversion as two possible conversion factors:

Set up the calculation by writing the supplied unit and then multiplying by the correct conversion factors. The correct conversion factor always has the unit that needs to be canceled in the denominator:

In a Nutshell: Energy Conversions

To convert between nutritional Calories and joules, it is best to use two conversion factors, one from Calorie to the calorie and then a second from calorie to joule.

Worked Example #11

How many joules of energy are there in 3.23×10^4 cal?

Identify the conversion(s):

1.000 cal = 4.184 J (exact)

Express each conversion as two possible conversion factors:

$$\frac{1.000 \text{ cal}}{4.184 \text{ J}} \; or \; \frac{4.184 \text{ J}}{1.000 \text{ cal}}$$

Set up the calculation by writing the supplied unit and then multiplying by the correct conversion factors. The correct conversion factor always has the unit that needs to be canceled in the denominator:

$$3.23 \times 10^4 \, \cancel{\text{cal}} \times \frac{4.184 \text{ J}}{1 \, \cancel{\text{cal}}} = 1.35 \times 10^5 \text{ J}$$

Try It Yourself #11

How many joules are in a banana that contains 121 Calories?

Identify the conversion(s):

Express each conversion as two possible conversion factors:

Set up the calculation by writing the supplied unit and then multiplying by the correct conversion factors. The correct conversion factor always has the unit that needs to be canceled in the denominator:

In a Nutshell: Dosage Calculations

For some medicines, the dosage must be adjusted according to the patient's weight. The dosage is itself a conversion factor between the mass or volume of the medicine and the weight of the patient.

Worked Example #12

Amoxil, an antibiotic, is prescribed for an infant who weighs 7.2 lbs. The dosage is 30 mg per kilogram of body weight per day. How many mg should one dose for this child contain?

Identify the conversion(s):

The conversion for the patient's weight from pounds to metric units is 1 kg = 2.205 lbs.

Express each conversion as two possible conversion factors:

English-to-metric units for the patient's weight are

$$\frac{1\,\text{kg}}{2.205\ \text{lbs}} \quad or \quad \frac{2.205\ \text{lbs}}{1\ \text{kg}}$$

The dosage is the conversion factor between the mass of medicine in mg and the weight of the patient in kg per dose:

$$\frac{30\ \text{mg}}{\text{kg}\cdot\text{dose}} \quad or \quad \frac{\text{kg}\cdot\text{dose}}{30\ \text{mg}}$$

Set up the calculation by writing the supplied unit and then multiplying by the correct conversion factors. The correct conversion factor always has the unit that needs to be canceled in the denominator:

$7.2\ \cancel{lbs} \times \frac{1\ \cancel{\text{kg}}}{2.205\ \cancel{\text{lbs}}} \times \frac{30\ \text{mg}}{\cancel{\text{kg}}\cdot\text{dose}} = 98\ mg/dose$, rounded to 100 mg/dose

Try It Yourself #12

A child who weighs 47 lbs needs to take Tylenol. The dosage is 10 mg per kilogram of body weight every 4 hours.

a. How many mg of Tylenol are in one dose for this child?
b. How much Tylenol would the child take in one day (24 hrs.)?

Identify the conversion(s):

Express each conversion as two possible conversion factors:

Set up the calculation by writing the supplied unit and then multiplying by the correct conversion factors. The correct conversion factor always has the unit that needs to be canceled in the denominator:

In a Nutshell: Density Calculations

Density represents a physical property. The density of a material is defined as its mass (*m*) divided by its volume (*V*). It can also be used as a conversion factor to calculate the mass of a substance whose volume and density are known, or the volume of a substance whose mass and density are known. The specific gravity is the ration of the density of a substance to the density of water at 4 °C.

Worked Example #13

What is the volume of 69.3 mL of alcohol? The density of alcohol is 0.79 g/mL.

Identify the conversion(s):

The conversion factor is the given in the physical quantity of density, so skip to the next step.

Express each conversion as two possible conversion factors:

$$\frac{0.79\ \text{g}}{1\ \text{mL}} \quad \text{or} \quad \frac{1\ \text{mL}}{0.79\ \text{g}}$$

Set up the calculation by writing the supplied unit and then multiplying by the correct conversion factors. The correct conversion factor always has the unit that needs to be canceled in the denominator:

$$69.3\text{mL} \times \frac{.79\text{g}}{1\text{mL}}$$

$$69.3\ \cancel{\text{mL}} \times \frac{0.79\ \text{g}}{1\ \cancel{\text{mL}}} = 54.7\ \text{g}, \text{ rounded to } 55\ \text{g}$$

Try It Yourself #13

What is the volume in mL of 10 lbs of ice? The density of ice is 0.9167 g/cm^3.

Identify the conversion(s):

Express each conversion as two possible conversion factors:

Set up the calculation by writing the supplied unit and then multiplying by the correct conversion factors. The correct conversion factor always has the unit that needs to be canceled in the denominator:

Worked Example #14

A patient supplies a urine sample that has a specific gravity of 1.015. What is the density of this patient's urine?

Since the specific gravity of the patient's urine is 1.015 and the density of water is 1 g/mL, the density of the patient's urine must be 1.015 g/mL.

$$\text{Density of substance} = \frac{\text{Sspecific gravity}}{\text{Ddensity of water}}$$

$$\text{Density of substance} = \frac{1.015}{1\ \text{g/mL}}$$

$$\text{Density of substance} = 1.015\ \text{g/mL}$$

Try It Yourself #14

What is density of a patient's urine that has a specific gravity of 1.027?

Density of substance =

In a Nutshell: Temperature

Temperature can be measured using a thermometer and can be reported in one of three temperature scales: Celsius, Fahrenheit, and Kelvin. The Fahrenheit and Celsius scales are relative scales, based on the freezing and boiling points of water. The Kelvin scale is used in chemistry and physics because it represents an absolute scale, one that assigns 0 K to zero kinetic energy. Equations are needed to convert between the temperature scales.

To convert from °C to °F, use the following equation:

$$°\mathrm{F} = 1.8 \times °\mathrm{C} + 32.0$$

To convert from °F to °C, use the following equation:

$$°\mathrm{C} = \frac{(°\mathrm{F} - 32)}{1.8}$$

To convert from °C to K, use the following equation:

$$\mathrm{K} = °\mathrm{C} + 273.15$$

To convert from K to °C, use the following equation:

$$°\mathrm{C} = \mathrm{K} - 273.15$$

Worked Example #15

It's snowing in New York City and the temperature is 28 °F. Your friend is visiting from Europe and would like to know what the temperature 28 °F corresponds to in °C. What do you tell him? What is the Kelvin temperature on this winter day?

Convert °F to °C by using the equation that has °C isolated on one side of the equal sign:

$$°\mathrm{C} = \frac{(°\mathrm{F} - 32)}{1.8}$$

Substitute the given value for °F into the equation and solve for °C:

$$°\mathrm{C} = \frac{(28 - 32)}{1.8} = -2.0\ °\mathrm{C}$$

A temperature of 28 °F corresponds to a temperature of −2.0 °C.

To find this temperature in Kelvin, use the equation that converts °C to K:

$K = °C + 273.15$

Substitute the given value for °C into the equation and solve for K:

$K = -2.0 + 273.15 = 271.2\ K$

Thus, 28 °F = −2.0 °C = 271.2 K.

Try It Yourself #15

It's a spring day in Paris, France, and the temperature is 19 °C. What is this temperature in °F and in K?

Convert °C to °F by using the equation that has °F isolated on one side of the equal sign. Substitute the given value for °C into the equation and solve for °C.

To find this temperature in Kelvin, use the equation that converts °C to K. Substitute the given value for °C into the equation and solve for K.

In a Nutshell: Specific Heat

Specific heat is the amount of heat required to raise the temperature of 1 g of substance by 1 °C. The specific heat is a physical property. It can be calculated using the following equation:

$$\text{Specific heat} = \frac{\text{Amount of heat } (q)}{\text{Mass} \times \text{change in temperature } (\Delta T)}$$

Heat refers to the amount of heat transferred to the substance and the change in temperature is the difference in temperature before and after heat has been transferred. Heat is usually measured in calories and mass in grams. A substance with a large specific heat requires a large amount of heat for a small increase in temperature.

Worked Example #16

Using Table 1-6, calculate the number of calories that must be added to warm 17.3 g of each of the following substances below from 15 °C to 28 °C. Which substance requires the most heat?

a. sand
b. water

First, rearrange the equation for specific heat to solve for q, *the amount of heat (the number of calories).*

$$\text{Amount of heat } (q) = \text{specific heat } (\frac{\text{cal}}{\text{g °C}}) \times \text{mass (g)} \times \Delta T \text{ (°C)}$$

Second, calculate the temperature difference:

ΔT = 28 – 15 = 13 °C

Third, substitute the supplied values into the equation and solve for the amount of heat.

a. *For sand:* $\text{Amount of heat} = 0.16 \frac{\text{cal}}{\cancel{\text{g}}\ \cancel{\text{°C}}} \times 17.2\ \cancel{\text{g}} \times 13\ \cancel{\text{°C}} = 36 \text{ cal}$

b. *For water:* $\text{Amount of heat} = 1.00 \frac{\text{cal}}{\cancel{\text{g}}\ \cancel{\text{°C}}} \times 17.2\ \cancel{\text{g}} \times 13\ \cancel{\text{°C}} = 2.2 \times 10^2 \text{ cal}$

Water requires the greater amount of heat to raise the temperature of 17.3 g by 13 °C.

Try It Yourself #16

Using Table 1-6, calculate the number of calories that must be added to warm 56 g of each of the following substances below from 31 °C to 68 °C. Which substance requires the most heat?

a. paraffin wax
b. lead

First, rearrange the equation for specific heat to solve for q, *the amount of heat (the number of calories):*

Second, calculate the temperature difference:

Third, substitute the supplied values into the equation and solve for the amount of heat:

Practice Problems for Using Dimensional Analysis

1. How many mg are there in 0.607 kg?

2. The Gateway Arch in St. Louis, Missouri, is 630 ft tall. What is its height in meters?

3. A granola bar has 105 Calories. How many calories and joules does it contain?

4. A 4.4-kg infant received a fentanyl infusion for pain relief at dosage of 5.7 μg/kg per hour. What was the total dose in a 24- hour period?

5. What is the mass of 150 mL of vegetable oil? The density of vegetable oil is 0.91 g/mL.
6. Dry ice has a temperature of 194.5 K. Convert this temperature to Fahrenheit and Celsius.

7. Calculate the number of calories required to raise 31 g of each of the following substances from 3 °C to 29 °C.
   a. ambient air
   b. ethanol
   c. aluminum
   d. copper

Chapter 1 Quiz

1. Indicate whether the following changes are physical changes or chemical changes.
   a. steam rising from a cup of coffee

   b. a cast-iron pan forming brown-colored rust (iron(II) oxide)

   c. a balloon deflating

   d. a candle burning

2. Indicate whether the following are examples of kinetic energy or potential energy.
   a. a can of gasoline

   b. a bird flying

   c. a granola bar

3. Rank the following from smallest to largest.
   a. 1 kg b. 1 mcg c. 1mg d. 1 ng e. 1 g

4. When a gold nugget is placed in a graduated cylinder containing 50.0 mL of water, the water level rises to 68.2 mL.
   a. What is the volume of the gold nugget in liters?

   b. What is the mass of the gold nugget in kilograms? Use the density supplied in Table 1-4.

5. How many nanometers are there in 728 cm?

6. How many mL are there in a 12-oz. cup of coffee? There are 32 oz. in a quart.

7. A thermometer in Europe reads 32 °C. Is it a winter day or a summer day? What temperature does this correspond to in degrees Fahrenheit and Kelvin?

8. A 37-lb child is prescribed Advil. The dosage is 10 mg/kg.
   How many mg should be given per dose?

9. A large bagel has 245 Calories. How many calories and joules does the bagel contain?

10. Using Table 1-6, calculate the number of calories that must be added to warm 68.4 g of paraffin wax from 21 °C to 43 °C.

Chapter 1

Answers to Additional Exercises

45 A bone-density scan measures the area and the mass of a section of bone. It estimates the third dimension to get the volume of the bone.

47 Her bone condition is normal.

49 A person with osteoporosis has less bone density than a person with osteopenia.

51 Gases and liquids conform to the shape of their container.

53 Temperature is a measure of the average kinetic energy of particles.

55 a. heat (energy is flowing from the fire) b. work (a body is moving up against gravity) c. heat (energy is flowing from the body to the ice pack) d. work (the gurney is being moved)

57 The rock sitting at the top of a hill has stored energy, so it has potential energy.

59 a. Kinetic energy (the water is moving) b. Potential energy (the skier is at the top of a hill) c. Potential energy (the water has stored energy due to its position) d. Potential energy (the atoms have stored energy in their bonds)

61 The faster moving molecules have more kinetic energy.

63 a. Molecules in the solid state have the least amount of kinetic energy.

65 Steam (The water molecules are in the gas phase.)

67 a. Physical change (The butter has undergone a phase change from a solid to a liquid.) b. Chemical change (The composition of hair has changed as it has changed color.) c. Physical change (The composition of the glass has not changed.) d. Chemical change (the composition of the paper has changed, observed by its color and texture.)

69 a. A chromium atom is on the atomic scale. (It is too small to be seen.) b. The human body is on the macroscopic scale. (You can see it.) c. A grain of sand is on the macroscopic scale. (You can see it.) d. A virus is on the atomic scale. (It is too small to be seen.)

71 *femto-* (There are 10^{15} in a base unit.), *milli-* (There are 10^{3} in a base unit.), *centi-* (There are 10^{2} in a base unit.), and *deci-* (There are 10 in a base unit.)

73 a. 7.6×10^{-6} b. 1×10^{-3} c. 1×10^{4} d. 1.4×10^{3}

75 a. 42,000 b. 13,700,000 c. 0.013

77 6.3×10^{-3} is the smaller number.

79 a. 10^3 b. 10^2 c. 2.5×10^{-3} d. the same number e. 0.00078

81 a. 10 m b. the same length c. the same length d. 1 nm

83 a. 1 g b. 1 μg c. the same mass d. 10 μg

85 c. 150 μg is equivalent to 1.5×10^{-7} kg.

87 V_{metal} = 203.5 mL – 200. mL = 3.5 mL

89 a. 0.25 cm^3 b. 80 L c. the same volume d. 0.05 μL

91 2 cm volume = length × width × height = 8 cm^3. $\sqrt[3]{8} = 2$.

93 The calorie and joule are the most common energy units used in science.

95 a. True b. False c. True d. False

97 The balance is precise and accurate.

99 a. three b. one c. three d. three

101 a. exact number b. not exact c. not exact

103 a. 1.8 b. 4.3 c. 28

105 b. 79.88

107 a. 2.26 + 8.1 = 10.4. There are three significant figures. b. 2.26 × 8.1 = 18. There are two significant figures.

109 a. 33,000. + 910. = 33,910.

b. 0.333 × 0.22 = 0.073

c. (37.55 mL + 22.2 mL) × 56.66 = 3,390 mL

111 a. $6.000\,\cancel{\text{L}} \times \frac{10^3\text{ mL}}{1\,\cancel{\text{L}}} = 6{,}000.\text{ mL}$

b. $2.0 \times 10^6\,\cancel{\text{cm}} \times \frac{1\,\cancel{\text{m}}}{10^2\,\cancel{\text{cm}}} \times \frac{10^{-3}\text{ km}}{1\,\cancel{\text{m}}} = 20.\text{ km}$

113 1 m = 10^{-3} km

115 $200.\,\cancel{\text{mg}} \times \frac{1\text{ g}}{10^3\,\cancel{\text{mg}}} = 0.200\text{ g}$

117 $2.000\,\cancel{\text{min}} \times \frac{60\text{ sec}}{1\,\cancel{\text{min}}} = 120.0\text{ sec}$

119 $150.\,\cancel{\text{kg}} \times \frac{2.205\text{ lb}}{1\,\cancel{\text{kg}}} = 331\text{ lb}$

121 $2.5\,\cancel{\text{cm}} \times \frac{1\text{ m}}{10^2\,\cancel{\text{cm}}} = 0.025\text{ m}$

123 b. $26.7\,\cancel{\text{mm}} \times \frac{1\,\cancel{\text{m}}}{10^3\,\cancel{\text{mm}}} \times \frac{39.37\text{ in}}{1\,\cancel{\text{m}}} = 1.05\text{ in}$

125 $2\,\cancel{\text{oz}} \times \frac{1\,\cancel{\text{cup}}}{8\,\cancel{\text{oz}}} \times \frac{1\,\cancel{\text{qt}}}{4\,\cancel{\text{cup}}} \times \frac{1\,\cancel{\text{L}}}{1.057\,\cancel{\text{qt}}} \times \frac{10^3\text{ mL}}{1\,\cancel{\text{L}}} = 60\text{ mL}$

127 a. $4.14\,\cancel{\text{J}} \times \frac{1\text{ cal}}{4.184\,\cancel{\text{J}}} = 0.989\text{ cal}$

b. $36.2\,\cancel{\text{kJ}} \times \frac{10^3\,\cancel{\text{J}}}{1\,\cancel{\text{kJ}}} \times \frac{1\text{ cal}}{4.184\,\cancel{\text{J}}} = 8.65 \times 10^3\text{ cal}$

c. $0.0587\ \cancel{\text{kJ}} \times \frac{10^3\ \cancel{\text{J}}}{1\ \cancel{\text{kJ}}} \times \frac{1\ \text{cal}}{4.184\ \cancel{\text{J}}} = 14.0\ \text{cal}$

d. $367\ \cancel{\text{J}} \times \frac{1\ \text{cal}}{4.184\ \cancel{\text{J}}} = 87.7\ \text{cal}$

129 $12\ \cancel{\text{lb}} \times \frac{1\ \cancel{\text{kg}}}{2.205\ \cancel{\text{lb}}} \times \frac{20\ \text{mg}}{\cancel{\text{kg}} \cdot \cancel{\text{day}}} \times \frac{1\ \cancel{\text{day}}}{3\ \text{dose}} = 36\ \text{mg/dose}$

131 a. A brick has the greater density. It has more mass than a loaf of bread for the same volume.

b. A bowling ball has the greater density. It has more mass than a soccer ball and they both have the same volume.

c. The bucket of concrete is denser. It has more mass than the bucket of water.

d. Normal bone is denser than a bone from a woman with osteoporosis.

e. A high-density lipoprotein has more density than a low-density lipoprotein.

133 $\text{Density} = \frac{3.8\ \text{g}}{2.0\ \cancel{\text{cm}^3}} \times \frac{1\ \cancel{\text{cm}^3}}{1\ \text{mL}} = 1.9\ \text{g/mL}$

135 A liquid with a density greater than 1.0 g/mL will sink in water.

137 Volume = (length of side)^3

Volume = (2.20 cm)^3

Volume = 10.6 cm^3

$10.6\ \cancel{\text{cm}^3} \times \frac{19.32\ \text{g}}{1\ \cancel{\text{cm}^3}} = 205\ \text{g}$

139 $\text{Specific gravity} = \frac{1.025\ \cancel{\text{g/mL}}}{1.0\ \cancel{\text{g/mL}}} = 1.025$

Yes, the specific gravity is within the normal range.

141 $\text{Density of urine} = 1.014 \times 1.0\ \text{g}\ \ \text{mL} = 1.014\ \text{g/mL}$

The specific gravity of the urine sample is within the normal range.

143 The freezing point of water is 0 °C and 32 °F.

145 b. in the refrigerator $1.8 \times 2\ °\text{C} + 32 = 36\ °\text{F}$

$1.8 \times 8\ °\text{C} + 32 = 46\ °\text{F}$

147 $20.\ °\text{C} + 273.15 = 293\ \text{Kelvin}$

$1.8 \times 20\ °\text{C} + 32 = 68\ °\text{F}$

You are wearing summer clothes.

149 $11\ °\text{C} + 273.15 = 284\ \text{K}$

$1.8 \times 11\ °\text{C} + 32 = 51\ °\text{F}$

You are wearing winter clothes.

151 $1.8 \times 38\ °\text{C} + 32 = 1.0 \times 10^2\ °\text{F}$

Yes, you have a fever.

153 $-78\ °C + 273.15 = 195\ K$

$1.8 \times -78\ °C + 32 = -110\ °F$

155 $\Delta T = 51 - 29 = 22\ °C$

a. For water: Amount of heat $= \frac{1.00\ \text{cal}}{\text{g}\ °\text{C}} \times 13.4\ \text{g} \times 22\ °\text{C} = 290\ \text{cal}$

b. For copper: Amount of heat $= \frac{0.093\ \text{cal}}{\text{g}\ °\text{C}} \times 13.4\ \text{g} \times 22\ °\text{C} = 27\ \text{cal}$

c. For sand: Amount of heat $= \frac{0.16\ \text{cal}}{\text{g}\ °\text{C}} \times 13.4\ \text{g} \times 22\ °\text{C} = 47\ \text{cal}$

Water requires the greatest input of heat to warm 13.4 g by 22°C. Water has the largest heat capacity.

157 $\Delta T = 45 - 30 = 15\ °C$

$$\text{Mass} = \frac{\text{Heat}}{\text{Sspecific heat} \times \Delta T} = \frac{412\ \text{cal}}{\frac{1.0\ \text{cal}}{\text{g}\ °\text{C}} \times 15\ °\text{C}} = 27\ \text{g}$$

159 $$\Delta T = \frac{\text{Heat}}{\text{Sspecific heat} \times \text{mass}} = \frac{137\ \text{cal}}{\frac{0.11\ \text{cal}}{\text{g}\ °\text{C}} \times 56.8\ \text{g}} = 21.9\ °C$$

161 First, calculate the amount of heat gained by the water. Use the temperature changes and mass related to water.

$\Delta T = 24 - 19 = 5\ °C$

Amount of heat $= \frac{1.00\ \text{cal}}{\text{g}\ °\text{C}} \times 81\ \text{g} \times 5\ °\text{C} = 400\ \text{cal}$

The amount of heat lost by the metal is 400 cal since that is the amount gained by the water. Use the specific heat equation to calculate the specific heat of the metal object. Use the temperature change and mass related to the metal object.

$\Delta T = 97 - 24 = 73\ °C$

$$\text{Specific heat} = \frac{\text{Heat}}{\text{Mass} \times \Delta T} = \frac{400\ \text{cal}}{23\ \text{g} \times 73\ °\text{C}} = 0.2\ \text{cal/g}\ °\text{C}$$

163 Glucose is the most important source of energy for the body. It is a sugar.

165 A person must consume the same amount of Calories that they expend in daily living to maintain their weight.

167 Glucose is a form of potential energy. The energy is stored in the chemical bonds.

Chapter 2

Atomic Structure and Nuclear Radiation

Chapter Summary

In this chapter, you learned about elements and the properties of the atom and how to read the periodic table. You also learned that the properties of the atom are determined by their valence electrons. Understanding valence electrons will help you understand properties of atoms, which in turn will help you understand the reactions of atoms and molecules. You also studied nuclear reactions, changes to the nucleus of an atom. You learned about different radioisotopes and different types of radiation emitted in a nuclear reaction. You were introduced to the different types of nuclear decay and their effects on biological tissue. The radiation produced in a nuclear reaction can be used in the diagnosis and treatment of disease.

2.1 Elements and the Structure of the Atom

An atom is the smallest intact component of all matter.

In a Nutshell: The Parts of an Atom

An atom has three subatomic particles: protons, electrons, and neutrons. A proton has a positive charge +1; an electron has a negative charge −1; and a neutron has no charge. An atom contains an equal number of protons and electrons. The mass of a subatomic particle is measured in units called atomic mass units, amu. The mass of a proton and the mass of a neutron is approximately 1 amu. The electron is much lighter.

In a Nutshell: The Structure of the Atom

Protons and neutrons are concentrated in a small volume at the center of the atom, known as the nucleus. Electrons occupy a larger volume of space. The probability of finding the electron in a particular volume is known as an electron orbital. The *s orbital* is spherical in shape, while the *p orbitals* are dumbbell shaped.

In a Nutshell: The Periodic Table

An element is defined by the number of protons that its atoms contain. The atomic number is the number of protons in an atom. Every element has a one- or two-letter atomic symbol. The atomic symbol for every element, its atomic number, and its atomic mass are given in the periodic table of elements. The elements are listed in the periodic table in order of their atomic number: from lowest atomic number to highest atomic number from left to right and from top to bottom across the table.

Worked Example #1

Using the periodic table, state the atomic number, element symbol, number of protons, and number of electrons for each of the following elements:

a. selenium
b. strontium
c. lithium

Solutions

Locate the element symbol on the periodic table. The atomic number will be located above the symbol, which is equal to the number of protons and the number of electrons. It will always be a whole number.

a. Selenium: atomic number = 34; symbol, Se; 34 protons, 34 electrons
b. Strontium: atomic number = 38; symbol, Sr; 38 protons, 38 electrons
c. Lithium: atomic number = 3; symbol, Li, 3 protons, 3 electrons

Try It Yourself #1

Using the periodic table, state the atomic number, element symbol, number of protons, and number of electrons for each of the following elements:

a. krypton
b. silicon
c. technetium

a. Locate the element symbol for krypton on the periodic table.
The element symbol is ____. The atomic number is ____________.
The number of protons is ____. The number of electrons is_____.

b. Locate the element symbol for silicon on the periodic table.
The element symbol is ____. The atomic number is ____________.
The number of protons is ____. The number of electrons is_____.

c. *Locate the element symbol for technetium on the periodic table.*

The element symbol is ____. The atomic number is ____________.

The number of protons is ____. The number of electrons is_____.

In a Nutshell: Isotopes of an Element

Isotopes are atoms with the same number of protons, but a different number of neutrons. Most elements have more than one naturally occurring isotope, and additional isotopes can be prepared artificially. Isotopes can be identified by their mass number. The mass number is the number of protons plus the number of neutrons. The symbol for an isotope often has the mass number written as a superscript to the left of the element symbol and the atomic number written as a subscript to the left of the element symbol.

Worked Example #2

How many protons and neutrons are there in potassium-41 with an atomic number of 19?

The number of protons is the same as the atomic number. The number of protons is 19. The neutrons can be calculated by subtracting the number of protons from its mass number (41).

number of neutrons = mass number − atomic number

number of neutrons = 41 − 19 = 22 neutrons

Try It Yourself #2

What are the mass number, number of protons, and number of neutrons in iron-57? The atomic number is 26.

The atomic number = ________

The mass number = ___________

The number of neutrons = _______________ − _____________

In a Nutshell: Average Atomic Mass

The mass of an atom is determined by the number of protons and neutrons it contains. The average atomic mass of an element is a weighted average of the mass of its isotopes based on their natural abundance. The term "average atomic mass" is often simplified to "atomic mass."

Worked Example #3

Germanium has five naturally occurring isotopes: germanium-70, germanium-72, germanium-73, germaniume-74, and germanium-76. The table below shows the natural occurrence of each of these isotopes as a percentage of all germanium isotopes.

Germanium isotope	Natural abundance
Germanium-70	21.234%
Germanium-72	27.663%
Germanium-73	7.731%
Germanium-74	35.942%
Germanium-76	7.442%

a. Which of these isotopes has the greatest mass and why?
b. Which of these isotopes is the most abundant isotope of germanium on Earth?
c. Which isotope has the least number of neutrons?

Solutions

a. *The isotope that has the greatest mass is germanium-76. It has the highest mass number of the isotopes listed.*
b. *The most abundant isotope is germanium-74. It has the highest percentage of natural abundance.*
c. *The isotope with the least number of neutrons is germanium-70. It has the smallest mass number. The mass number is the number of protons and the number of neutrons. Since all the isotopes have the same number of protons, the isotope with the smallest mass number will have the smallest number of neutrons.*

Try It Yourself #3

Tungsten has five naturally occurring isotopes: tungsten-180, tungsten-182, tungsten-183, tungsten-184, and tungsten-186.

The table below shows the natural occurrence of each of these isotopes as a percentage of all tungsten isotopes.

Tungsten isotope	Natural abundance
Tungsten-180	0.134%
Tungsten-182	26.32%
Tungsten-183	14.31%
Tungsten-184	30.6715%
Tungsten-186	28.62%

a. Which of these isotopes has the smallest mass and why?

b. Which of these isotopes is the least abundant isotope of tungsten on Earth?

c. Which isotope has the greatest number of neutrons?

Practice Problems for Elements and the Structure of the Atom

1. For the following elements, determine the atomic number, the element symbol, the number of protons, and the number of electrons.
   a. silver

   b. magnesium

   c. xenon

   d. iodine

2. Chromium has four naturally occurring isotopes: chromium-50, chromium-52, chromium-53, and chromium-54. Complete the table below:

Isotope	Mass number	Atomic number	Number of protons	Number of neutrons
Chromium-50		24		
Chromium-52		24		
Chromium-53		24		
Chromium-54		24		

a. What do all chromium isotopes have in common?

b. How are the isotopes of chromium different?

c. Which isotope of chromium has the greatest mass? Explain.

2.2: Navigating the Periodic Table of Elements

In a Nutshell: Groups and Periods

Elements in the same column of the periodic table are known as a group or family of elements. Each group is identified by a group number. Elements in the first two and last six columns, Groups 1A to 8A, are known as main group elements. The middle section (columns numbered 1B to 8B) contains the transition metal elements. Four of the groups have common names: 1) group 1A are the alkali metals; 2) group 2A are the alkaline earth metals; 3) group 7A are the halogens; 4) group 8A are the inert gases, or noble gases. Elements within the same group exhibit similar chemical and physical properties.

Rows of elements are known as periods. There are seven periods in the table. Sections of periods 6 and 7 are always separated from the rest of the periodic table and appear below the main table in two rows. These two rows actually belong in the gap after the elements La and Ac, and they are known as the lanthanides and actinides, respectively.

In a Nutshell: Metals, Nonmetals, and Metalloids

The elements can also be classified as being metals, nonmetals, or metalloids. Metals appear to the left of the bold zigzag that runs diagonally from boron to polonium. Nonmetals appear to the right of the bold line. The elements located along the zigzag line are called metalloids. Metalloids display characteristics of both metals and nonmetals. Hydrogen is classified as a nonmetal. Metals and nonmetals have very different physical and chemical properties, whereas metalloids display characteristics of both metals and nonmetals.

In a Nutshell: Important Elements in Biochemistry and Medicine

Some of the elements, classified as building-block elements, macronutrients, micronutrients, and radioisotopes in nuclear medicine, play an important role in medicine and biochemistry. The building-block elements include carbon, hydrogen, nitrogen, oxygen, phosphorus, and sulfur. These are the elements that make up the chemical structure of the majority of molecular compounds found in living organisms. Essential nutrients (macronutrients and micronutrients) must be supplied through the diet. Macronutrients are required in large quantities (more than 100 mg a day) and micronutrients are required in less than 100 mg a day.

Worked Example #4

For the following elements, determine the atomic number, the element name, the group number, and the family name (if it exists). Also state if it is a nonmetal, metal, or metalloid.

a. K
b. Sb

Solutions

a. *Locate the element symbol on the periodic table. The atomic number will be located above the element symbol. K: potassium, atomic number 19. Potassium is in group 1A—the alkali metals. Potassium is a metal because it falls to the left of the line.*
b. *Locate the element symbol on the periodic table. The atomic number will be located above the element symbol. Sb: antimony, atomic number 51. Antimony is in group 5A. It does not have a family name. Antimony is a metalloid because it falls along the line.*

Try It Yourself #4

For the following elements, determine the atomic number, the element name, the group number, and the family name (if it exists). Also state if it is a nonmetal, metal, or metalloid.

a. Rn

b. Co

Solutions

a. Rn

Locate the element symbol on the periodic table.

The element name is ____. The atomic number is ____________.

Rn is in ____________. The family name is ______________.

Rn lies to the (left or right) of the gray line.

Rn is a ________________.

b. Co

Locate the element symbol on the periodic table.

The element name is ____. The atomic number is ____________.

Co is in ____________. The family name is ______________.

Co lies to the (left or right) of the gray line.

Co is a ________________.

Practice Problems for Navigating the Periodic Table of Elements

For the following elements, determine the atomic number, the element name, the group number, and the family name (if it exists). Also state if it is a nonmetal, metal, or metalloid.

a. Ba

b. At

c. B

d. Pd

2.3: Electrons

The current model of the electron is based on quantum mechanics. An electron can behave as a particle and a wave. An electron can exist in only certain allowed energy levels and no energies in between. Electron levels are designated n = 1, 2, 3 … where the greater the value of n, the higher the energy of the electron. The lower the value of n, the closer the electrons are to the nucleus.

In a Nutshell: Ground-State Electron Arrangements

The maximum number of electrons in a given energy level, n, is $2n^2$ electrons, where n = *1, 2, 3, etc.* Therefore, the n = 1 level can accommodate only two electrons, the n = 2 level can accommodate up to eight electrons. Every element differs in the number of protons that its atoms contain, and that number of protons is equal to the number of electrons. To determine the electron arrangement for a particular element, fill its n = 1 level first, up to two electrons, before filling the n = 2 energy level, up to eight electrons, and so forth.

In a Nutshell: Valence Electrons and Periodicity

Valence electrons are the outermost electrons found in the highest energy level. The valence electrons are farthest from the nucleus. Main-group elements in the same group have the same number of valence electrons that corresponds to the group number.

Worked Example #5

Fill in the table below.

Element	Symbol	Atomic number	Group number	Number of n = 1 electrons	Number of n = 2 electrons	Number of n = 3 electrons	Number of valence electrons
Phosphorus	P			2	8	3	

Solutions

The periodic table shows the atomic number for each element. Table 2-7 provides electron arrangements for the first 20 elements. The number of valence electrons equals the number of electrons in the outermost shell.

Element	Symbol	Atomic number	Group number	Number of n = 1 electrons	Number of n = 2 electrons	Number of n = 3 electrons	Number of valence electrons
Phosphorus	P	**15**	**5**	**2**	**8**	**5**	**5**
Aluminum	**Al**	**13**	**3**	2	8	3	**3**

Try It Yourself #5

Fill in the table below.

Element	Symbol	Atomic Number	Group Number	Number of n = 1 electrons	Number of n = 2 electrons	Number of n = 3 electrons	Number of valence electrons
	Li			2	8	0	

Practice Problems for Electrons

Fill in the table below.

Element	Symbol	Atomic number	Group number	Number of n = 1 electrons	Number of n = 2 electrons	Number of n = 3 electrons	Number of valence electrons
	S	17		2	8	1	

2.4: Ions

An ion is an atom that has lost or gained one or more valence electrons, giving an ion a net charge. Therefore, an ion has an unequal number of protons and electrons, which give it a positive (+) or a negative (−) charge. The magnitude of the charge on an ion is equal to the difference between the number of protons and electrons in the ion. When the number of electrons is greater than the number of protons, the ion is negatively charged and is known as an anion. When the number of electrons is less than the number of protons, the ion is positively charged and known as a cation. As a general rule, metals tend to lose electrons to become cations, while nonmetals tend to gain electrons to become anions or share electrons to become molecules.

When a main-group metal loses all of its valence electrons, the ion produced has a full outermost shell—an arrangement of electrons similar to the noble gas in the period above it. Transition metals can lose a variable number of electrons. Nonmetal elements in groups 5A through 7A have a tendency to gain electrons and form anions. These elements gain the number of electrons needed to achieve a full valence shell—typically eight electrons (although hydrogen achieves a full valence shell with two electrons).

Worked Example #6

Write the ions that can be formed from the following elements, and indicate whether the elements are metals or nonmetals. Indicate the number of protons and electrons in each ion.

a. magnesium
b. sulfur
c. mercury

Solutions

a. *Looking at the periodic table, you see that magnesium is in group 2A. Group 2A elements lose two electrons (all of the valence electrons) to become an ion with a full outermost shell of electrons. Magnesium forms* Mg^{2+} *as an ion. It is a metal.* Mg^{2+} *has 12 protons and 10 electrons.*
b. *Sulfur is in group 6A. Group 6A elements gain two electrons to become an ion with a full outermost shell of electrons. Sulfur forms* S^{2-} *as an ion. It is a nonmetal.* S^{2-} *has 16 protons and 18 electrons.*

c. *Mercury is a transition metal. Figure 2-5 shows that mercury loses two electrons or one electron. Mercury can form Hg^{2+} or Hg^{+} as ions. Hg^{2+} has 82 protons and 80 electrons. Hg^{+} has 81 protons and 80 electrons.*

Try It Yourself #6

Write the ions that can be formed from the following elements, and indicate whether the elements are metals or nonmetals. Indicate the number of protons and electrons in each ion.

a. rubidium
b. arsenic
c. aluminum

a. *Rubidium is in group: ________*
Rubidium will (lose or gain) __________ electrons.
How many electrons will be lost or gained? ____________
Ion formed: ___________
Metal or nonmetal: __________________
Number of protons: __________________
Number of electrons: _________________

b. *Arsenic is in group: _______*
Arsenic will (lose or gain) ___________ electrons.
How many electrons will be lost or gained? __________
Ion formed: ___________
Metal or nonmetal: __________________
Number of protons: __________________
Number of electrons: _________________

c. *Aluminum is in group: _________*
Aluminum will (lose or gain) _________ electrons.
How many electrons will be lost or gained? __________
Ion formed: ___________
Metal or nonmetal: __________________

Number of protons: ___________________

Number of electrons: ___________________

Practice Problems for Ions

Write the ions that can be formed from the following elements, and indicate whether the elements are metals or nonmetals. Indicate the number of protons and electrons in each ion.

a. strontium

b. gallium

c. bromine

d. iron

<u>**2.5 Radioisotopes**</u>

Some isotopes are unstable due to an imbalance in the ratio of neutrons to protons or as a result of too many protons and neutrons in the nucleus. Unstable isotopes are known as radioactive isotopes or radioisotopes. All isotopes with an atomic number greater than 82 are unstable because they have too many protons and neutrons. In addition to naturally occurring radioisotopes, there are man-made isotopes, known as artificial radioisotopes.

The nucleus of a radioactive isotope undergoes a natural process known as radioactive decay to become a more stable nucleus. Radioactive decay typically produces an isotope with a different atomic number, and hence a different element. Radioactive decay is always accompanied by a release of a form of energy called radiation.

Worked Example #7

Fill in the empty cells in the table below for the given radioisotopes.

Radioisotope	Atomic number	Mass number	Number of neutrons	Number of protons
Lead-197				
	85	209		

Solutions

Tools: Periodic Table

Radioisotope	Atomic number	Mass number	Number of neutrons	Number of protons
Lead-197	***82***	***197***	***115***	***82***
Astatine-209	85	209	***124***	***85***

Remember the atomic number is equal to the number of protons, and the mass number is equal to the sum of the number of protons and the number of neutrons.

Try It Yourself #7

Fill in the empty cells in the table below for the given radioisotopes.

Radioisotope	Atomic number	Mass number	Number of neutrons	Number of protons
Americium-242				
	59	125		

In a Nutshell: Radiation

Radiation from the decay of radioisotopes appears in either of two forms: (1) high-energy electromagnetic radiation, such as x-rays or γ-rays; or (2) high-energy particles, such as α particles, β particles, or positrons. Electromagnetic radiation is a form of energy that travels through space as a wave at the speed of light. The wavelength of electromagnetic radiation is defined as the distance between wave crests. Longer wavelengths correspond to lower frequencies, while shorter wavelengths correspond to higher frequencies. Electromagnetic radiation is classified into regions according its wavelength and frequency. The electromagnetic spectrum is the range of electromagnetic radiation spanning all possible wavelengths. The different forms of electromagnetic radiation include γ-rays, x-rays, ultraviolet, visible, infrared, microwaves, and radio waves. γ-rays and x-rays have short wavelengths, whereas radio waves have long wavelengths.

The energy associated with a particular type of electromagnetic radiation depends on its wavelength or frequency. The shorter the wavelength (higher frequency), the higher in energy the electromagnetic radiation is. Longer wavelength, lower frequency corresponds to lower energy, while, shorter wavelength, higher frequency corresponds to higher energy. High-energy electromagnetic radiation can damage biological tissue. The most damaging forms of electromagnetic radiation are γ-rays and x-rays.

Worked Example #8

Refer to Figure 2-27 and answer the questions below by choosing from among the following three forms of electromagnetic radiation.

i. visible ii. γ-rays iii. infrared

a. Which form has the longest wavelength?
b. Which form is the highest in energy?
c. Which form causes the most biological damage?

a. *Of these three, infrared has the longest wavelength.*
b. *γ- rays are the highest in energy.*
c. *γ-rays cause the most biological damage.*

Try It Yourself #8

Refer to Figure 2-27 and answer the questions below by choosing from among the following three forms of electromagnetic radiation.

i. radio waves ii. ultraviolet iii. x-rays

a. Which form has the shortest wavelength?
b. Which form is the lowest in energy?

c. Which form causes the most biological damage?

a. *The shortest wavelength:* ____________________

b. *The lowest in energy:* ____________________

c. *The most biological damage:* ________________________

In a Nutshell: α Decay

Some radioisotopes undergo radioactive decay by emitting α particles. An α particle is a slow-moving, high-energy particle consisting of two protons and two neutrons. Its nuclear symbol is ${}^{4}_{2}\alpha$ or ${}^{4}_{2}He$. The symbol for helium is sometimes used because the composition of an α particle is the same as a helium nucleus. Alpha particles are extremely dense and have a +2 charge. After emission of an α particle, the mass number of the isotope decreases by four and the atomic number decreases by two.

Radioactive decay can be depicted by writing a nuclear equation. The convention is to write the radioisotope to the left of the arrow and the new isotope as well as the emitted radiation to the left of the arrow. The atomic number and mass number for all species in the equation are included. The radioisotope undergoing decay is often referred to as the parent nuclide. When a different element is formed, it is referred to as the daughter nuclide. In a balanced equation, the sum of the atomic numbers (subscripts) of all species on the left side of the arrow must equal the sum of the atomic numbers of all species on the right side of the arrow; likewise for the mass numbers (superscripts).

Worked Example #9

Predict the daughter nuclide for α decay of Ra-219.

Write the radioisotope with its atomic mass numbers to the left of the reaction arrow, and write the α particle and the daughter nuclide on the right of the reaction arrow. Determine the identity of the daughter nuclide, ${}^{A}_{Z}X$, by subtracting 4 from the mass number and subtracting 2 from the atomic number of the parent nuclide. In this case:

$A = 219 - 4 = 215$; and

$Z = 88 - 2 = 86.$

Thus, the daughter nuclide is ${}^{215}_{86}X$. *Use the periodic table to find the element that corresponds to atomic number 86; the element is radon. The balanced nuclear equation is:*

$${}^{219}_{88}Ra \rightarrow {}^{215}_{86}Rn + {}^{4}_{2}\alpha$$

Try It Yourself #9

Predict the daughter nuclide for α decay of Po-209.

Write the radioisotope with its atomic mass numbers to the left of the reaction arrow, and write the α *particle and the daughter nuclide on the right of the reaction arrow.*

Determine the identity of the daughter nuclide.

Mass number of daughter nuclide: ___________

Atomic number of daughter nuclide: __________

Element that corresponds to atomic number of daughter nuclide: _________

Balanced nuclear equation: ______________________

In a Nutshell: β Decay

Some radioisotopes undergo radioactive decay by emitting a β particle. A β particle is a high-energy electron whose nuclear symbol is ${}^{0}_{-1}\beta$, indicating that it contains no protons or neutrons, it is just an electron. The daughter nuclide will have the same mass number as the parent nuclide, but its atomic number is increased by one; therefore, it is the element with the next higher atomic number in the periodic table.

Worked Example #10

Predict the daughter nuclide for β decay of Am-247 and write the balanced nuclear equation for the reaction.

The daughter nuclide will be the next element on the periodic table. Alternatively, you can write the daughter nuclide as X, then insert the values for Z and A after you balance the nuclear equation, and finally look up the element symbol on the periodic table that corresponds to the atomic number calculated.

$$^{247}_{95}Am \rightarrow\ ^{247}_{96}Cm +\ ^{0}_{-1}\beta$$

Try It Yourself #10

Predict the daughter nuclide for β decay of U-237 and write the balanced nuclear equation for the reaction.

Atomic number of daughter nuclide: __________

Mass number of daughter nuclide: __________

Element that corresponds to atomic number of daughter nuclide: _________

Balanced nuclear equation: ______________________

In a Nutshell: X-rays and γ Radiation

X-ray and γ radiation are short-wavelength, high-energy forms of electromagnetic radiation. X-rays accompany some forms of radioactive decay. Gamma-ray (γ) emission accompanies almost all forms of radioactive decay. However, the γ emission is not usually shown in the nuclear equation because it does not affect the atomic number or the mass number.

After radioactive decay has occurred, the daughter nuclide is often in an excited state, a condition in which the nucleus still contains excess energy. An isotope in its excited state is referred to as a metastable isotope, and is notated by the abbreviation m following the mass number of an isotope. The daughter nuclide releases its excess energy in returning to the ground state by releasing a pulse of γ radiation.

Worked Example #11

Show the nuclear reaction that Fr-214m undergoes to become Fr-214.

Since no particle is emitted, the atomic number, mass number, and element symbol for the daughter nuclide is the same.

$$^{214m}_{87}Fr \rightarrow\ ^{214}_{87}Fr + \gamma$$

Try It Yourself #11

Show the nuclear reaction that Cs-119m undergoes to become Cs-119.

Atomic number of daughter nuclide: __________

Mass number of daughter nuclide: __________

Element that corresponds to atomic number of daughter nuclide: __________

Balanced nuclear equation: ____________________

In a Nutshell: Half-Life

The time it takes a macroscopic sample of the radioisotope to decay to one-half its original mass is known as its half-life. Half-lives range from a few seconds to billions of years.

Worked Example #12

For a 50.0 g sample of gallium-67:

a. How much Ga-67 is left after four half-lives?
b. How much time has elapsed after four half-lives?

a. *Divide the original amount of material, 50.0 g, in half four consecutive times (four half-lives):*

$$50.0\ g \xrightarrow{1} 25.0\ g \xrightarrow{2} 12.5\ g \xrightarrow{3} 6.25\ g \xrightarrow{4} 3.1\ g$$

Therefore, 3.1 g of the original sample is left after four half-lives.

b. *Table 2-10, shows that the half-life for Ga-67 is 78 hr. Determine how many hours four half-lives correspond to, given that one half-life is 78 hours.*

$$4\ \cancel{\text{half-lives}} \times \frac{78\ \text{hr}}{1\ \cancel{\text{half-life}}} = 312\ \text{hr}\text{, rounded to 310 hr}$$

Try It Yourself #12

For a 150. g sample of gold-198:

a. How much Au-198 is left after three half-lives?
b. How much time has elapsed after three half-lives?

a. *Initial amount of Au-198:* ______________________
Amount after one half-life: __________________
Amount after two half-lives: _________________
Amount after three half-lives: ____________________

b. *Length of one half-life for Au-198:* ____________
Length of three half-lives: ________________

In a Nutshell: Biological Effects of Nuclear Radiation

The radiation emitted from a nuclear reaction is classified as ionizing radiation because it has the energy to dislodge an orbital electron from an atom creating an ion. When an atom in a molecule is ionized, it changes the molecule in a significant way. In living organisms, this change can be quite destructive. The biological effects of nuclear radiation depend in large part on the type of radiation produced. The energy of the radiation and the penetrating power of the radiation determine the biological effects of the radiation.

Penetrating power varies with the type of radiation. An α particle is relatively large, slow moving, and high in energy. Due to its size and slow speed, an α particle has little penetrating power, though its high-energy makes it very destructive to human tissue. A piece of paper or light clothing is sufficient protection against α particles. On the other hand, inhalation or ingestion of an α emitter can cause major damage to internal organs.

Beta particles have slightly less energy than α particles and much more penetrating power because they are substantially lighter. Specialized heavy clothing or a thick piece of aluminum is required for protection against β particles.

The energy of γ-rays and x-rays is less than or equal to the energy of β particles. However, γ-rays have the most penetrating power of all forms of radiation. Several inches of lead are required to protect against γ radiation. A thin sheet of lead is sufficient protection against x-rays.

Worked Example #13

Which form of radiation has greater penetrating power, an α particle or an x-ray?

Solution

The x-ray has more penetrating power; it is lighter.

Try It Yourself #13

Which form of radiation has greater energy, an α particle or an x-ray?

Solution:

In a Nutshell: Measuring Radiation

There are different ways to measure radiation. The Geiger counter is an inexpensive instrument used in the field that can detect all forms of radiation. A radiation badge is used to monitor radiation exposure of personnel who work in areas or use instruments that produce radiation.

Several units of measurement are used to measure radiation. These units may indicate the number of radioactive emissions, the amount of energy absorbed (absorbed dose) or the biological effectiveness of the energy absorbed (effective dose).The amount of radioactive decay can be measured in becquerel (Bq) or curie (Ci), which indicate the number of emissions per second. These units do not distinguish between the different types of radiation.

An absorbed dose measurement indicates the energy of radiation absorbed per mass of tissue. The Gray (Gy) unit is the most common unit for absorbed dose used in medicine; another unit is the Rad. Absorbed dose measurements do not account for the difference in penetrating power of the different forms of radiation.

The effective dose encompasses both the penetrating power and the amount of energy to give the actual biological effect. The effective dose is calculated by multiplying the absorbed dose by a quality factor, Q, which varies for the different types of radiation. When the unit of absorbed dose is the Gray, the unit of effective dose is the sievert (Sv). Another common unit for the effective dose is rem.

Worked Example #14

An ionizing smoke detector contains a small amount of americium-241. As long as the americium stays in the detector, you receive about 0.01 mrem of radiation. The americium emits 1×10^{-6} Ci. What type of information does mrem and Ci convey?

Solution

A mrem is a unit for measuring the relative biological effectiveness of radiation exposure, while Ci is a unit for measuring the amount of radiation emitted by the americium.

Try It Yourself #14

A chest x-ray produces about 0.1 mSv per x-ray image. What type of information does the Sv convey?

Solution:

In a Nutshell: Radiation Sickness

Exposure to radiation can occur as a single dose (acute exposure) or as smaller doses over a longer period of time (chronic exposure). Radiation sickness results from acute exposure to radiation. The severity of the symptoms from radiation sickness is directly proportional to the effective dose received. The effective dose is often measured in values of LD_x. LD_x refers to the lethal dose of the radiation in *x*% of the population after 30 days.

Worked Example #15

Identify the following situations as either acute or chronic exposure to radiation.

a. A patient with hyperthyroidism (an overactive thyroid) who is treated with a single dose of I-131
b. An airline pilot, who is exposed to cosmic radiation on each flight

Solutions

a. *Acute exposure. It is a single exposure to radiation.*
b. *Chronic exposure. It is exposure to smaller doses of radiation over longer periods of time.*

Try It Yourself #15

Identify the following situations as either acute or chronic exposure to radiation.

a. A hospital worker who carries out CT scans on patients regularly.
b. A patient receiving a chest x-ray.

Solution:

a. *Radiation exposure is:* ________________
b. *Radiation exposure is:* ________________

Worked Example #16

Using Table 2-12, what does it mean if the LD_{100} is 6–10 Sv?

An LD_{100} signifies that 100% of the population would die in 30 days after exposure to 6–10 Sv of radiation.

Try It Yourself #16

Using Table 2-12, what does it mean if the LD_{10} is 1.0–2.0 Sv? What are the symptoms of this dose of radiation?

Solution:

An LD_{10} signifies: ________________________________

The symptoms of exposure to 1.0–2.0 Sv of radiation are: ________________________

Practice Problems for Radioisotopes

1. Refer to Figure 2-27 and answer the questions below by choosing from among the following three forms of electromagnetic radiation.

   i. visible ii. ultraviolet iii. radio waves

   a. Which form has the shortest wavelength?

   b. Which form is the lowest in energy?

   c. Which form causes the most biological damage?

2. Write the balanced nuclear equation for the following:

   a. α decay of Sg-261

   b. Formation of Sm-134 through β decay

3. How much of a 20.0 g sample of fluorine-18 remains after 327 minutes? How many half-lives does 327 minutes represent? The half-life of fluorine-18 is 109 minutes.

4. Which form of radiation has greater penetrating power?

   a. α particle or γ-ray

b. positron or x-ray

c. γ-ray or β particle

5. Which form of radiation has greater energy?

a. γ-ray or positron

b. β particle or α particle

c. x-ray or α particle

6. A bone scan with 600 MBq (megabequerel) of technetium-99m produces 3 mSv. What do the units Bq and Sv convey?

7. Using Table 2-12, what does it mean if the LD_{60} is 4–6 Sv? What are the symptoms of this dose of radiation?

Chapter 2 Quiz

1. Fill in the following table.

Element	Symbol	Atomic number	Mass number	Group number	Number of protons	Number of neutrons	Number of valence electrons	Metal, nonmetal, or metalloid
	O							
		35						
					5	6		

2. How many valence electrons does group 2A have? ________________

3. Answer the following questions about iron.
   a. What is its atomic number? ____________________________
   b. Is it a metal or nonmetal? ____________________________
   c. Is it a building block element, micronutrient, macronutrient, or a radioisotope used in nuclear medicine? ____

4. Answer the following questions about the element potassium. Potassium has three isotopes: potassium-39 (with an abundance of 93.258144%), potassium-40 (with an abundance of 0.01171%), and potassium-41 (with an abundance of 6.730244%).
   a. What is the symbol for the element? ________________________________
   b. What is its atomic number? ______________________________________
   c. What is its mass number? ____
   d. Which isotope is more abundant on Earth? __________________________
   e. How many neutrons are present in each isotope of potassium? ____________

f. Is potassium a metal or nonmetal? ______________________________

g. Identify the group number and family for potassium. __________________

h. How many valence electrons does potassium have? __________________

5. Fill in the empty cells in the table below for the given radioisotopes.

Radioisotope	Atomic number	Mass number	Number of neutrons	Number of protons
		269		108
	87		113	

6. Refer to Figure 2-27 and answer the questions below by choosing from among the following three forms of electromagnetic radiation.

   i. ultraviolet   ii. γ-rays   iii. microwaves

   a. Which form has the shortest wavelength?

   b. Which form is the lowest in energy?

   c. Which form causes the most biological damage?

7. Y-90 is used in treatment of cancer. It undergoes β decay. Identify the daughter nuclide produced in this process. Write the nuclear equation for the β decay of Y-90.

8. Thallium-179 undergoes α decay. Write the nuclear equation for this process. What is the daughter nuclide produced in the process?

9. How much of a 15.0 g sample of P-32 is left after three half-lives? How much time has elapsed in three half-lives? The half-life of P-32 is 14.3 days.

10. What type of radiation has a large penetrating power but a small amount of energy? Which units are used to measure effective doses of radiation? Which units are used to measure absorbed doses of radiation?

Chapter 2

Answers to Additional Exercises

53 Anemia is a condition that occurs when there is a shortage of red blood cells. Red blood cells transport oxygen to cells throughout the body.

55 Iron deficiency anemia can be caused by poor absorption of iron, lack of iron in the diet, chronic blood loss or pregnancy. Also, some medications reduce the body's ability to absorb iron.

57 atom

59 An element is a substance that is composed of only one type of atom. A compound is a substance that is composed of two or more different types of atoms held together by chemical bonds. A compound can be broken down into its elements by chemical means.

61 A proton has a +1 charge, a neutron has no charge, and an electron has a −1 charge.

63 The proton and the neutron have more mass than the electrons, so the nucleus is denser.

65 the electron

67 a. 23 protons and electrons in vanadium b. 16 electrons and protons in sulfur c. 12 protons and electrons in magnesium

69 The atomic number is determined by the number of protons in an element.

71 The mass of a proton is 1 amu and 1.66×10^{-24} g.

73 $^{16}_{8}O$, $^{17}_{8}O$, $^{18}_{8}O$

75

Isotope	Mass number	Atomic number	Number of protons	Number of neutrons
Sulfur-32	**32**	**16**	**16**	**16**
Sulfur-33	**33**	**16**	**16**	**17**
Sulfur-34	**34**	**16**	**16**	**18**
Sulfur-36	**36**	**16**	**16**	**20**

77 The lightest isotope is sulfur-32. It has the smallest mass number.

79 a. Iron-54 is the lightest. b. Iron-58 is the least abundant. c. Iron-56 is the most abundant. d. Iron-54 has the fewest neutrons. e. Iron-56 is present in the greatest amount, but there are three other isotopes of iron that contribute the mass of iron.

81 a. boron, atomic number 5 b. magnesium, atomic number 12 c. osmium, atomic number 76 d. silver, atomic number 47 e. mercury, atomic number 80 f. americium, atomic number 95 g. cesium, atomic number 55

83 a. beryllium, Be b. manganese, Mn c. palladium, Pd d. thorium, Th e. seaborgium, Sg

85 Titanium is located between scandium and vanadium.

87 A family or group of elements has elements in the same column in the periodic table.

89 A row of elements is called a period.

91 a. Group 1A, alkali metals b. Group 4A c. Group 8B d. Group 3A e. Group 8A, noble gases

93 a. Oxygen is a nonmetal. b. Germanium is a metalloid. c. Carbon is a nonmetal.

95 a. Potassium, K b. Radon, Rn

97 The building block elements make up the structure of the majority of compounds found in living organisms. They are: carbon, 4 valence electrons; hydrogen, 1 valence electron; nitrogen 5 valence electrons; oxygen, 6 valence electrons; phosphorus, 5 valence electrons; and sulfur, 6 valence electrons.

99 Building-block elements are nonmetals.

101 Iron carries oxygen through the blood stream and delivers it to cells throughout the body. Iron is part of the structure of hemoglobin.

103 Zinc is important in the immune system. Good sources of zinc are oysters, breakfast cereal, beef, pork, chicken, yogurt, baked beans, and nuts.

105 A helium atom has the smaller diameter because it has fewer electrons. As the number of electrons increases, the outermost electrons spend more of their time in larger orbitals that extend farther from the nucleus, thereby increasing the diameter of the atom.

107 The group number equals the number of valence electrons for the elements in that group.

109

Element	**Symbol**	**Atomic number**	**Group number**	**Number of $n = 1$ electrons**	**Number of of $n = 2$ electrons**	**Number of $n = 3$ electrons**
Oxygen	*O*	*8*	*6A*	**2**	**6**	**0**
Beryllium	*Be*	**4**	*2A*	*2*	*2*	*0*
Argon	*Ar*	*18*	*8A*	*2*	*8*	*8*
Fluorine	**F**	*9*	*7A*	*2*	*7*	*0*

111 Group 4A

113 The ground state

115 Boron has three valence electrons in the $n = 2$ level, while aluminum has three valence electrons in the $n = 3$ level.

117 2 electrons

119 The magnitude of the charge is equal to the difference between the number of protons and electrons in the ion.

121 A cation has a positive charge, while an anion has a negative charge. The metals form cations.

123 a. Ca^{2+} b. Sn^{2+} and Sn^{4+} c. N^{3-} d. Ag^{+}

125 Carbon achieves stability by sharing electrons with other nonmetal atoms.

127 Cu^{+} has one more electron than Cu^{2+}. Both ions have the same number of protons.

129 a. magnesium ion, 12 protons, 10 electrons. There are two more protons than electrons. b. iron ion, 26 protons, 24 electrons. There are two more protons than electrons. c. chloride, 17 protons, 18 electrons. There is one more electron than protons. d. fluoride, 9 protons, 10 electrons. There is one more electron than protons. e. oxide, 8 protons, 10 electrons. There are two more electrons than protons.

131 Radioisotopes are unstable isotopes of an element. They have an imbalance in the ratio of neutrons to protons in the nucleus.

133 Background radiation is radiation that comes from natural sources such as cosmic rays from the Sun or radon gas from the ground.

135 Electromagnetic radiation or high-energy particles are released when a nucleus undergoes radioactive decay.

137 All types of electromagnetic radiation travel at the speed of light.

139 a. Radio wave has the longer wavelength. b. X-ray has the longer wavelength. c. Visible has the longer wavelength.

141 a. γ-rays are more damaging to biological tissue because they are higher in energy. b. Ultraviolet radiation is more damaging to biological tissue because it is higher in energy.

143 ${}^{213}_{83}\mathrm{Bi} \rightarrow {}^{209}_{81}\mathrm{Tl} + {}^{4}_{2}\alpha$

145 ${}^{192}_{77}\mathrm{Ir} \rightarrow {}^{192}_{78}\mathrm{Pt} +. {}^{0}_{-1}\beta$ Platinum-192 is the daughter ion produced.

147 a. ${}^{26}_{11}\mathrm{Na} \rightarrow {}^{26}_{12}\mathrm{Mg} + {}^{0}_{-1}\beta$ b. ${}^{210}_{86}\mathrm{Rn} \rightarrow {}^{206}_{84}\mathrm{Po} + {}^{4}_{2}\alpha$

149 Thirty-two days represent four half-lives.

$$32\ \cancel{\text{days}} \times \frac{1\ \text{half-life}}{8\ \cancel{\text{days}}} = 4\ \text{half-lives}$$

18.0 g (initial amount), 9.0 g (one half-life), 4.5 g (two half-lives) 2.25 g (three half-lives), 1.1 g (four half-lives)

151 It has a very short half-life and won't be left in the body for a long time.

153 100 g (initial amount), 50 g (one half-life)

$1 \cancel{\text{half lives}} \times \frac{66\ hours}{1\ \cancel{\text{half life}}} = 33\ hours$

155 80.0 g (initial amount), 40.0 g (one half-life), 20.0 g (two half-lives), 10.0 g (three half-lives), 5.0 g (four half-lives), 2.5 g (five half-lives), 1.25 g (six half-lives), 0.625 g (seven half-lives), 0.313 g (eight half-lives)

$0.03 \times 80.0\ \text{g} = 2.4\ \text{g}$ More than five half-lives have elapsed when 3 percent of the material remains.

157 Ionizing radiation damages tissue because it has sufficient energy to dislodge a valence electron from an atom, forming a cation. It can cause mutations in DNA, the genetic material of our cells.

159 Ionizing radiation damages tissue because it has sufficient energy to dislodge a valence electron from an atom, forming a cation.

161 X-rays have more penetrating power than radio waves. Therefore, when you go to the dentist you need to wear a lead apron,

163 The lead apron protects your neck and chest from the penetrating power of the x-rays.

165 a. β particles have greater penetrating power. b. γ-rays have greater penetrating power. c. γ-rays have greater penetrating power.

167 α particles, β particles, x-rays are stopped by a thin piece of lead.

169 The bequerel and curie measure the same property of radiation. They measure the rate of radiation emissions from a sample. The abbreviations are Bq for bequerel and Ci for curie. A millicurie would be abbreviated with mCi.

171 An absorbed dose measures the energy of radiation absorbed per mass of tissue but does not take into account the penetrating power of the radiation. The effective dose takes into account both the penetrating power of radiation and the amount of energy to give a biological effect. The units for absorbed dose are the Gray and the Rad. The effective dose is measured in sieverts and rem.

173 An LD_{50} indicates a level of exposure that would result in death in 50% of the population in 30 days.

175 The CT scan exposed the man to a higher dose of radiation.

177 a. After one half-life, there will be half as much radiation, so there will be 10mCi. b. The Tc-99m emits γ-rays.

179 The quarter is higher in density than the tissue in the esophagus. The quarter absorbed more x-rays and is lighter in color than the tissue of the esophagus.

181 The MRI is best for imaging soft tissue areas of the body. The technique uses low-energy radio waves that do not pose a risk of biological damage. An MRI is not ideal for imaging denser tissues in the body such as bones and joints.

Chapter 3
Compounds and Molecules

Chapter Summary

In this chapter, you have learned about compounds and molecules and the differences between ionic compounds and covalent compounds. You have learned to distinguish the six basic shapes of simple molecules. The molecular shape and the individual bond dipoles determine the polarity of the molecule. The polarity of the molecule determines how the molecule interacts with other molecules. You learned about the intermolecular forces of attraction between molecules. These intermolecular forces impact the physical properties of a compound and play an important role in the structure of biological molecules.

3.1 Ionic Compounds

A compound is substance composed of two or more different atoms in a defined whole-number proportion. There are two basic types of compounds: ionic and covalent. An ionic compound contains ionic bonds, while a covalent compound contains covalent bonds.

An ionic compound is composed of metal cations and nonmetal anions held together by the strong attraction between oppositely charged ions, known as an ionic bond, a type of electrostatic attraction. Ionic bonds are the glue that holds ions together to form an ionic compound. Most ionic compounds are called salts and are brittle solids at room temperature. In a salt, the ions form a crystal lattice on the atomic scale. In a lattice structure, every cation is surrounded by anions and every anion is surrounded by cations.

An ion formed from a single atom is called a monoatomic ion. Recall from Chapter 2 that a monoatomic cation is formed when a metal loses one or more electrons and a monoatomic cation is formed when a nonmetal gains one or more electrons. A polyatomic ion is an ion formed when a molecule, rather than a single atom, gains or loses one, two, or three electrons. In a polyatomic ion, the bonds between the atoms are primarily covalent bonds, but an imbalance in the number of protons and electrons overall creates a net charge on the molecule, hence it is an ion. The charge can be localized on one atom or it can be spread over several atoms in the polyatomic ion.

In a Nutshell: The Formula Unit

An ionic compound is identified by its formula unit, the ratio of cations to anions in the lattice, indicated by the subscripts in the formula unit. The lowest whole number ratio of ions that gives an electrically neutral ionic compound determines the subscripts. The formula unit of an ionic compound is written with atomic symbol for the cation followed by the atomic symbol for the anion, excluding their charges. The ratio of cation to anion is indicated with subscripts following the symbols. A subscript is understood to be 1 when none is shown. The sum of all the positive and negative charges in the formula unit of an ionic compound must add up to zero because an ionic compound is electrically neutral. When the magnitude of the charge on the cation and the anion are different, the ratio of ions is not 1:1 and subscripts must be added to indicate the ration that gives a neutral ionic compound. Writing the formula unit for an ionic compound containing a polyatomic ion is similar to the process described for monoatomic ions, except if there is a subscript following the polyatomic ion, in which case the entire polyatomic ion must be enclosed in parentheses.

The following guidelines can determine the formula unit: 1) Determine the charge on the cation; 2) determine the charge on the anion; 3) insert subscripts; 4) if the subscripts can be divided by a common divisor, do so; and 5) double check that the formula unit indicates a neutral compound.

In determining the charge on a cation, for monatomic cations derived from main-group elements in groups 1A and 2A, the charge on the cation corresponds to its group number. If the monatomic cation is derived from a transition metal (or a main-group metal in groups 3A–5A), there is often more than one possible charge for the cation. For polyatomic cations, either memorize the charge associated with the ion or look it up in Table 3-1.

For determining the charge on the anion, the magnitude of the charge on a monatomic anion corresponds to 8 minus its group number. For a polyatomic anion, memorize the charge associated with a particular ion or look it up in Table 3-1.

To determine the subscripts, insert subscripts so that the sum of all the charges on the ions in the formula unit is equal to zero. If the subscript is 1, it is implied and should not be written in. The subscript on each ion is equal to the numerical value of the charge on the other ion.

If a subscript is needed for a polyatomic ion, parentheses must be used to enclose the formula unit of the polyatomic ion, and the subscript placed after the parentheses.

In a Nutshell: Naming Ionic Compounds

To name an ionic compound, write the name of the cation first followed by the name of the anion. A cation derived from a main group element has the same name as the main group metal. For metals that have variable charged forms, the charge on the cation is indicated by placing a Roman numeral, corresponding to the magnitude of the charge, within parentheses immediately following the name of the cation. To name a monoatomic anion, change the ending on the element name to "ide." For a polyatomic ion, the name given in Table 3-1 is used.

In a Nutshell: Ions in Health and Consumer Products

Ionic compounds can be found in many foods and other consumer products. Some examples are zinc oxide, iron(III) oxide, sodium fluoride, hypochlorite ion, sodium nitrate, and sodium nitrite. These compounds are used in sunscreens, toothpaste, lotion, bleach and preservatives.

Worked Example #1

Write the formula unit for gallium chloride.

Solution

Determine the charge on the cation.

Gallium is in group 3; the charge on gallium is +3. The ion formed is Ga^{3+}.

Determine the charge on the anion.

Chlorine is in group 7A; the charge on chlorine is −1.

Insert the subscripts.

$Ga^{3+} \quad Cl^{-} \rightarrow GaCl_3$

If the subscripts can be divided by a common divisor, do so.

No common divisor exists for the subscripts 1 and 3.

Double check that the formula gives a neutral compound.

(cation charge × cation subscript) + (anion charge × anion subscript) = 0

(+3 × 1) + (−1 × 3) =

(+3)..+..(−3) = 0

Try It Yourself #1

Write the formula unit for titanium(II) selenide.

Determine the charge of the cation.

Determine the charge of the anion.

Insert subscripts.

If the subscripts can be divided by a common divisor, do so.

Double check that the formula gives a neutral compound.

Worked Example #2

What is the name of the compound with the formula unit PbF_4?

Solution

First, we must determine the charge on the cation. You can determine the charge of the cation from the formula unit, knowing that the compound must be neutral overall:

(? × 1) + (−1 × 4) = 0

(? × 1) + (−4) = 0

? × 1 = 4

? = +4

Second, name the cation first followed by the anion: lead(IV) fluoride.

Try It Yourself #2

What is the name of the compound with the formula unit Cu_3P?

Determine the charge on the cation.

Name the compound.

Worked Example #3

Answer the questions below about the carbonate ion:

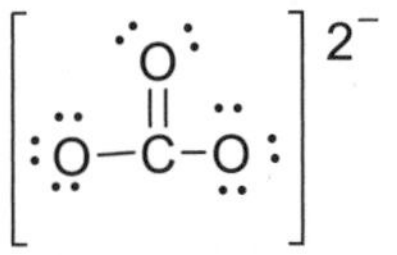

Carbonate ion

a. The charge on the carbonate ion is spread over all three oxygen atoms. What is the total charge on the carbonate ion?
b. Why is the carbonate ion classified as a polyatomic ion and not a monatomic ion?
c. How is the carbonate ion different from a molecule?
d. What is the formula unit for sodium carbonate?
e. What is the formula unit for calcium carbonate?

Solutions

a. *The total charge is −2, as indicated outside the brackets.*
b. *Carbonate is a polyatomic ion because there are several atoms joined by covalent bonds, and collectively they carry a net −2 charge.*
c. *Carbonate is an ion because there is a charge on the collection of atoms.*
d. *Determine the charge on the cation.*
   The cation is the sodium ion. The sodium ion has a +1 charge.
   Determine the charge on the anion.
   The anion is the carbonate ion. The carbonate ion has a −2 charge.

Insert the subscripts.

Na^{+} CO_3^{2-} → Na_2CO_3.

If the subscripts can be divided by a common divisor, do so.

No common divisor exists for the subscripts 1 and 2.

Double check that the formula gives a neutral compound.

(cation charge × cation subscript) + (anion charge × anion subscript) = 0

$(+1 \times 2) + (-2 \times 1) = 0$

$+2 + (-2) = 0$

e. *Determine the charge on the cation.*

The cation is the calcium ion. The calcium ion has a +2 charge.

Determine the charge on the anion.

The anion is the carbonate ion. The carbonate ion has a −2 charge.

Insert the subscripts.

Ca^{2+} CO_3^{2-} → $CaCO_3$.

If the subscripts can be divided by a common divisor, do so.

Double check that the formula gives a neutral compound.

(cation charge × cation subscript) + (anion charge × anion subscript) = 0

$(+2 \times 1) + (-2 \times 1) = 0$

$+2 + (-2) = 0$

Try It Yourself #3

Answer the questions below about the cyanide ion:

$[:C\equiv N:]^-$

Cyanide ion

a. What is the total charge on the cyanide ion?

b. Why is the cyanide ion classified as a polyatomic ion and not a monatomic ion?

c. How is the cyanide ion different from a molecule?

d. What is the formula unit for sodium cyanide?

e. What is the formula unit for calcium cyanide?

Solutions

a. *The total charge:* ___________

b. *Number of different atoms in cyanide:* ______________

c. *The cyanide ion is different from a molecule because*

______________________________.

d. *Determine the charge on the cation.*

Determine the charge on the anion.

Insert the subscripts.

If the subscripts can be divided by a common divisor, do so.

Double check that the formula gives a neutral compound.

(cation charge × cation subscript) + (anion charge × anion subscript) = 0

e. *Determine the charge on the cation.*

Determine the charge on the anion.

Insert the subscripts.

If the subscripts can be divided by a common divisor, do so.

Double check that the formula gives a neutral compound.

(cation charge × cation subscript) + (anion charge × anion subscript) = 0

Practice Problems for Ionic Compounds

1. Write the formula unit for the following compounds.
   a. potassium sulfide
   b. aluminum oxide
   c. lead(IV) oxide
   d. copper(II) chloride
   e. sodium phosphide

2. Write the names of the following compounds.
   a. FeF_2
   b. $CaCl_2$
   c. $CuCl$
   d. BCl_3
   e. Mg_3P_2

3. Write the formula unit for the following ionic compounds:
   a. sodium hydrogen carbonate (also known as baking soda)
   b. ammonium hydroxide
   c. sodium sulfate
   d. calcium phosphate

4. Name the following ionic compounds:
   a. Na_2HPO_4

   b. $Fe(NO_3)_3$

   c. NaOCl

   d. $Al(OH)_3$

3.2 Covalent Compounds

Covalent compounds are composed of identical molecules. A molecule is a discrete entity composed of two or more nonmetal atoms held together by covalent bonds. Most covalent compounds are composed of molecules assembled from the building block elements: carbon, hydrogen, nitrogen, oxygen, phosphorus, sulfur, and halogens.

Ionic and covalent compounds have very different physical and chemical properties. Ionic compounds are solids at room temperature, while covalent compounds can be found in all three states of matter at room temperature. Covalent compounds require less energy to melt or vaporize than covalent compounds.

In a Nutshell: The Covalent Bond

Nonmetal atoms can achieve a full valence energy level by sharing some or all of their valence electrons with another nonmetal. A shared pair of electrons, between two nonmetal atoms, is known as a single covalent bond. The distance between two nuclei joined by a bond is known as the bond length. The driving force for covalent bond formation is the lower potential energy of a covalent bond than two separate atoms.

In a Nutshell: The Molecular Formula

A covalent compound is composed of molecules with the same unique composition, described in part by its molecular formula. In a molecular formula, each atom type is listed, usually in alphabetical order, followed by a subscript indicating how many atoms of that type are in the molecule. Molecular formulas differ from formula units in that they do not represent a ratio but the actual number of each type of atom in the molecule. A molecular formula cannot be altered without also changing the identity of the compound.

Covalent compounds containing only two different types of nonmetal elements—binary compounds—can be named using a simple set of rules. Construct the name of the binary compound by naming the first atom in the formula according to its element name followed by the element name of the second atom in the formula, with a change to the ending of its name to *-ide*. If more than one atom of a given type is present, as indicated by subscripts, insert a prefix before the element name to indicate the number of atoms of that type in the molecule. Sometimes, the prefix *mono-* is used, but normally it is assumed that if an atom has no prefix, only one such atom is present.

Worked Example #4

Name the compound PBr_3.

Phosphorus tribromide. There is one phosphorus atom, named after the element: phosphorus. There are three bromine atoms, so the prefix "tri-" is inserted before the element name and the ending is changed to "-ide."

Try It Yourself #4

Name the compound H_2S.

Number of hydrogen atoms: _________
Prefix for hydrogen: _________
Number of sulfur atoms: ___________
Prefix for sulfur: ___________
Name of compound: _______________

Worked Example #5

Write the molecular formula for diphosphorus pentasulfide.

P_2S_5. The prefix "di-" indicates that there are two phosphorus atoms; the prefix "penta-" indicates that there are five sulfur atoms.

Try It Yourself #5

Write the molecular formula for carbon tetrabromide.

The two elements are: ____________

Number of atoms of each element: ________ *and* _______

Molecular formula: _______________

In a Nutshell: Writing Lewis Dot Structures

Lewis dot structures are a simple and convenient method to show the arrangement of atoms and covalent bonds in a molecule. Lewis dot structures are created by combining Lewis dot symbols according to a set of rules. A Lewis dot symbol for an element is a way of representing an atom and its valence electrons by writing the atomic symbol surround on up to four sides with its valence electrons represented as dots.

To write a Lewis dot structure, arrange the valence electrons so that each atom in period 2 and higher is surrounded by eight electrons, known as the octet rule. In a Lewis dot structure, a pair of shared electrons, a single bond, is usually represented as a line. In order for some atoms to achieve an octet, they must share four electrons, known as a double bond. A double bond will be represented by two lines. Other atoms must share six electrons, known as a triple bond, which will be represented by three lines.

The group-5A, -6A, and -7A elements do not share all of their valence electrons. The electrons that are not shared are known as nonbonding electrons or lone-pair electrons because they spend their time around only one nucleus. Nonbonding electrons are always represented as pairs of dots, written on the atom to which they belong, but with no second atom attached.

The periodicity of nonmetal elements is seen in the number of covalent bonds that elements with a group form. Halogens typically form one covalent bond. The group-6A elements form two bonds; the group-5A elements form three bonds; and the group-4A elements form four bonds.

You can determine the Lewis dot structure from a simple molecular formula using the following guidelines:

1) Add up the total number of valence electrons from all the atoms in molecule.
2) Determine which atom in the molecule is the central atom and which atoms are on the periphery of the molecule and bonded to the central atom. Place a single bond between each pair of atoms. Hydrogen and halogens are usually on the periphery.
3) Determine the number of valence electrons remaining from step 1, after step 2 has been completed, and distribute these remaining electrons as nonbonding pairs. Do this by placing pairs of dots around each atom so that each atom has an octet (8), except hydrogen, which should have two electrons. If you run out of electrons before every atom has an octet, proceed to step 4.
4) If any atoms are short of an octet after all the valence electrons have been distributed in step 3, turn one nonbonding pair of electrons from an adjacent atom into a double bond or turn two nonbonding pairs of electrons into a triple bond.
5) Double check each atom in the molecule against Table 3-3 to make sure that each atom is surrounded by the expected number of bonding and nonbonding electrons.

Worked Example #6

Write the Lewis dot structure for NBr_3.

Step 1: Add up the total number of valence electrons from all the atoms in the molecule. The total number of valence electrons is nitrogen—5, bromine—7: 5 + (3 × 7) = 26 total valence electrons.

Step 2: Determine which atom in the molecule is the central atom and which atoms on the periphery. Place a single bond between each pair of atoms.

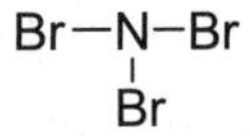

```
Br—N—Br
   |
   Br
```

Nitrogen is the central atom because it is the atom closest to the center of the Periodic Table. Br is a halogen and usually on the periphery.

Step 3: Out of the 26 total valence electrons calculated in step 1, six electrons have been used in the N-Br single bonds.

Step 4: Determine the number of valence electrons remaining from step 1, after step 2 has been completed and distribute these remaining electrons as nonbonding pairs by placing pairs of dots around each atom until it has an octet (8), except hydrogen (2).

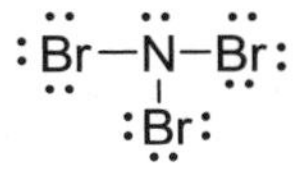

```
 ..   ..  ..
:Br — N — Br:
 ..   |   ..
     :Br:
      ..
```

Turn nonbonding electrons into multiple bonds if any atoms are short of an octet.

All of the atoms have an octet; therefore, multiple bonds are not needed.

Double check each atom in the molecule against Table 3-3 to check that each atom is surrounded by the expected number of bonding and nonbonding electrons.

The nitrogen atom has three bonds and one nonbonding pair, and the bromine atoms have one bond and three nonbonding pairs.

Try It Yourself #6

Write the Lewis dot structure for $CHCl_3$.

Step 1: Add up the total number of valence electrons from all the atoms in the molecule.

Step 2: Determine which atom in the molecule is the central atom and which atoms are on the periphery. Place a single bond between each pair of atoms.

Step 3: Determine the number of valence electrons remaining from step 1, after step 2 has been completed, and distribute these remaining electrons as nonbonding pairs by placing pairs of dots around each atom until it has an octet (8), except hydrogen (2).

Step 4: Turn nonbonding electrons into multiple bonds if any atoms are short of an octet.

Step 5: Double check each atom in the molecule against Table 3-3 to check that each atom is surrounded by the expected number of bonding and nonbonding electrons.

In a Nutshell: Extension Topic: Expanded Octets

The key building-block elements, carbon, nitrogen, and oxygen will always have an octet in a molecule. Atoms in period 3, phosphorus and sulfur, are sometimes found with an expanded octet; these atoms may be surrounded by more than eight electrons in the outermost shell.

Worked Example #7

Evaluate each of the following molecules to determine if it contains an atom with an expanded octet. Indicate the number of electrons surrounding each expanded octet.

a.
O
H ||
H–C—S—Cl:
H ||
O

b.
O
H ||
H–C—P—O–H
H :O:
H

a. *The sulfur atom has an expanded octet; there are 12 electrons around the sulfur atom.*
b. *The phosphorus atom has an expanded octet; there are 10 electrons around the phosphorus atom.*

Try It Yourself #7

Evaluate each of the following molecules to determine if it contains an atom with an expanded octet. Indicate the number of electrons surrounding each expanded octet.

a.

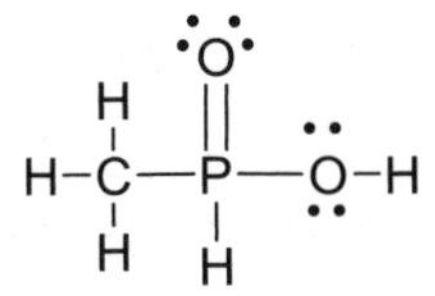

b.

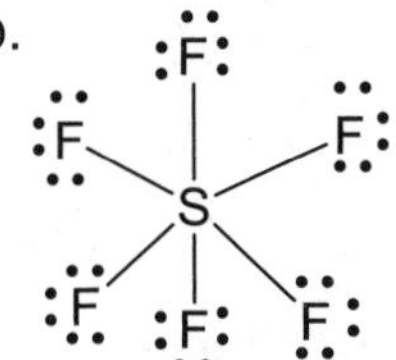

a. *Atom with expanded octet:* _________

Number of electrons surrounding the expanded octet: _________

b. *Atom with expanded octet:* _________

Number of electrons surrounding the expanded octet: _________

Practice Problems for Covalent Compounds

1. Write the Lewis dot structure for the following molecules:
   a. HBr

   b. ClF

   c. BrCN

2. Write the Lewis dot structure for nitrogen triiodide.

3. Write the Lewis dot structure for phosphorus pentabromide.

4. Write the Lewis dot structure for CS_2. What is the name of this compound?

5. Write the Lewis dot structure for COS.

3.3 Three-Dimensional Shapes of Molecules

Molecules are three-dimensional structures. The shape of a molecule containing a central atom surrounded by two or more atoms is defined by the spatial arrangement of the atoms surrounding the central atom, which in turn is determined from the number and type (bonding and nonbonding) of electrons on the central atom. The arrangement of electrons around the central atom is referred to as the electron geometry of the molecule, while the arrangement of atoms around the central atom is the molecular shape of the molecule.

In a Nutshell: Molecular Models

Molecular models are a common tool used for visualizing the three-dimensional shapes of molecules. Two types of molecular models are routinely used: the ball-and-stick model and the space-filling model. The ball-and-stick models depict molecules by representing atoms as balls and bonds as sticks and are used primarily to show molecular shape. Space-filling models are used to show the relative space occupied by the atoms in a molecule. Occasionally, a tube model will be used in place of a ball-and-stick model, particularly when the molecule is large. In a tube model, both bonds and atoms appear as part of a tube, where end points represent the atoms.

In a Nutshell: Using Lewis Dot Structures to Predict Electron Geometry

The process for determining the molecular shape of a molecule can be described in three steps. Step 1) Write the Lewis dot structure for the molecule. Step 2) From the Lewis dot structure, determine the electron geometry. Step 3) From the electron geometry, determine the molecular shape.

Predicting electron geometry is based on valence shell electron pair repulsion theory (VSEPR). This theory is based on the premise that groups of valence electrons (bonding and nonbonding) surrounding a central atom repel because electrons have like charges. Therefore, bonding and nonbonding groups of electrons around a central atom adopt a geometry that places them as far apart from one another as possible, while still remaining attached to the central atom. To determine the electron geometry, simply count the number of "groups" of electrons surrounding the central atom in a Lewis dot structure. A group of electrons refers to the six electrons in a triple bond, the four electrons in one double bond, the two electrons in one single bond, or the two electrons in one nonbonding pair of electrons.

VSEPR theory gives us three electron geometries that are derived from the number of electron groups around the central atom: four groups of electrons—tetrahedral electron geometry; three groups of electrons—trigonal planar electon geometry; two groups of electrons—linear electron geometry. The linear geometry is one-dimensional and has the shape of a straight line, with both groups of electrons directed 180° from each other. The trigonal planar geometry is two-dimensional: Each of the three groups of electrons points to the three corners of an equilateral triangle, separated by 120°. The tetrahedral geometry is three-dimensional: Each of the four groups of electrons points to the four corners of a tetrahedron, creating angles of 109.5°.

In a Nutshell: Using Electron Geometries to Determine Molecular Shape

In the third step of the three-step process for determining molecular shape, the electron geometry is used to predict the shape of the molecule. When the central atom in a molecule has only bonding electrons and no nonbonding electrons, its molecular shape is the same as its electron geometry. Thus, four bonding groups around a central atom produce a molecule with a tetrahedral shape, three bonding groups produce a molecule with a trigonal planar shape, and two bonding groups of electrons around a central atom produce a molecule with linear shape. When a central atom has one or more nonbonding pairs of electrons, the molecular shape is different from its electron geometry because shape is defined by the relative position of the atoms only.

Thus, the shape of a molecule depends on both its electron geometry and the relative number of bonding and nonbonding groups of electrons on the central atom. There are three different molecular shapes derived from a tetrahedral electron geometry depending on the number of nonbonding electrons on the central atom and there are two molecular shapes derived from a trigonal planar geometry depending on the number of nonbonding electrons on the central atom.

There are three molecular shapes derived from a tetrahedral electron geometry: tetrahedral, trigonal pyramidal, and bent shapes. The tetrahedral molecular shape arises when a central atom is surrounded by four bonding groups of electrons and zero nonbonding electrons. The trigonal pyramidal molecular shape arises when there are three bonding groups and one nonbonding group of electrons around a central atom with tetrahedral electron geometry. A bent molecular shape arises when there are two bonding groups and two nonbonding groups of electrons around a central atom with a tetrahedral electron geometry.

A bond angle describes the angle between the central atom and any two of the atoms to which it is bonded. Since the tetrahedral, trigonal pyramidal, and bent molecular shapes are all derived from the same electron geometry, they all have approximately the same bond angles of 109.5°.

Three bonding groups and no nonbonding groups form a trigonal planar electron geometry. Two molecular shapes are derived from a trigonal planar electron geometry. The trigonal planar geometry gives rise to molecules with bond angles of approximately 120°.

When all three groups of electrons around the central atom are bonding in a molecule with a trigonal planar electron geometry, then the molecular shape is also trigonal planer. If one of the electron groups in a molecule with a trigonal planar electron geometry is a nonbonding pair of electrons, the molecular shape is bent.

When a central atom is surrounded by two groups of electrons, they will always be bonding groups; therefore, the molecule has a linear electron geometry and a linear molecular shape. The bond angle in a linear molecule is 180°.

Most molecules are easily drawn in a way that accurately represents their molecular shape. To draw the structure of a tetrahedral center in three dimensions, write the central atom with two of its four bonds in the plane of the paper, using a normal line to represent the bond. The bonds should form an angle greater than 90° to represent the 109.5° bond angle. Next, draw the third bond attached to the central atom projecting toward you by writing a wedged line. Finally, write the fourth bond attached to the central atom by projecting away from you as a dashed line.

Worked Example #8

Predict the electron geometry, molecular shape, and bond angles for following compounds.

a. $CHCl_3$

b. OCS (C is the central atom)

a. *First, write the Lewis dot structure.*

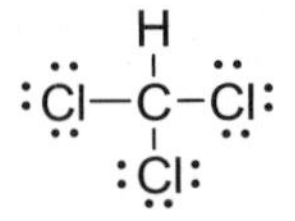

```
      H
      |
:Cl—C—Cl:
      |
    :Cl:
```

Second, determine the number of electron groups around the central atom.

The carbon has four single bonds, or four electrons groups, around it.

Third, determine the electron geometry from the number of electron groups.

Four electron groups indicate a tetrahedral geometry. The molecular shape is tetrahedral and the bond angles are 109.5°.

b. *First, write the Lewis dot structure.*

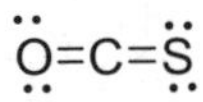

```
O=C=S
```

Second, determine the number of electron groups around the central atom.

The carbon has two double bonds, or two electron groups, around it.

Third, determine the electron geometry from the number of electron groups.

Two electron groups indicate a linear geometry. The molecular shape is linear and the bond angles are 180°.

Try It Yourself #8

Predict the electron geometry, molecular shape, and bond angles for following compounds.

a. CF_2Cl_2

b. HCN

a. *Write the Lewis dot structure.*

Determine the number of electron groups around the central atom.

Central atom: ______

Number of electron groups: ________

Determine the electron geometry.

Geometry: ________________________

b. *Write the Lewis dot structure.*

Determine the number of electron groups around the central atom.

Central atom: ______

Number of electron groups: ________

Determine the electron geometry.

Geometry: ________________________

Worked Example #9

Predict the electron geometry, molecular shape, and bond angles for following compounds.

a. OCl_2
b. PCl_3
c. BCl_3 (an exception to the octet rule)

a. *Write the Lewis dot structure.*

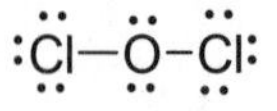

$$:\underset{..}{\ddot{Cl}}-\underset{..}{\ddot{O}}-\underset{..}{\ddot{Cl}}:$$

Determine the number of electron groups around the central atom.

The central atom is oxygen and it has two nonbonding groups and two single bonds, so four electron groups around it.

Determine the electron geometry.

The electron geometry is tetrahedral.

Determine the number of nonbonding groups and bonding groups.
There are two nonbonding groups and two bonding groups; therefore, the molecular shape is bent. The bond angles are approximately 109.5°.

b. *Write the Lewis dot structure.*

```
 ..   ..  ..
:Cl—P—Cl:
 ..   |   ..
     :Cl:
      ..
```

Determine the number of electron groups around the central atom.
Phosphorus is the central atom; there are three single bonds and one nonbonding pair; therefore, there are four electron groups around phosphorus.

Determine the electron geometry.
The electron geometry is tetrahedral.

Determine the number of nonbonding groups and the number of bonding groups.
There are one nonbonding group and three bonding groups; therefore, the molecular shape is trigonal pyramidal. The bond angles are approximately 109.5°.

c. *Write the Lewis dot structure.*

```
 ..      ..
:Cl—B—Cl:
 ..   |   ..
     :Cl:
      ..
```

Determine the number of electron groups around the central atom.
The central atom is boron and there are three single bonds around it; therefore, there are three electron groups around it.

Determine the electron geometry.
The electron geometry is trigonal planar.

Determine the number of nonbonding groups and the number of bonding groups.

There are three bonding groups and no nonbonding groups; therefore, the molecular shape is trigonal planar. The bond angles are approximately 120°.

Try It Yourself #9

Predict the electron geometry, molecular shape, and bond angles for following compounds.

a. PH_3
b. SF_2

a. *Write the Lewis dot structure.*

Determine the number of electron groups around the central atom.

Central atom: ______

Number of electron groups: ________

Determine the electron geometry.

Geometry: _______________

Determine the molecular shape.

Number of bonding groups: __________

Number of nonbonding groups: ________

Geometry: ______________

b. *Write the Lewis dot structure.*

Determine the number of electron groups around the central atom.

Central atom: ______

Number of electron groups: ________

Determine the electron geometry.

Geometry: _______________

Determine the molecular shape.

Number of bonding groups: __________

Number of nonbonding groups: ________

Geometry: ______________

In a Nutshell: Shapes of Larger Molecules

The six different molecular shapes described for smaller molecules can also be used to evaluate the shapes of larger molecules having more than one central atom. The shape of a larger molecule can be viewed as a combination of the shapes of the individual atom centers.

Worked Example #10

Predict the shape of ethanol, CH_3CH_2OH.

Write the Lewis dot structure.

```
    H  H
    |  |  ..
  H-C--C--O-H
    |  |  ..
    H  H
```

Evaluate the geometry around each central atom.

The two carbons have four bonding groups; therefore, they are tetrahedral. The oxygen atom has two bonding groups and two nonbonding groups; therefore, it is bent. The overall structure has a zigzag shape.

Try It Yourself #10

Predict the shape of hydrogen peroxide, H_2O_2.

Write the Lewis dot structure.

Central atoms: ______________

Number of bonding groups around each central atom: ___________

Number of nonbonding groups around each central atom: _________

Geometry of each central atom: _______________

Practice Problems for Three-Dimensional Shapes of Molecules

1. Predict the electron geometry, molecular shape, and bond angles for following compounds.
   a. H_2CO_3

   b. $COCl_2$

   c. CS_2

2. Predict the shapes of the central atoms in the following compounds.
   a. N_2H_4

   b. CH_3SH

   c. C_2Cl_4

3.4 Molecular Polarity

Polarity, a characteristic of molecules, determines how they interact with other molecules. Polarity accounts for the formation of cell membranes and determines whether or not drug molecules can cross the blood-brain barrier.

In a Nutshell: Electronegativity

When two atoms form a covalent bond, the electrons do not necessarily spend equal time around both nuclei. Some atoms are better able to draw electrons toward their nuclei than other atoms. Electronegativity is the measure of an atom's ability to draw electrons toward itself in a covalent bond. The elements toward the bottom left of the periodic table are the least electronegative and elements toward the top right part of the periodic table are the most electronegative. The most electronegative element is fluorine.

Several factors affect the electronegativity of an element. One is proximity of the valence electrons to the nucleus. The closer the valence electrons are to the positively charged nucleus, the more electronegative the atom. Within a group of elements, electronegativity increases as you go from the bottom of a group to the top of a group. Within a period of elements, electronegativity increases as you go from left to right.

In a Nutshell: Bond Dipoles

A polar covalent bond is created between two atoms with a significant difference in electronegativity, such that the bonding electrons spend more time around the more electronegative atom. A partial negative charge (□−) is formed on the more electronegative atom, and a partial positive charge (□+) develops on the less electronegative atom. A polar covalent bond is defined as a covalent bond between two atoms with an electronegativity value difference greater than 0.5 but less than 2. A dipole arrow can be placed alongside the bond with the tail of the arrow shown next to the less electronegative atom and the head of the arrow pointing toward the more electronegative atom. An electron-density diagram is a space filling model that indicates regions of higher and lower concentrations of electrons by color.

When covalent bonds are formed between atoms with comparable electronegativity values (electronegativity < 0.5), the bonding electrons are evenly distributed between both atoms, and the bond is known as a nonpolar covalent bond.

Worked Example #11

Identify the more electronegative atom in each pair.

a. sodium and rubidium
b. chlorine and silicon

a. *Figure 3-21 shows that sodium is above rubidium in the same group; therefore, sodium is the more electronegative atom.*

b. *Figure 3-21 shows that chlorine is to the right of silicon in the same period; therefore, chlorine is more electronegative.*

Try It Yourself #11

Identify the more electronegative atom in each pair.

a. sulfur and copper

b. platinum and bromine

Tools: Figure 3-21

a. *The more electronegative atom is* ________________.

b. *The more electronegative atom is* ________________.

Worked Example #12

Indicate which of the following represent polar covalent bonds by showing a bond dipole arrow.

a. P—F

b. O—H

c. N—O

a. $\overset{+\!\!\longrightarrow}{\text{P–F}}$ *Figure 3-21 shows that fluorine is more electronegative than phosphorus.*

b. $\overset{\longleftarrow\!\!+}{\text{O–H}}$ *Figure 3-21 shows that oxygen is more electronegative than hydrogen.*

c. *N—O is a nonpolar covalent bond. They have similar electronegativities.*

Try It Yourself #12

Indicate which of the following represent polar covalent bonds by showing a bond dipole arrow.

a. S—O

b. C—Br

c. Si—C

Tools: Figure 3-21

a. S—O

b. C—Br

c. Si—C

In a Nutshell: Polar and Nonpolar Molecules

A molecule with more than one polar bond can be either polar or nonpolar, depending on its molecular shape. A polar molecule has a separation of charge—a positive end and a negative end of the molecule. In contrast, a nonpolar molecule has an even distribution of electrons throughout the molecule. Polar and nonpolar molecules have very different physical properties.

If a molecule has only nonpolar covalent bonds, the molecule is nonpolar. For a molecule with two or more polar covalent bonds, the shape of the molecule determines whether the molecule is polar or nonpolar. Bond dipoles will cancel when there are two identical bond dipoles in a linear molecule or when there are three identical bond dipoles in a molecule with a trigonal planar shape or four identical bond dipoles in a molecule with tetrahedral shape. When the bond dipoles cancel each other, the molecule will be nonpolar. In general, when the covalent bonds around a central atom are identical and the electron geometry is the same as the molecular shape (tetrahedral, trigonal planar and linear), the bond dipoles will cancel and the molecule is nonpolar.

Molecules with one or more bond dipoles and a trigonal pyramidal or bent molecular shape are polar because the bond dipoles cannot cancel as a result of the molecular shape of the molecule.

Worked Example #13

Which of the following molecules are nonpolar? Explain why.

a. O_2
b. CF_4
c. CH_3Cl
d. HBr

a. O_2 is nonpolar. The electrons in the covalent bond are shared equally by the identical atoms that form the covalent bond.

b. Carbon tetrafluoride is nonpolar. CF_4 is tetrahedral and has four C—F bonds, which are polar covalent bonds. The dipoles are equivalent and are directed into the four opposite corners of the tetrahedron; therefore, the dipoles cancel each other, causing the molecule to be nonpolar.

c. CH_3Cl is a polar molecule. There is one polar covalent bond (C—Cl) in the molecule.

d. H—Br is a polar molecule.

Try It Yourself #13

Which of the following molecules are nonpolar? Explain why.

a. CF_3Cl
b. NBr_3
c. BCl_3
d. C_4H_{10}

Solution

____________________ *are nonpolar.*

Practice Problems for Molecular Polarity

1. Which of the following elements is less electronegative?
   a. fluorine and hydrogen

   b. arsenic and nitrogen

   c. sodium and bromine

2. Indicate which of the following represent a polar covalent bond by showing a bond dipole arrow.
   a. C—H

   b. N—H

   c. P—Cl

   d. C—O

3. Which of the following molecules are nonpolar?
   a. CH_2ClF

   b. C_2H_4

   c. PF_3

   d. HCN

3.5 Intermolecular Forces of Attraction

Molecules interact with each other through intermolecular forces of attraction—those that exist between molecules rather than within molecules. Intermolecular forces of attraction determine the physical properties of a substance such as boiling point, melting point, and solubility.

Intermolecular forces are much weaker than covalent bonding forces. Covalent bonds are found where relatively permanent connections between atoms are required, and intermolecular are found where temporary connections between molecules are needed. Intermolecular forces arise from electrostatic interactions between partial charges resulting from bond dipoles in one molecule being attracted to the opposite partial charges in other molecules. A dipole can be either a permanent dipole or a temporary dipole. The three basic types of intermolecular forces of attraction between molecules listed in order of increasing strength are dispersion forces, dipole-dipole forces, and hydrogen-bonding forces.

In a Nutshell: Dispersion Forces

All molecules and elements interact through dispersion forces, also known as London forces. Dispersion forces are the only intermolecular force of attraction between nonpolar molecules. In nonpolar molecules, the electrons can shift temporarily toward one end of the molecule, creating a temporary dipole. A temporary dipole induces a corresponding dipole in adjacent molecules, causing them to shift in a way that brings opposite partial charges together. Since the induced dipoles are short-lived, dispersion forces are the weakest of the three types of intermolecular forces of attraction.

Since many electrons are distributed throughout a molecule, there are many opportunities for temporary dipoles to form. It follows, therefore, that dispersion forces of attraction are greater in molecules that have a greater number of electrons or a greater surface area.

In a Nutshell: Dipole-Dipole Forces

In addition to dispersion forces, polar molecules are capable of interacting through dipole-dipole forces, created by the electrostatic attraction that occurs between the opposite partial charges of the permanent dipoles of polar molecules. Polar molecules arrange themselves in a way that brings together opposite charges. Because the dipoles are permanent, dipole-dipole forces are much stronger than the temporary dipoles that are responsible for dispersion forces.

In a Nutshell: Hydrogen Bonding Forces

Hydrogen bonding is the strongest intermolecular force of attraction between molecules. A hydrogen bond is not a covalent bond, but rather a type of dipole-dipole interaction that exists in molecules that have one of the three most polar covalent bonds: an H—F bond, an H—O, or an H—N bond. In these bonds, because of the differences in electronegativity, the hydrogen atom has the greatest partial positive charge of any atom in a covalent bond, and F, O, and N have the greatest partial negative charge of any atom in a covalent bond. Therefore, a molecule containing an O—H or N—H bond should exhibit the strongest force of attraction between the positive pole—hydrogen—of the bond and a nitrogen or oxygen atom in another molecule—the negative pole. The presence of hydrogen bonding impacts the boiling points of molecules.

Deoxyribonucleic acid (DNA) is probably the most important biomolecule because it contains an organism's genetic information. It is a composed of two extremely large molecules-strands-wound together in the shape of a double helix. Hydrogen bonds hold the two strands of DNA together throughout the length of the helix.

Worked Example #14

Which type of intermolecular force of attraction is formed between temporary dipoles? What types of molecules have temporary dipoles?

Dispersion forces are formed between temporary dipoles of different molecules. All molecules and elements have temporary dipoles.

Try It Yourself #14

How do dipole-dipole forces differ from dispersion forces? How are they the same?

Solution

Worked Example #15

What type of intermolecular force will be exerted by the following molecules?

a. CH_3OH (methanol)
b. C_4H_8

a. *Follow the flow chart in Figure 3-27. Are there polar molecules present? Yes, CH_3OH is polar. It has a polar covalent bond, O—H. Are H—O or H—N bonds present? Yes, therefore it has hydrogen bonding.*

b. *Follow the flow chart in Figure 3-27. Are there polar molecules present? No, C_4H_8 has only C—H bonds, which are nonpolar covalent bonds. Therefore, dispersion forces are present.*

Try It Yourself #15

What type of intermolecular force will be exerted by the following molecules?

a. CH_3Cl
b. CH_3NH_2

Solution

Tools: Figure 3-27

a. *CH_3Cl has ____________________________.*
b. *CH_3NH_2 has ____________________________.*

Worked Example #16

Which molecule in each pair exhibits stronger intermolecular forces of attraction with like molecules? Explain your choice. State the type of intermolecular force.

a. C_4H_{10} or $C_{10}H_{22}$
b. NH_3 or PH_3

a. *$C_{10}H_{22}$ exerts stronger dispersion forces than C_4H_{10} because there are more atoms in the molecule and therefore more electrons. Dispersion forces increase with the number of electrons in a molecule.*
b. *NH_3 exhibits the stronger force because it interacts with other NH_3 molecules through hydrogen bonding, the strongest force of attraction. PH_3 takes part in dipole-dipole interactions because it has a permanent dipole.*

Try It Yourself #16

Which molecule in each pair exhibits stronger intermolecular forces of attraction with like molecules? Explain your choice. State the type of intermolecular force.

a. CH_4 or H_2O

b. H₂C=CH₂ (H, H / C=C / H, H) or H–C(=O)–H (O / ‖ / C / H, H)

Solution

a. ____________________ exhibits stronger intermolecular forces of attraction. The reason for the stronger intermolecular forces of attraction is ____________________________. The intermolecular force is ____.

b. ____________________ exhibits stronger intermolecular forces of attraction. The reason for the stronger intermolecular forces of attraction is ____________________________. The intermolecular force is ____.

Practice Problems for Intermolecular Forces of Attraction

1. What type of intermolecular force will be exhibited by the following molecules?
   a. CH_3F

   b. C_2H_6

   c. CH_3CH_2OH

   d. H–C(=O)–H

2. Draw the hydrogen bonds that can be formed between three water molecules.

3. Which molecule in each pair is capable of hydrogen bonding?

    a. CHF_3 or CH_3OH

    b. CH_3OH or CH_3OCH_3

Chapter 3 Quiz

1. Identify the compounds shown below as either an ionic compound or a covalent compound:

   a. NF_3

   b. $FeCl_2$

   c. CH_3CH_2OH

   d. NaBr

   e. COS

   f. SO_2

2. Name the following compounds:

   a. V_2O_5

   b. HBr

   c. PbI_2

   d. SO_3

   e. RbCl

3. Draw the Lewis dot structure for the following compounds:

   a. HI

   b. OCS

c. acetylene, C_2H_2

d. CF_2Cl_2

e. CH_3NH_2

f. ClF

g. CH_3Cl

h. propane, C_3H_8

4. Predict the electron geometry, molecular shape, and bond angles for the following compounds.

   a. $CHClF_2$

   b. CF_4

c. C_2H_2

d. C_2H_4

e. H_2CO

f. $CH_3CH_2CH_2OH$

5. Which molecules in Question 4 are polar molecules? Which molecules in Question 4 are three-dimensional?

6. Which molecules in Question 4 exert only dispersion forces as an intermolecular force of attraction? Explain. Which molecules in Question 4 have hydrogen bonding? Which molecules in Question 4 exert dipole-dipole interactions?

7. Predict the electron geometry, molecular shape, and bond angles for the following compounds.
   a. CH_3CN

   b. $SiCl_4$

   c. C_3H_6

   d. H_2S

8. Which molecules in Question 7 are polar molecules?

9. Which molecules in Question 7 are three-dimensional?

10. What types of intermolecular forces of attraction are exerted by the molecules in Question 7?

Chapter 3

Answers to Additional Exercises

33 Adenosine, the brain to prepare for sleep

35 Epinephrine stimulates the production of dopamine. Dopamine makes you feel good.

37 There are ionic compounds and covalent compounds. Ionic compounds have ionic bonds and covalent compounds have covalent bonds.

39 An ionic bond is a strong force of attraction. Electrostatic attraction is another name for this kind of attraction.

41 A salt is an ionic compound.

43 A monoatomic ion is formed from a single atom, while a polyatomic ion is formed from a molecule.

45 Na^+. The sodium cation is a monoatomic ion.

47 a. polyatomic b. polyatomic c. monoatomic d. monoatomic e. polyatomic

49 a. Na^+, sodium cation, charge +1; O^{2-}, oxide, charge −2 b. Cu^{2+}, cupric ion, charge +2; O^{2-}, oxide, charge −2 c. Ag^+, silver cation, charge +1; Cl^-, chloride ion, charge −1 d. Zn^{2+}, zinc ion, charge +2; S^{2-}, sulfide, charge −2 e. Ga^{3+}, gallium ion, charge +3; As^{3-}, arsenide ion, charge −3

51 a. Ba^{2+}, charge +2; SO_4^{2-}, charge −2 b. Li^+, charge +1; CO_3^{2-}, charge −2 c. Mg^{2+}, charge +2; SO_4^{2-}, charge −2 d. K^+, charge +1; $H_2PO_4^-$, charge −1 e. Na^+, charge +1, NO_2^-, charge −1

53 a LiI, lithium (cation), iodide (anion)
b. RbF, rubidium (cation), fluoride (anion)
c. $CaBr_2$, calcium (cation), bromide (anion)
d. BaI_2, barium (cation), iodide (anion)
e. FeS, iron (cation), sulfide (anion)
f. Al_2O_3, aluminum (cation) oxide (anion)

55 a. strontium oxide b. potassium iodide c. sodium iodide d. lithium fluoride e. gallium oxide

57 b. potassium hydrogen phosphate

59 a. Na_3PO_4 b. NH_4Cl c. $Mg(OH)_2$ d. $NaHCO_3$

61 $Pb_3(PO_4)_2$

63 A molecule is a discrete entity composed of two or more nonmetal atoms held together by covalent bonds.

65 Ionic compounds are solids at room temperature and it takes more energy to melt or vaporize an ionic compound than covalent compounds. Covalent compounds can be found in all three states of matter at room temperature.

67 a. covalent b. ionic c. ionic

69 c. two nonmetals

71 When two atoms share electrons, the atoms achieve a full valence energy level. The shared electrons are in an orbital that encompasses both nuclei.

73 The octet rule states that, when writing a Lewis dot structure, the valence electrons are arranged so that each atom in period 2 and higher is surrounded by eight electrons.

75 a. ·Ċ· b. H· c. ·Ö· d. ·P̈·

77 a. The carbon atoms contain four pairs of bonding electrons and zero pairs of nonbonding electrons. b. The nitrogen atom has three pairs of bonding electrons and one pair of nonbonding electrons. c. Yes, every atom has an octet. d. There are single and double bonds in this molecule.

79 :Br̈—Br̈: This molecule is an element.

81 Selenium would have two bonding pairs and two nonbonding pairs. It is in the same family/group as oxygen and sulfur, so it has the same number of valence electrons.

83 a. :F̈–P̈–F̈: with :F̈: bonded below P

b. H–S̈–H

c. :O: double-bonded (=) to C, with H–C–H

d. H–Cl̈:

e. :O: double-bonded (=) to C, with H–Ö–C–Ö–H

85

```
    H
    |
H - C - C≡N :
    |
    H
```

87 sulfur trioxide

89 a. nitrogen oxide b. nitrogen dioxide c. nitrous oxide d. dinitrogen trioxide e. dinitrogen tetraoxide f. dinitrogen pentaoxide

91 a.

```
    H
    |
H - C - H
    |
    H
```

The model shown is a space-filling model.

b. H−C≡N: The model shown is a ball-and-stick model. The bond angle is 180°.

93 The molecule is linear if there are only two electron groups surrounding the central atom.

95 The electron geometry for four electron groups around a central atom is tetrahedral. The possible molecular shapes are tetrahedral, trigonal pyramidal, and bent. The bond angles for the three molecular shapes are 109.5°.

97 a. trigonal pyramidal

99 The bond angles in a bent shape can be either 120° or 109.5°. The bond angles of 120° come from a trigonal planar electron geometry, while the bond angles of 109.5° come from a tetrahedral electron geometry.

101 a.

```
    H   ..
    |
H - C - Cl:
    |   ..
   :Cl:
    ..
```

The electron geometry is tetrahedral. The molecular shape is tetrahedral and the bond angles are 109.5°.

b.

```
 ..  ..  ..
: I — N — I :
 ..   |   ..
     : I :
      ..
```

The electron geometry is tetrahedral. The molecular shape is trigonal pyramidal. The bond angles are 109.5°.

103 c. 180°, linear

105 a. The molecular shape is linear. The bond angle around the carbon atom is 180°. b. The molecular shape is tetrahedral. The bond angles around the phosphorus atom are 109.5°.

107 The electron geometry of this molecule is trigonal planar; there are three bonding groups and no nonbonding groups. The O—N—O bond angle is 120°.

109 a. H–P–H (with H below P; one lone pair on P) The molecular shape is trigonal pyramidal; there are three bonding groups and one nonbonding group; the bond angles are 109.5°.

b. :Cl: / :Cl–C–Cl: / :Cl: The molecular shape is tetrahedral; there are four bonding groups and no nonbonding groups; the bond angles are 109.5°.

c. :F–B–F: / :F: The molecular shape is trigonal planar; there are three bonding groups and no nonbonding groups; the bond angles are 120°.

d. :F–S–F: The molecular shape is bent; there are two bonding groups and two nonbonding groups; the bond angles are 109.5°.

e. O=C=S The molecular shape is linear; there are two bonding groups; the bond angle is 180°.

111 a. F F / F–C–C–O–C–F / Cl H H F b. The bond angle around each carbon atom is 109.5°; there are four bonding groups around each carbon atom. c. The bond angle around the oxygen atom is 109.5°; there are two bonding groups and two nonbonding groups around the oxygen atom. d. The molecular shape at each carbon center is tetrahedral. e. The molecular shape at the oxygen atom is bent.

113 The molecular shape is trigonal planar when there are three bonding groups and no nonbonding groups surrounding a central atom; the bond angles are 120°. The molecular shape is trigonal pyramidal when there are three bonding groups and one nonbonding group surrounding a central atom; the bond angles are 109.5°.

115 Based on VSEPR, the bond angles should be 109.5° because each atom is tetrahedral. Based on geometry, the angles within an equilateral triangle are 60°. This molecule is not commonly found in nature because the bond angles predicted by VSEPR and defined by geometry do not match.

117
1. There are three groups of electrons that surround the central carbon atom.
2. They are bonding electrons.
3. There are no nonbonding electrons around the central atom.
4. The electron geometry is trigonal planar.

5. No, they are not. The carbon on the left is tetrahedral; it has four bonding groups around it. The carbon on the right is trigonal planar; it has three bonding groups around it.

6. The C—C—C bond angle around the central carbon is 120°.

119 a. Oxygen is the more electronegative element. Oxygen lies farther to the right of carbon within the same period.

b. Oxygen is the more electronegative element. Oxygen and sulfur are in the same group, and oxygen lies above sulfur.

c. Fluorine is the more electronegative element. Fluorine lies farther to the right of lithium within the same period.

121 Electronegativity increases as you move from bottom to top in a group of elements.

123 The halogens are the most electronegative. They are the furthest to the right in the periodic table.

125 If the electronegativities of the two atoms in a covalent bond are similar, then the bond will be nonpolar. If the electronegativities are different, then the covalent bond will be a polar covalent bond.

127 a. Water is polar. H O H

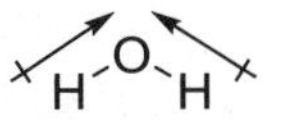

b. Ethanol is polar.

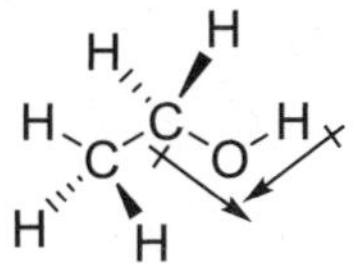

H H H C C O H H H

c. C_2H_4 is not polar. Carbon and hydrogen have similar electronegativities.

129 a. I_2 is nonpolar. The two atoms in the covalent bond are the same.

b. CH_4 is nonpolar. Carbon and hydrogen have similar electronegativities.

c. H—Br is polar. Hydrogen and bromine have different electronegativities.

131 A covalent bonding force occurs between two atoms within the same molecule and is much stronger than an intermolecular force of attraction. An intermolecular force of attraction occurs between atoms on different molecules.

133 The three types of intermolecular forces of attraction are: dispersion forces, dipole-dipole interactions, and hydrogen bonds. The strongest forces of attraction are hydrogen bonds. The weakest forces of attraction are dispersion forces.

135 Nonpolar molecules are attracted to each other by dispersion forces.

137 Dipole-dipole interactions are due to the interaction between permanent dipoles, while dispersion forces are due to the interaction between temporary dipoles.

139 H—O, H—N, and H—F bonds are capable of hydrogen bonding.

141 a. C_5H_{12} is nonpolar, so it should only exert dispersion forces.

b. Acetone is polar, so it should exert dispersion forces and dipole-dipole interactions. The dipole-dipole interactions are stronger.

c. Water is polar and has O—H bonds, so it should exert hydrogen bonding, as well as dispersion forces, and dipole-dipole interactions. The hydrogen bonding is the strongest intermolecular force.

143 Hydrogen bonding gives DNA a helical shape.

145 Estrogen binds to the estrogen receptor and activates several genes. This gene activation also stimulates the proliferation of breast-cancer cells.

147 The estrogen receptor is a large protein molecule. Estradiol fits the estrogen-binding site perfectly because it has a complimentary shape to the binding site—the cavity within the receptor.

149 When Tamoxifen binds to the receptor, it changes the shape of the receptor, preventing gene activation from occurring.

Chapter 4

Chemical Quantities and Chemical Reactions

Chapter Summary

In this chapter, you learned about the relationship between energy and chemical reactions, reaction rates and equilibrium. Fundamental to this relationship is understanding how to quantify matter. You observed how the composition of a substance is changed in a chemical reaction where bonds are broken and new bonds are formed. You learned how to write and balance chemical equations to reflect these changes. You discovered that energy transfer is intimately connected with chemical reactions. When chemical bonds are broken and formed in chemical reactions, energy is transferred to or from the surroundings. You also considered how fast chemical reactions occur, known as chemical kinetics. You learned about reversible reactions and what it means when a chemical reaction is at equilibrium. Understanding chemical quantities and chemical reactions will give you insight into the reactions in our bodies that produce the substances and the energy that make it possible for us to walk, talk, breathe, grow, and reproduce.

4.1 The Mole: Counting and Weighing Matter

The mole is an important counting unit in chemistry that gives us information about the number of atoms, ions, or molecules in a sample of matter. There is a quantitative relationship between the number of particles (atoms, ions, or molecules) and their mass.

In a Nutshell: Formula Mass and Molecular Mass

In Chapter 2, you learned that the average atomic mass of an element is given on the periodic table of elements, in atomic mass units (amu). The mass of one molecule, known as molecular mass, is the sum of all the individual atomic masses of its component atoms (given in the molecular formula). Similarly, the mass of one formula unit of an ionic compound, known as a formula mass, is calculated from the atomic masses of its component ions.

Worked Example #1

Calculate the molecular mass of acetaminophen (Tylenol), $C_8H_9NO_2$.

The substance is a covalent compound, so the terms "molecular mass" and "molecular formula" are used.

Atom type	*Number of atoms*		*Atomic mass (from Periodic Table)*		*Total*
C	*8*	×	*12.01*	=	*96.08 amu*
H	*9*	×	*1.008*	=	*9.072 amu*
N	*1*	×	*14.01*	=	*14.01 amu*
O	*2*	×	*16.00*	=	*32.00 amu*

The molecular mass = 96.08 + 9.072 + 14.01 + 32.00 = 151.16 amu/molecule. Remember to avoid rounding errors; do not round the atomic mass values derived from the Periodic Table until the end of the calculation.

Try It Yourself #1

Naproxen is a nonsteroidal anti-inflammatory drug. Calculate the molecular mass of naproxen, $C_{14}H_{14}O_3$.

Tools: The Periodic Table

Fill in the table below.

Atom type	*Number of atoms*		*Atomic mass (from Periodic Table)*		*Total*
		×		=	
		×		=	
		×		=	

The molecular mass: _______________

Worked Example #2

Calculate the formula mass of calcium phosphate, $Ca_3(PO_4)_2$.

The substance is an ionic compound, so the terms "formula mass" and "formula unit" are used. Remember, the atomic mass of atoms within parentheses must be multiplied by the subscript outside of the parentheses, as well as the subscript within the parentheses. In this example, there are eight oxygen atoms, as indicated by the subscript "4" following the symbol for oxygen and the subscript "2" after the parentheses.

Atom type	*Number of atoms*		*Atomic mass (from Periodic Table)*		*Total*
Ca	*3*	×	*40.08*	=	*120.24 amu*
P	*2*	×	*30.97*	=	*61.94 amu*
O	*8*	×	*16.00*	=	*128.00 amu*

The formula mass = 120.24 + 61.94 + 128.00 = 310.18 amu/formula unit.

Try It Yourself #2

Calculate the formula mass of calcium hydroxyapatite, $Ca_5(PO_4)_3OH$.

Tools: The Periodic Table

Fill in the table below.

Atom type	*Number of atoms*		*Atomic mass (from Periodic Table)*		*Total*
		×		=	
		×		=	
		×		=	
		×		=	

The formula mass: ________________________________

In a Nutshell: Molar Mass and Counting Particles

There is a relationship between the number of atoms in a sample and the mass of the sample, we can calculate the number of particles in the sample by weighing it. The chemist's counting unit is the mole, abbreviated mol, which represents 6.02×10^{23} atoms, ions, or molecules. The number 6.02×10^{23}, known as Avogadro's number, is the number in one mole. Since 1 mole is equal to 6.02×10^{23} (Avogadro's number), we can use it as a conversion between units of moles and number of particles. The mass of one mole of any element or compound is known as its molar mass, given in units of grams per mole. The molar mass of a compound has the same numerical value as its formula mass or molecular mass except the units are g/mol rather than amu.

Worked Example #3

Which has a greater mass: 1 mol of helium or 1 mol of oxygen? Explain why.

Using the Periodic Table, we see that a mole of helium has a mass of 4.00 g/mol and a mole of oxygen has a mass of 16.00 g/mol. Therefore, a mole of oxygen has a greater mass than a mole of helium. Oxygen has a greater average atomic mass.

Try It Yourself #3

Which has a greater mass: 1 mol of titanium or 1 mol of platinum? Explain why.

Tools: The Periodic Table

Average atomic mass titanium: ______________________

Average atomic mass platinum: ______________________

Molar mass titanium: __________________________

Molar mass platinum: __________________________

To interconvert between moles and ions, atoms, or molecules, we use the conversion $6.02 \times 10^{23} = 1$ mole.

Worked Example #4

How many silver atoms are there in 0.45 mol of silver?

Use dimensional analysis to set up the calculation using the correct from of the conversion factor (Avogadro's number) that will allow moles to cancel.

$$0.45\ \cancel{\text{mol}} \times \frac{6.02 \times 10^{23}\ \text{atoms}}{1\ \cancel{\text{mol}}} = 2.7 \times 10^{23}\ \text{atoms}$$

Try It Yourself #4

How many carbon dioxide molecules are there in 2.4 mol of CO_2?

Use dimensional analysis to set up the calculation using the correct from of the conversion factor (Avogadro's number) that will allow moles to cancel.

Interconverting between mass and moles is one of the most routine calculations performed in the chemical laboratory. The key to this calculation is recognizing that the molar mass of an element or compound is a conversion factor between the mass and the moles of the substance. The following steps are used to convert between mass and moles or moles and mass. 1) If the substance is an element, look up the molar mass of the element on the Periodic Table. If the substance is a compound, calculate the molar mass from the atomic masses of its constituents. 2) Use dimensional analysis to set up the calculation using the correct form of the conversion factor (molar mass) that will allow the units to cancel. Double check that the final answer has been rounded to the correct number of significant figures.

Worked Example #5

How many moles of sucrose are in 4.2 grams (1 teaspoon)? The molecular formula for sucrose is $C_{12}H_{22}O_{11}$.

For a compound, calculate the molar mass from its constituent masses, using the Periodic Table.

Calculate the molar mass of sucrose, $C_{12}H_{22}O_{11}$, from the atomic masses of its constituent atoms:

12 C + 22 H + 11 O = (12 × 12.01) + (22 × 1.008) + (11 × 16.00) = 342.30 g/mol

Use dimensional analysis to set up the calculation using the correct form of the conversion factor (molar mass) that will allow grams to cancel:

$$4.2\ \cancel{g} \times \frac{1\text{ mol}}{342.30\ \cancel{g}} = 0.012\text{ mol}$$

Try It Yourself #5

How many moles of ibuprofen, $C_{13}H_{18}O_2$, are in 400 mg?

Tools: The Periodic Table

Calculate the molar mass of ibuprofen from its constituent mass.

Use dimensional analysis to set up the calculation using the correct form of the conversion factor (molar mass) that will allow grams to cancel. You will also need to convert mg to g.

Worked Example #6

How many milligrams of gold are there in 0.15 mol of gold?

Look up the molar mass of the element on the Periodic Table.

The molar mass is 197.0 g/mol.

Use dimensional analysis to set up the calculation using the correct form of the conversion factor (molar mass) that will allow moles to cancel. We will also need to convert grams to milligrams using the metric conversion.

$$0.15\ \cancel{\text{mol Au}} \times \frac{197.0\ \cancel{\text{g}}}{1\ \cancel{\text{mol Au}}} \times \frac{10^3\text{mg}}{1\ \cancel{\text{g}}} = 3.0 \times 10^4\ \text{mg}$$

Try It Yourself #6

How many grams of copper are in 2.5 mol of copper?

Tools: The Periodic Table

Look up the molar mass of the element in the Periodic Table.

Use dimensional analysis to set up the calculation using the correct form of the conversion factor (molar mass) that will allow moles to cancel.

In a Nutshell: Converting Between Mass and Number of Particles

Sometimes, you may need to know how many atoms, ions, or molecules are present in a given sample. This type of calculation requires two steps, involving two conversion factors: Avogadro's number, which serves as a conversion factor between moles and the number of particles, and the molar mass the conversion between moles and mass of a substance.

Worked Example #7

How many sucrose ($C_{12}H_{22}O_{11}$) molecules are there in 25 g of sucrose?

You first need to convert g to moles and then moles to molecules.

Calculate the molar mass of sucrose, $C_{12}H_{22}O_{11}$, from the atomic masses of its constituent atoms:

12 C + 22 H + 11 O = (12 × 12.01) + (22 × 1.008) + (11 × 16.00) = 342.30 g/mol

There are two conversions here: g to moles (use the molar mass) and moles to molecules (use Avogadro's number).

1st conversion: $\frac{342.30\text{ g}}{1\text{ mol}}$ or $\frac{1\text{ mol}}{342.30\text{ g}}$

2nd conversion: $\frac{6.02\times10^{23}\text{ molecules}}{1\text{ mol}}$ or $\frac{1\text{ mol}}{6.02\times10^{23}\text{ molecules}}$

Using dimensional analysis, multiply the supplied unit by the appropriate form of the conversion factors (molar mass and Avogadro's number) that allow grams and moles to cancel. This calculation may be done in two sequential steps or in one step multiplying the appropriate conversion factors.

$$25\ \cancel{\text{g}}\times\frac{1\ \cancel{\text{mol}}}{342.30\ \cancel{\text{g}}}\times\frac{6.02\times10^{23}\text{ molecules}}{1\ \cancel{\text{mol}}}=4.4\times10^{22}\text{ molecules}$$

Try It Yourself #7

How many $CaCO_3$ molecules are there in 32.3 g of calcium carbonate?

Calculate the formula mass of $CaCO_3$ from its constituent atoms.

Express the conversions as two possible conversion factors.

Using dimensional analysis, multiply the supplied unit by the appropriate form of the conversion factors (molar mass and Avogadro's number, that allow grams and moles to cancel. This calculation may be done in two sequential steps or in one step multiplying the appropriate conversion factors.

Practice Problems for The Mole: Counting and Weighing Matter

1. A patient's blood test shows that she has 130 mg/dL of cholesterol ($C_{27}H_{46}O$) in her blood. How many moles of cholesterol are there in every deciliter of her blood?

2. How many grams of $Al(OH)_3$ are there in 1.7 mol $Al(OH)_3$?

3. Which has more atoms: half a mole of zinc or half a mole of chromium?

4. How many platinum atoms are in 2.34 mmol of platinum?

5. How many molecules of Na_3PO_4 are there in 8.23 g of Na_3PO_4?

6. What is the mass of 3.7×10^{20} carbon dioxide (CO_2) molecules?

4.2 The Law of Conservation of Mass and Balancing a Chemical Equation

In a Nutshell: The Law of Conservation of Mass

On the macroscopic scale, what we typically observe in a chemical reaction is one or more of the following: a change in color, formation of a precipitate, a change in temperature, the evolution of a gas (bubbles), a change in state, or a noticeable smell or aroma. On the atomic scale, particles are in constant motion, colliding with the walls of their container and occasionally colliding with one another. In most collisions, they bounce of one another unchanged. Occasionally, however, the kinetic energy of two colliding molecules or atoms is sufficient and occurs in the proper orientation for a chemical reaction to take place. In a chemical reaction, a change in the chemical structure occurs. The reacting substances are known as reactants, and the new substance(s) formed are known as products. In all chemical reactions, the number and types of atoms in the product(s) are the same as in the reactant(s). No new atoms have been created, and no atoms have been destroyed. This fundamental principle of chemical reactions is known as the law of conservation of mass: Matter can be neither created nor destroyed.

In a Nutshell: Writing a Chemical Equation

A chemical reaction is described by a chemical equation, which conveys the identity of the reactants and the products and their relative proportions. Other pertinent information about the reaction, such as the physical state of the reactants and products, may also be shown. In a chemical equation, the reactants and products are represented by their chemical formulas or chemical structure. A reaction arrow (→) is used to separate the reactants on the left side of the arrow from products on the right side of the arrow. A "+" sign is used to separate the individual reactants and the individual products when more than one is present. Whole number coefficients are placed before each chemical formula to indicate the molar ratio of the individual reactants and products. When no number is shown, the coefficient is assumed to be 1. The physical state of each compound or element is sometimes included in parentheses following the formula of the compound: solid (s), liquid (l), gas (g), or aqueous solution (aq).

Worked Example #7

For the equation below:

$C_6H_{12}O_6$ (aq) → 2 C_2H_6O (aq) + 2 CO_2 (g)

a. Indicate the reactant(s).
b. Indicate the product(s).
c. Indicate the ratio in which the reactants and products react.
d. If 2 moles of $C_6H_{12}O_6$ react, how many moles of C_2H_6O and how many moles of CO_2 are produced?
e. Indicate the physical state (s, l, g, aq) of each reactant and product.

Solutions

a. *The reactant is $C_6H_{12}O_6$. The reactant(s) appear to the left of the reaction arrow.*
b. *The products are C_2H_6O and CO_2. The products appear to the right of the reaction arrow.*
c. *One mole of $C_6H_{12}O_6$ reacts to produce 2 moles of C_2H_6O and 2 moles of CO_2.*
d. *If two moles of $C_6H_{12}O_6$ reacts, then 4 moles of C_2H_6O and 4 moles of CO_2 are produced. There is twice as much C_2H_6O and twice as much CO_2 formed as the number of moles of $C_6H_{12}O_6$, based on the 1, 2, and 2 coefficients.*
e. *$C_6H_{12}O_6$ and C_2H_6O are in the aqueous state, and CO_2 is in the gas state.*

Try It Yourself #7

For the equation below:

$Ca(OH)_2$ (aq) + 2 HCl (aq) → $CaCl_2$ (aq) + 2 H_2O (l)

a. Indicate the reactant(s).
b. Indicate the product(s).
c. Indicate the ratio in which the reactants and products react.
d. If 2 moles of $Ca(OH)_2$ react, how many moles of $CaCl_2$ are produced?
e. Indicate the physical state (s, l, g, aq) of each reactant and product.

Solutions

a. *The reactants:* ____________________________

b. *The products:* ____________________________

c. *Ratio of reactant to products:* ______________________

d. *Number of moles of $CaCl_2$ produced:* __________________

e. *Physical state $Ca(OH)_2$:* ____________

Physical state HCl: __________________

Physical state $CaCl_2$: ______________

Physical state H_2O: ____________________

In a Nutshell: Balancing a Chemical Equation

The law of conservation of mass requires that the total number of each type of atom on the reactant side of a chemical equation must always equal the total number of each type of atom on the product side of an equation. A chemical equation containing the proper coefficients such that there are an equal number of each type of atom on both sides of the equation is known as a balanced chemical equation. To balance a chemical equation, the correct number of coefficients must be inserted in the equation. To balance a chemical equation, follow these guidelines: 1) balance the equation one atom type at a time by inserting whole-number coefficients; and 2) check that the coefficients cannot be divided by a common factor (divisor). To balance the equation one atom at a time, systematically place coefficients in front of each reactant and product as necessary to arrive at an equal number of each type of atom on both sides of the equation. Then repeat the process with the next atom type. Never change the numerical subscripts in a chemical formula because this changes the identity of the compound. Remember that adding a coefficient alters the number of every type of atom in the formula. Multiply the coefficient by the subscript for a particular atom type to determine the number of that atom type. The balancing process is simplified if you balance last any atom type that appears in more than one compound or element on one side of the equation.

Worked Example #8

Balance the following combustion equation:

_____$C_7H_6O_2$ (s) + _____O_2 (g) → _____CO_2 (g) + _____H_2O (g)

Balance the equation one atom type at a time by inserting coefficients.

Balance carbon or hydrogen first. Balance oxygen last because it is present in both compounds on the product side. If you begin with carbon, insert coefficient 7 in front of CO_2.

____$C_7H_6O_2$ (s) + ____O_2 (g) → 7 CO_2 (g) + ____H_2O (g) *(carbon is now balanced)*

Balance hydrogen by inserting the coefficient 3 in front of H_2O.

____$C_7H_6O_2$ (s) + ____O_2 (g) → 7 CO_2 (g) + 3 H_2O (g) *(carbon and hydrogen are now balanced)*

To balance oxygen insert the coefficient, 15/2 or 7 ½, in front of oxygen as a temporary coefficient. Seventeen oxygen atoms are required on the left hand side to balance the seventeen total oxygen atoms on the right side of the equation (fourteen from 7 CO_2 and three from 3 H_2O). Only fifteen oxygen atoms are used for O_2 because there are two oxygen atoms in $C_7H_6O_2$.

$C_7H_6O_2$ (s) + 15/2 O_2 (g) → 7 CO_2 (g) + 3 H_2O (g) *(oxygen is now balanced)*

Turn the fraction into a whole number by multiplying every coefficient in the equation by 2:

2 $C_7H_6O_2$ (s) + 15 O_2 (g) → 14 CO_2 (g) + 6 H_2O (g)

Check that the coefficients cannot be divided by a common factor (divisor).

There is no common factor for the coefficients 2, 15, 14, and 6. The lowest set of whole number coefficients is in the balanced equation. The values in the second column now equal the values in the third column.

Kind of atom	*Number on reactant side*	*Number on product side*
C	*14*	*14*
H	*12*	*12*
O	*34*	*34*

Try It Yourself #8

Balance the following equation:

____ H_2O_2 (l) → ____H_2O (l) + ____O_2 (g)

Solution

Balance the equation one atom type at a time by inserting coefficients.

Check that the coefficients cannot be divided by a common factor (divisor).

From a balanced equation, it is possible to calculate the number of moles of product produced if the number of moles of each reactant is known. Since there is a relationship between mass and moles, we can determine the mass of products formed if we know the mass of reactants. This type of calculation is known as a stoichiometry calculation. Reaction stoichiometry calculations always begin with a balanced chemical equation because the coefficients in a chemical equation represent the molar ratio of each reactant and product. In order to perform a stoichiometry calculation to determine the number of moles of products formed, two steps are followed: Step 1) select the appropriate coefficients from the balanced chemical equation to construct a conversion factor between the substances of interest; and step 2) use dimensional analysis to convert moles of reactant to moles of product.

Worked Example #9

How many moles of hydrogen, H_2, would be required to react completely with 4.2 moles of nitrogen, N_2?

$$N_2(g) + 3\ H_2(g) \rightarrow 2\ NH_3(g)$$

Select the appropriate coefficients from the balanced equation to construct a conversion factor between the substances of interest.

$$\frac{1\ \text{mol nitrogen}}{3\ \text{mol hydrogen}}\ \text{or}\ \frac{3\ \text{mol hydrogen}}{1\ \text{mol nitrogen}}$$

Using dimensional analysis, multiply the supplied moles of nitrogen by the conversion factor from step 1 that allows moles of nitrogen to cancel:

$$4.2\ \cancel{\text{mol nitrogen}} \times \frac{3\ \text{mol hydrogen}}{1\ \cancel{\text{mol nitrogen}}} = 13\ \text{mol hydrogen}$$

Try It Yourself #9

How many moles of ammonia, NH_3, will be produced in Worked Example #9?

Select the appropriate coefficients from the balanced equation to construct a conversion factor between the substances of interest.

Using dimensional analysis, multiply the supplied moles of nitrogen by the conversion factor from step 1 that allows moles of nitrogen to cancel:

Practice Problems for The Law of Conservation of Mass and Balancing a Chemical Equation

1. For the following equations, indicate the reactants, the products, the ratio in which the reactants and products react, and the physical state of the reactants and products.

   a. $4\ Al\ (s) + 3\ O_2\ (g) \rightarrow 2\ Al_2O_3\ (s)$

   b. $CO_2\ (g) + 4\ H_2\ (g) \rightarrow CH_4\ (g) + 2\ H_2O\ (g)$

   c. $Zn\ (s) + 2\ HCl\ (aq) \rightarrow ZnCl_2\ (aq) + H_2\ (g)$

2. Balance the following equations.

   a. $Fe(OH)_3\ (aq) \rightarrow Fe_2O_3\ (aq) + H_2O\ (l)$

   b. $C_5H_{12}\ (l) + O_2\ (g) \rightarrow CO_2\ (g) + H_2O\ (g)$

   c. $Mg\ (s) + Mn_2O_3\ (aq) \rightarrow MgO\ (aq) + Mn\ (s)$

3. Balance the following equation and then determine how many grams of iron oxide are produced when 12.7 g of iron reacts with an excess of oxygen.
   $Fe (s) + O_2 (g) \rightarrow Fe_2O_3 (s)$

Extension Topic 4-1: Gram-A to Mole-A to Mole-B to Gram-B Stoichiometry Calculations

The coefficients in a chemical equation represent mole ratios of reactants and products, not mass ratios. However, since there is a relationship between moles of a substance and its mass, any number of moles can be converted into a mass using the molar mass of the substance as a conversion factor. Start with a balanced chemical equation. Step 1) convert the mass of the supplied substance into moles of supplied substance; step 2) convert the moles of supplied substance to moles of requested substance; and step 3) convert moles of requested substance, calculated in step 2, to mass of requested substance.

Worked Example #10

How many grams of carbon dioxide are produced when 5.0 g of butane reacts with an excess of oxygen?

$$2\ C_4H_{10} (g) + 13\ O_2 (g) \rightarrow 8\ CO_2 (g) + 10\ H_2O (g)$$

Butane

Convert the mass of the supplied substance into moles of supplied substance.
In this problem, butane is the supplied substance. Use the molar mass of butane as the conversion factor.

$$5.0\ \cancel{\text{g butane}} \times \frac{1\ \text{mol butane}}{58.11\ \cancel{\text{g butane}}} = 0.086\ \text{mol butane}$$

mass of butane molar mass of butane

Convert the moles of supplied substance to moles of requested substance.

Convert moles of butane to moles of carbon dioxide.

Use the coefficients in the balanced equation to set up a conversion factor between mol butane and mol carbon dioxide. Remember, when no number is shown, the coefficient is 1:

$$0.086\ \cancel{\text{mol butane}} \times \frac{8\ \text{mol carbon dioxide}}{2\ \cancel{\text{mol butane}}} = 0.34\ \text{mol carbon dioxide}$$

Convert moles of requested substance, calculated in step 2, to mass of requested substance.

Convert moles of carbon dioxide to grams of carbon dioxide.

Use the molar mass of carbon dioxide as a conversion factor.

$$0.34\ \cancel{\text{mol carbon dioxide}} \times \frac{44.01\ \text{g carbon dioxide}}{1\ \cancel{\text{mol carbon dioxide}}} = 15\ \text{g carbon dioxide}$$

mol of carbon dioxide molar mass of carbon dioxide

Therefore, 5.0 g of butane produces 15 g of carbon dioxide.

Try It Yourself #10

How many grams of water will be produced in Worked Example #10?

Convert the mass of the supplied substance into moles of supplied substance.

Molar mass of butane: ______________________________

Convert moles of requested substance, calculated in step 2, to mass of requested substance.

Convert the moles of supplied substance to moles of requested substance.

Convert moles of butane to moles of water.

Ratio of coefficients for butane and water: ____________________

Convert moles of water to grams of water.

Molar mass of water: __________________________

Grams of water produced: __________________________

Practice Problems for Gram-A to Mole-A to Mole-B to Gram-B Stoichiometry Calculations

1. Balance the following equation and then determine how many grams of iron oxide are produced when 12.7 g of iron reacts with an excess of oxygen.

   $Fe\ (s) + O_2\ (g) \rightarrow Fe_2O_3\ (s)$

2. How many grams of mercury are produced when 144.8 g of mercury (II) sulfide is heated in the presence of excess oxygen?

   $HgS\ (s) + O_2(g) \rightarrow Hg\ (l) + SO_2(g)$

3. How many grams of sulfur dioxide are produced when 144.8 g of mercury (II) sulfide is heated in the presence of excess oxygen in the reaction shown in Question 2?

4.3 Energy and Chemical Reactions

Recall that energy is defined as the capacity to do work or transfer heat. Work is the act of moving an object over a distance against an opposing force. Bioenergetics is the field of study concerned with the transfer of energy in reactions occurring in living cells.

In a Nutshell: Heat Energy

Energy is found in several forms—chemical, mechanical, and heat energy. The energy transferred during a chemical reaction commonly appears in the form of heat energy. Chemical reactions transfer heat energy by either absorbing from or releasing heat to the surroundings. In a chemical reaction, not all of the energy is transferred in the form of heat.

In chemistry, some reactions release heat and others absorb heat. Breaking a bond requires an input of energy, whereas making a bond results in an output of energy. Whether heat is absorbed or released in a chemical reaction depends on the difference in potential energy between the products and the reactants. The difference can appear in the form of heat on either the reactant or product side of the reaction. Thus, in a chemical reaction, energy is conserved, as stated in the first law of thermodynamics: Energy can be neither created nor destroyed.

The difference in potential energy between reactants and products can be measured and is known as the change in enthalpy, ΔH, where the Greek letter Δ in science always means "change." When the value for ΔH is negative (less than 0), it means that the reaction releases energy, which is known as an exothermic reaction. When the value for ΔH is positive (greater than 0), it means that the reaction absorbs energy, which is known as an endothermic reaction.

Exothermic and endothermic reactions can be illustrated in an energy diagram. In an energy diagram, the *y*-axis represents energy and the *x*-axis represents the progress of the reaction from reactants to products. In an exothermic reaction, products are lower in energy than reactants. In an endothermic reaction, the products are higher in energy than the reactants. The difference in energy, represented by a vertical arrow spanning the distance between the energy of reactants and the products, represents the magnitude of ΔH. In both endothermic and exothermic reactions, energy is never lost and never created. Energy can only be transformed.

In an endothermic reaction, the reactants are lower in energy than the products; therefore, energy is absorbed from the surroundings (the surroundings become cooler). The change in enthalpy is positive ($\Delta H > 0$). In an endothermic reaction, energy must be continuously supplied to sustain the reaction. In an exothermic reaction, the reactants are higher in energy than the products; therefore, heat is released to the surroundings (the surroundings become warmer). The change in enthalpy is negative ($\Delta H < 0$). It follows that if a reaction is exothermic, the reverse reaction is endothermic, by the same numerical value of ΔH, only the sign of ΔH (+ or −) changes.

Worked Example #11

For each of the following reactions, indicate whether it is an endothermic or exothermic reaction. Indicate whether the reactants or products are lower in energy.

a. $C_6H_8O_7(aq) + 3\ NaHCO_3(s) \rightarrow 3\ CO_2(g) + 3\ H_2O(l) + Na_3C_6H_5O_7(aq)$; energy is absorbed
b. $2\ SO_2\ (g) + O_2\ (g) \rightarrow 2\ SO_3\ (g)$ + heat

a. *Energy is absorbed, so the reaction is an endothermic reaction. The reactants are lower in energy than the products.*
b. *Energy is released; it is on the right hand side of the chemical equation. The products are lower in energy than the reactants.*

Try It Yourself #11

For each of the following reactions, indicate whether it is an endothermic or exothermic reaction. Indicate whether the reactants or products are lower in energy.

a. $N_2\ (g) + 2\ O_2\ (g)$ + heat $\rightarrow 2\ NO_2\ (g)$
b. $H_2O\ (l) \rightarrow H_2O\ (s)$ + heat

a. *Energy is:* ____________
   Reaction is: __________________

b. *Energy is:* _____________
   Reaction is: ______________________

In a Nutshell: Calorimetry

Calorimetry is a technique used to measure enthalpy changes (ΔH) in chemical reactions, using an apparatus known as a calorimeter. A bomb calorimeter is designed to measure heat released in a combustion reaction. The caloric content of food can be measured using a bomb calorimeter. The Caloric content of the three basic food molecules are as follows: Carbohydrates have 4 Cal/g, proteins have 4 Cal/g, and fats have 9 Cal/g. If we know both the mass and type of food (carbohydrate, protein, and fat), we can calculate the total caloric content of a particular food sample. Calculate the number of calories from each food type by multiplying the mass of each type of food by its caloric content and then add up these values to obtain the total caloric content of the food item. Food labels generally round the Calories to the nearest tens place.

Worked Example #12

A cup of 1% milk has 2.5 g of fat, 13 g of carbohydrates, and 8 g of protein. How many total Calories does a cup of 1% milk contain?

Determine the Calories supplied by each type of food.

Calories provided by fat: $2.5\ \cancel{g} \times \frac{9\ \text{Cal}}{\cancel{g}} = 23\ \text{Cal}$

Calories provided by carbohydrate: $13\ \cancel{g} \times \frac{4\ \text{Cal}}{\cancel{g}} = 52\ \text{Cal}$

Calories provided by protein: $8\ \cancel{g} \times \frac{4\ \text{Cal}}{\cancel{g}} = 32\ \text{Cal}$

Sum these values to obtain the total Calories supplied by milk:

23 Cal + 52 Cal + 32 Cal = 107 Cal or 110 Cal

Try It Yourself #12

A serving (1/2 cup) of oatmeal has 3 g of fat, 27 g of carbohydrate, and 5 g of protein. How many total Calories does a serving of oatmeal contain?

Determine the Calories supplied by each type of food.

Calories provided by fat

Calories provided by carbohydrate:

Calories provided by protein:

Total Calories supplied by oatmeal: ______________________

In a Nutshell: An Overview of Metabolism and Energy

The chemical reactions of the cell are referred to as biochemical reactions and typically occur as a sequence of reactions. A biochemical pathway is defined by a particular sequence of reactions. There are two basic types of biochemical pathways: catabolic and anabolic. Catabolic and anabolic pathways together are known as metabolism. Catabolic pathways convert large biomolecules, such as carbohydrates, proteins, and fats, to smaller molecules. Anabolic pathways build larger molecules, such as proteins, lipids, and DNA, from smaller molecules. Catabolic pathways release energy overall; while anabolic pathways absorb energy overall. The energy required to build molecules via anabolic pathways comes from the energy released in catabolic processes.

Practice Problems for Energy and Chemical Reactions

1. Are the following reactions endothermic or exothermic? Indicate whether ΔH greater or less than zero for each reaction.
   a. H_2O (s) + heat → H_2O (l)

   b. $C_2H_4 + 3\ O_2 \rightarrow 2\ CO_2 + 2\ H_2O$ + heat

2. One egg has 6 g of protein, 4.5 g of fat, and 1 g of carbohydrate. How many total Calories does one egg contain?

3. Which type of metabolic pathways are exothermic, releasing energy overall?

4.4 Reaction Kinetics

Another important aspect of chemical reactions is how fast they occur, a topic known as kinetics. The reaction rate—how fast a reaction proceeds—is determined by measuring either the consumption of reactants, or the formation of products over time.

In a Nutshell: Activation Energy, E_A

The minimum amount of energy that must be attained initially by the reactants for a reaction to proceed is known as the activation energy, E_A. Until the reactants acquire the necessary activation energy, they will only bounce off each other without reacting. The activation energy can be illustrated in an energy diagram. The activation energy is represented as a "hill" that reactants must climb in order to become products. The activation energy, E_A, for a reaction is proportional to the rate of a reaction. The lower the activation energy, E_A, the faster the reaction will become. Reactions that have low activation energies proceed faster than reactions that have high activation energies. An energy diagram with a greater activation energy (E_A) represents a slower reaction. The activation energy, E_A, for a reaction has no effect on the potential energy of either the reactants or the products and therefore no effect on ΔH. The activation energy, E_A, is related to the rate of a reaction, whereas the change in enthalpy, ΔH, is related to the difference in potential energy between the reactants and the products.

Worked Example #13

Does this energy diagram represent endothermic or exothermic reactions? Which reaction would you predict to proceed faster, a or b?

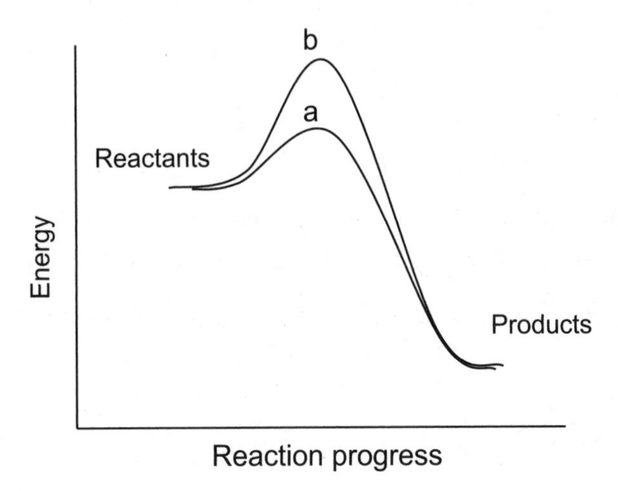

Solution

The energy diagram represents exothermic reactions. The products are lower in energy than the reactants. Reaction a would proceed faster because it has a lower activation energy, E_A.

Try It Yourself #13

Does this energy diagram represent endothermic or exothermic reactions? Which reaction would you predict to proceed faster, a or b?

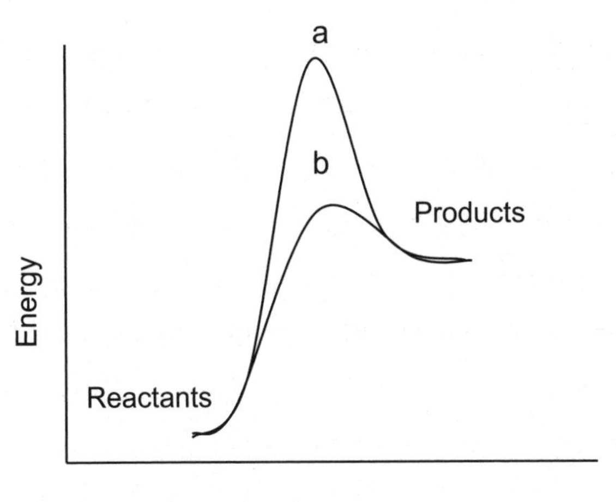

Solution

Products are (higher or lower) in energy than the reactants.

The reaction is: ____________________

The reaction with lower activation energy, E_A*:* ______________

The faster reaction is: ______________________________

In a Nutshell: Factors Affecting the Rate of a Reaction

The rate of a reaction is most influenced by three factors: concentration, temperature, and whether or not a catalyst is present. Generally, the rate of a reaction decreases as the concentration of reactants decreases because, as reactants are consumed, there are fewer reactant molecules that can collide, reducing the rate of product formation. Therefore, the rate of a reaction generally decreases over time.

As heat is added to a substance, its particles move faster. The greater kinetic energy of the particles allows reactant molecules or atoms to collide more frequently and with greater energy, increasing the likelihood that the reactants will attain the activation energy. For every 10 °C increase in temperature, the rate of reaction approximately doubles Conversely, if the temperature of a reaction is lowered, by removing heat, the rate decreases because the molecules have less kinetic energy and thus fewer collisions and fewer collisions with sufficient energy to attain the activation energy

A catalyst is a substance that increases the rate of a reaction by lowering the activation energy for the reaction but is itself not a reactant. A catalyst does not undergo a chemical change and can be reused over and over. In a chemical equation, the name of the catalyst is written above the reaction arrow. Coefficients, therefore, do not apply to catalysts. A catalyst, however, does not influence the change in enthalpy, ΔH, for a reaction. A catalyst only alters the pathway leading from reactants to products but not the potential energy of the reactants and products. Catalysts facilitate the collision process, making it easier for the reactants to come together. Biological catalysts are molecules are called enzymes.

Most enzymes are proteins, large biomolecules made up of carbon, hydrogen, nitrogen, oxygen, and sulfur atoms. Enzymes are capable of remarkable rate enhancements, ranging from 10^5 to 10^{17} faster than in the absence of the enzyme. An enzyme contains a pocket or groove, known as the active site, where the reactant molecule(s) bind. The enzyme reduces the freedom of motion available to the reactant(s), thereby forcing reactants into a spatial orientation conducive to reaction and lowering the activation energy for the reaction. You can usually recognize a reference to an enzyme because it ends in *-ase*. By convention, many enzyme names are italicized.

Worked Example #14

For the following reactions, state whether the indicated change in conditions would increase or decrease the rate of the reaction. Explain why.

a. $Zn\ (s) + 2\ HCl\ (aq) \rightarrow ZnCl_2\ (aq) + H_2\ (g)$

   More HCl is added to the reaction.

b. $I_2\ (s) + H_2\ (g) \rightarrow 2\ HI\ (g)$

   The temperature is decreased from 130 °C to 25 °C.

c. $2\ KClO_3\ (s) \rightarrow 2\ KCl\ (s) + 3\ O_2(g)$

   A catalyst, manganese dioxide, MnO_2 is added.

a. *The reaction rate increases because the concentration of one of the reactants was increased.*

b. *The reaction rate decreases because the temperature is lowered. At low temperatures, atoms and molecules have less kinetic energy, so there are fewer collisions between them.*

c. *The reaction rate increases because a catalyst has been added. The catalyst lowers the activation energy.*

Try It Yourself #14

For the following reactions, state whether the indicated change in conditions would increase or decrease the rate of the reaction. Explain why.

a. $2\ N_2O\ (g) \rightarrow 2\ N_2\ (g) + O_2\ (g)$

   A catalyst, gold, is added to the reaction.

b. $Zn\ (s) + 2\ HCl\ (aq) \rightarrow ZnCl_2\ (aq) + H_2\ (g)$

   The temperature of the reaction is increased.

a. *Reaction rate increases or decreases.*

b. *Reaction rate increases or decreases.*

Practice Problems for Reaction Kinetics

1. Which part of an energy diagram determines the rate of a reaction?

2. Name three ways to increase the rate of a reaction.

3. *Hexokinase* is an enzyme that catalyzes one of the steps in the metabolic pathway, glycolysis. Is *hexokinase* chemically altered when it catalyzes the reaction?

4.5 Chemical Equilibrium

When a reaction "goes to completion," all of the reactants are completely converted into products. At the end of the reaction, only products are present; no reactants remain. Many reactions are reversible, which means that both the forward and the reverse reactions occur simultaneously. In other words, reactants combine to form products—the forward reaction—and products combine to form reactants—the reverse reaction. At the end of the reaction, both reactants and products are present.

At equilibrium, the forward and reverse reactions continue to proceed, but since the rate of the forward reaction is the same as the rate of the reverse reaction, the concentration of reactants and products no longer changes. Hence, equilibrium is often called a dynamic equilibrium.

In a Nutshell: Le Châtelier's Principle

A reversible reaction at equilibrium can be disturbed by adding or removing one of the reactants or products or by adding or removing heat (a temperature change) from the reaction. Le Châtelier's principle predicts that a reversible reaction at equilibrium will respond to a disturbance in such a way as to counteract the disturbance and establish a new equilibrium. When a reaction at equilibrium is disturbed, the reaction responds by shifting in a direction that restores equilibrium: either the forward direction (to the right) or the reverse direction (to the left.) For example, if more reactant is added to a reversible reaction, the rate of the forward reaction will increase until a new equilibrium is established.

If the temperature of a reaction at equilibrium is changed by adding or removing heat (cooling), the direction of the shift depends on whether the reaction is exothermic or endothermic. In an exothermic reaction, heat is released and appears a product; therefore, adding heat to the reaction will cause it to shift to the left, so as to consume the added heat. If an exothermic reaction is cooled, the reaction will shift to the right, so as to restore some of the heat that has been removed. The results are the opposite for an endothermic reaction, for the same reasons.

Worked Example #15

Consider the reversible reaction shown below, which occurs in ethylene glycol poisoning:

$$\underset{\text{oxalic acid}}{HO{-}\overset{\overset{O}{\|}}{C}{-}\overset{\overset{O}{\|}}{C}{-}OH} + \underset{\text{water}}{H_2O} \rightleftharpoons \underset{\text{oxalate ion}}{HO{-}\overset{\overset{O}{\|}}{C}{-}\overset{\overset{O}{\|}}{C}{-}O^-} + \underset{\text{hydronium ion}}{H_3O^+}$$

a. What substances are present at equilibrium?
b. At equilibrium, are the concentrations of oxalic acid and oxalate ion constant or changing?
c. How will the equilibrium shift if oxalic acid is removed from the solution?
d. How the equilibrium shift if more water is added to the solution?

a. *Oxalic acid, water, oxalate ion, and hydronium ion are all present at equilibrium.*

b. *The concentrations of oxalic acid and oxalate ion are constant at equilibrium.*

c. *If oxalic acid is removed, the concentration of one of the reactants decreases; therefore, the equilibrium will shift to the left.*

d. *If more water is added, the concentration of one of the reactants increases; therefore, the equilibrium will shift to the right.*

Try It Yourself #15

Phenylacetic acid builds up in the blood of people with phenylketonuria. In aqueous solution, phenylacetic acid undergoes the following reversible reaction:

$$C_6H_5-CH_2-COOH + H_2O \rightleftharpoons C_6H_5-CH_2-COO^- + H_3O^+$$

phenylacetic acid | water | phenylacetate ion | hydronium ion

a. What substances are present at equilibrium?

b. At equilibrium, are the concentrations of phenylacetic acid and phenylacetate ion constant or changing?

c. How will the equilibrium shift if phenylacetate ion is removed from the solution?

d. How will the equilibrium shift if water is removed from the solution?

a. *The substances present at equilibrium are:* ____________________

b. *The concentrations are:* ____________________

c. *Phenylacetate ion is a reactant or a product?* ____________________

The concentration of phenylacetate: __________*; therefore, the reaction shifts:*

____________________.

d. *Water is a reactant or a product?* ____________________

The concentration of water: __________*; therefore, the reaction shifts:*

____________________.

Chapter 4 Quiz

1. How many milligrams are in 0.017 mol of celecoxib, $C_{17}H_{14}F_3N_3O_2S$, the active ingredient of Celebrex?

2. How many moles are in 250 mg of the potassium salt of penicillin V, $C_{16}H_{17}KN_2O_5S$?

3. For the following unbalanced chemical equations, identify the reactants and the products and the physical state of the reactants and the products. Balance the chemical equations.
   a. Mg (s) + O_2 (g) → MgO (s)

   b. C_8H_{18} (l) + O_2 (g) → CO_2 (g) + H_2O (g)

   c. $Mg(OH)_2$ (aq) + HCl (aq) → $MgCl_2$ (aq) + H_2O (l)

   d. P_4O_{10} (s) + H_2O (l) → H_3PO_4 (l)

4. Balance the following chemical equation and then determine how many grams of calcium oxide are formed when excess oxygen reacts with 4.35 g of calcium.
   $Ca\ (s) + O_2\ (g) \rightarrow CaO\ (s)$

5. Calculate the total calories in the following foods:
   a. A slice of apple pie containing 19.4 g of fat, 57.5 g of carbohydrate, and 3.7 g of protein

   b. A slice of pizza containing 9.8 g of fat, 33.6 g of carbohydrate, and 12.3 g of protein

   c. A blueberry bagel containing 1.5 g of fat, 33.6 g of carbohydrate, and 9 g of protein

6. Are the following reactions exothermic or endothermic?
   a. $H_2O\ (l) + \text{heat} \rightarrow H_2O\ (g)$

   b. $2\ Na\ (s) + Cl_2\ (g) \rightarrow 2\ NaCl\ (s) + \text{heat}$

7. Draw energy diagrams for the reactions shown in Questions 6a and 6b and include labels for reactants, products, and ΔH.

8. Which of the following will increase the rate of a reaction?
   a. Adding a catalyst

   b. Decreasing the concentration of a reactant

   c. Decreasing the temperature

9. Reaction "X" is an exothermic reaction that occurs twice as fast in the presence of a catalyst. Draw side-by-side energy diagrams for reaction "X" occurring with and without a catalyst.

10. Which metabolic pathways create larger molecules?

Chapter 4

Answers to Additional Exercises

43 b and c

45 Acetaldehyde is produced in the first step when ethanol is metabolized in the liver.

47 *Alcohol dehydrogenase* catalyzes the conversion of ethanol into acetaldehyde.

49 Alcohol contains “empty” calories because it contains no nutrients.

51 The formula mass of a compound is the sum of the mass of all the ions in a formula unit.

53 25 C + 34 H + 6 O = (25 ×12.01) + (34 × 1.008) + (6 × 16.00) = 430.52 amu/molecule

55 1 Mg + 1 S + 4 O = (1 × 24.31) + (1 × 32.07) + (4 ×16.00) = 120.38 amu/formula unit

57 Avogadro’s number is the number of items in one mole.

59 1×10^{100} eggs in a googol of eggs

61 One mole of nickel atoms is 58.69 g.

63 a. 29 C + 30 H + 6 N + 6 O = (29 × 12.01) + (30 × 1.008) + (6 × 14.01) + (6 × 16.00) = 558.59 g/mol b. 16 C + 14 H + 3 F + 3 N + 2 O + 1 S = (16 × 12.01) + (14 × 1.008) + (3 × 19.00) + (3 × 14.01) + (2 × 16.00) + (1 × 32.07) = 369.37 g/mol

65 a. 1 Al + 3 O + 3 H = 26.99 + (3 × 16.00) + (3 × 1.008) = 78.00 g/mol

67 One mole of feathers has the smaller mass. They have an equal number.

69 One mole of zinc has greater mass.

71 The molar mass of a compound is the same as the numerical value as its formula mass except that the units are g/mol rather than amu/formula unit.

73 Use dimensional analysis to set up the calculation using the correct form of the conversion factor (Avogadro’s number) that will allow molecules to cancel:

$$8.65 \times 10^{18} \cancel{\text{crestor molecules}} \times \frac{1 \text{ mol}}{6.02 \times 10^{23} \cancel{\text{crestor molecules}}} = 1.44 \times 10^{-5} \text{ mol}$$

75 Use dimensional analysis to set up the calculation using the correct form of the conversion factor (Avogadro’s number) that will allow formula units to cancel:

$$1.45 \times 10^{20} \cancel{\text{MgSO}_4 \text{ formula units}} \times \frac{1 \text{ mol}}{6.02 \times 10^{23} \cancel{\text{MgSO}_4 \text{ formula units}}} = 2.41 \times 10^{-4} \text{ mol}$$

77 For parts a, b, c, and d, first, calculate the molar mass of H_2O_2. Then use dimensional analysis to set up the calculation using the correct form of the conversion factor (molar mass) that will allow moles to cancel.

For a compound, calculate the molar mass from its constituent masses, using the Periodic Table. Calculate the molar mass of hydrogen peroxide, H_2O_2, from the atomic masses of its constituent atoms:

2 H + 2 O = (2 × 1.008) + (2 × 16.00) = 34.02 g/mol

Set up the calculation so that the supplied units cancel:

a. $1.0 \cancel{\text{mol}} \times \frac{34.02\text{ g}}{1 \cancel{\text{mol}}} = 34\text{ g}$

b. $0.50 \cancel{\text{mol}} \times \frac{34.02\text{ g}}{1 \cancel{\text{mol}}} = 17\text{ g}$

c. $10. \cancel{\text{mol}} \times \frac{34.02\text{ g}}{1 \cancel{\text{mol}}} = 340\text{ g}$

d. $0.678 \cancel{\text{mol}} \times \frac{34.02\text{ g}}{1 \cancel{\text{mol}}} = 23.1\text{ g}$

79 First, find the molar mass of calcium on the periodic table. The molar mass of calcium is 40.08 g/mol.

Then use dimensional analysis to set up the calculation using the correct form of the conversion factor (molar mass) that will allow grams to cancel. The molar mass of is in units of *grams* per mole, but the supplied unit is given in *milli*grams. So the calculation will also require dimensional analysis to convert milligrams to grams:

$$8.8 \cancel{\text{mg calcium}} \times \frac{1 \cancel{\text{g calcium}}}{10^3 \cancel{\text{mg calcium}}} \times \frac{1\text{ mol calcium}}{40.08 \cancel{\text{g calcium}}} = 2.2 \times 10^{-4}\text{ mol calcium}$$

81 First, find the molar mass of sodium on the periodic table. The molar mass of sodium is 22.99 g/mol.

Then use dimensional analysis to set up the calculation using the correct form of the conversion factor (molar mass) that will allow moles to cancel. The molar mass of is in units of grams per *mole*, but the supplied unit is given in *milli*mole. So the calculation will also require dimensional analysis to convert millimole to mole:

$$145 \cancel{\text{mmol sodium}} \times \frac{1 \cancel{\text{mol sodium}}}{10^3 \cancel{\text{mmol sodium}}} \times \frac{22.99\text{ g sodium}}{1 \cancel{\text{mol sodium}}} = 3.33\text{ g sodium}$$

83 This type of calculation requires two steps using two conversion factors: Avogadro's number and molar mass. The first step will be to convert the mass of acetaminophen into moles of acetaminophen and then convert the moles of acetaminophen into the number of acetaminophen molecules.

First, calculate the molar mass of acetaminophen, $C_8H_9NO_2$, from the atomic masses of its constituent atoms:

8 C + 9 H + 1 N + 2 O = (8 × 12.01) + (9 × 1.008) + (1 × 14.01) + (2 × 16.00) = 151.16 g/mol

Then, using dimensional analysis, multiply the two conversion factors so that the units cancel. The molar mass of is in units of *grams* per mole, but the supplied unit is given in *milli*grams. So the calculation will also require dimensional analysis to convert milligrams to grams:

$$325\ \cancel{\text{mg}} \times \frac{1\ \cancel{\text{g}}}{10^3\ \cancel{\text{mg}}} \times \frac{1\ \cancel{\text{mol}}}{151.16\ \cancel{\text{g}}} \times \frac{6.02 \times 10^{23}\,\text{molecules}}{1\ \cancel{\text{mol}}} = 1.29 \times 10^{21}\ \text{molecules}$$

85 This type of calculation requires two steps using two conversion factors: Avogadro's number and molar mass. The first step will be to convert the mass of silver nitrate into moles of silver nitrate and then convert the moles of silver nitrate into the number of silver nitrate formula units.

First, calculate the molar mass of silver nitrate, $AgNO_3$, from the atomic masses of its constituent atoms:

1 Ag + 1 N + 3 O = (1 × 107.9) + (1 × 14.01) + (3 × 16.00) = 169.91 g/mol

Then, using dimensional analysis, multiply the two conversion factors so that the units cancel:

$$0.70\ \cancel{\text{g AgNO}_3} \times \frac{1\ \cancel{\text{mol AgNO}_3}}{169.91\ \cancel{\text{g AgNO}_3}} \times \frac{6.02 \times 10^{23}\ \text{formula units}}{1\ \cancel{\text{mol AgNO}_3}} = 2.5 \times 10^{21}\ \text{formula units}$$

87 The calculations in this problem will use two conversion factors: Avogadro's number and molar mass. The first step will be to convert the mass of lead into moles of lead and then convert the moles of lead into the number of lead atoms.

First, find the molar mass of lead, Pb, in the periodic table. The molar mass of lead is 207.2 g/mol:

Then use dimensional analysis to set up the calculation using the correct form of the conversion factor (molar mass) that will allow grams to cancel. The molar mass of is in units of *grams* per mole, but the supplied unit is given in *micro*grams. So the calculation will also require dimensional analysis to convert micrograms to grams:

$$5.0\ \cancel{\text{mcg lead}} \times \frac{1\ \cancel{\text{g lead}}}{10^{6}\ \cancel{\text{mcg lead}}} \times \frac{1\ \text{mol lead}}{207.2\ \cancel{\text{g lead}}} = 2.4\ \times\ 10^{-8}\ \text{mol lead}$$

Then use dimensional analysis to set up the calculation using the correct form of the conversion factor (Avogadro's number) that will allow moles to cancel

$$2.4\ \times\ 10^{-8}\ \cancel{\text{mol lead}} \times \frac{6.02\ \times\ 10^{23}\text{atoms}}{1\ \cancel{\text{mol lead}}} = 1.4\ \times\ 10^{16}\ \text{atoms}$$

89 a, b, c, and d. Changes in color, changes in temperature, the formation of a precipitate, and bubbles forming with an evolution of a gas can all be observed on a macroscopic scale when a chemical reaction occurs.

91 The abbreviation (s) indicates that the substance is in the solid state. The abbreviation (g) indicates that the substance is in the gaseous state. The abbreviation (aq) indicates that the substance is in the aqueous state.

93 Matter cannot be created nor destroyed in a chemical reaction. It is an application of the law of conservation of matter.

95 a. Assess the equation.

Kind of atom	Number on reactant side	Number on product side
C	6	1
H	14	2
O	2	3

The equation is unbalanced because the values in the reactant column do not equal the values in the product column for all types of atoms.

Balance the equation one atom type at a time by inserting coefficients.
Begin by balancing carbon or hydrogen first. Balance oxygen last because it is present in both compounds on the product side. If you begin with carbon, insert the coefficient 6 in front of CO_2.

C_6H_{14} (l) + O_2 (g) → 6 CO_2 (g) + H_2O (l) (carbon is now balanced)
Balance hydrogen by inserting the coefficient 7 in front of H_2O.
C_6H_{14} (l) + O_2 (g) → 6 CO_2 (g) + 7 H_2O (l) (carbon and hydrogen are now balanced)
To balance oxygen, insert the coefficient 19/2 in front of oxygen as a temporary coefficient.
C_6H_{14} (l) + 19/2 O_2 (g) → 6 CO_2 (g) + 7 H_2O (l) (oxygen is now balanced)
Turn the fractions into a whole number by multiplying every coefficient in the equation by 2:
2 C_6H_{14} (l) + 19 O_2 (g) → 12 CO_2 (g) + 14 H_2O (l)

Check that the coefficients cannot be divided by a common factor (divisor).
There is no common factor for the coefficients 2, 19, 12, and 14. The lowest set of whole number coefficients is in the balanced equation. The values in the second column now equal the values in the third column.

Kind of atom	Number on reactant side	Number on product side
C	12	12
H	28	28
O	38	38

b. Assess the equation.

Kind of atom	Number on reactant side	Number on product side
C	2	1
H	6	2
O	1	3

The equation is unbalanced because the values in the reactant column do not equal the values in the product column for all types of atoms.

Balance the equation one atom type at a time by inserting coefficients.
Begin by balancing carbon or hydrogen first. Balance oxygen last because it is present in both compounds on the product side. If you begin with carbon, insert the coefficient 2 in front of CO_2.

C_2H_6O (g) + O_2 (g) → 2 CO_2 (g) + H_2O (l) (carbon is now balanced)
Balance hydrogen by inserting the coefficient 3 in front of H_2O.
C_2H_6O (g) + O_2 (g) → 2 CO_2 (g) + 3 H_2O (l) (carbon and hydrogen are now balanced)
To balance oxygen, insert the coefficient 3 in front of oxygen.
C_2H_6O (g) + 3 O_2 (g) → 2 CO_2 (g) + 3 H_2O (l) (oxygen is now balanced)

Check that the coefficients cannot be divided by a common factor (divisor).
There is no common factor for the coefficients 1, 3, and 2. The lowest set of whole number coefficients is in the balanced equation. The values in the second column now equal the values in the third column.

Kind of atom	Number on reactant side	Number on product side
C	2	2
H	6	6
O	7	7

c. Assess the equation.

Kind of atom	Number on reactant side	Number on product side
C	3	1
H	6	2
O	2	3

The equation is unbalanced because the values in the reactant column do not equal the values in the product column for all types of atoms.

Balance the equation one atom type at a time by inserting coefficients.

Begin by balancing carbon or hydrogen first. Balance oxygen last because it is present in both compounds on the product side. If you begin with carbon, insert the coefficient 3 in front of CO_2.

C_3H_6 (g) + O_2 (g) → 3 CO_2 (g) + H_2O (l) (carbon is now balanced)

Balance hydrogen by inserting the coefficient 3 in front of H_2O.

C_3H_6 (g) + O_2 (g) → 3 CO_2 (g) + 3 H_2O (l) (carbon and hydrogen are now balanced)

To balance oxygen, insert the coefficient 9/2 in front of oxygen as a temporary coefficient.

C_3H_6 (g) + 9/2 O_2 (g) → 3 CO_2 (g) + 3 H_2O (l) (oxygen is now balanced)

Turn the fractions into a whole number by multiplying every coefficient in the equation by 2:

2 C_3H_6 (g) + 9 O_2 (g) → 6 CO_2 (g) + 6 H_2O (l)

Check that the coefficients cannot be divided by a common factor (divisor). There is no common factor for the coefficients 2, 9, and 6. The lowest set of whole number coefficients is in the balanced equation. The values in the second column now equal the values in the third column.

Kind of atom	Number on reactant side	Number on product side
C	6	6
H	12	12
O	18	18

97 c. For every 12 moles of iron that react, 6 moles of iron(III) oxide are produced.

E4.5 Assess the equation.

Kind of atom	Number on reactant side	Number on product side
Al	1	2
Cu	1	1
O	1	3

The equation is unbalanced because the values in the reactant column do not equal the values in the product column for all types of atoms.

Balance the equation one atom type at a time by inserting coefficients.

Begin by balancing aluminum, copper, or oxygen first. If you begin with aluminum, insert the coefficient 2 in front of Al.

2 Al (s) + CuO (s) → Al_2O_3 (s) + Cu (s) (aluminum is now balanced)

Balance oxygen by inserting the coefficient 3 in front of CuO.

2 Al (s) + 3 CuO (s) → Al_2O_3 (s) + Cu (s) (aluminum and oxygen are now balanced)

To balance copper, insert the coefficient 3 in front of copper.

2 Al (s) + 3 CuO (s) → Al_2O_3 (s) + 3 Cu (s) (copper is now balanced)

Check that the coefficients cannot be divided by a common factor (divisor).

There is no common factor for the coefficients 2, 3, and 1. The lowest set of whole number coefficients is in the balanced equation. The values in the second column now equal the values in the third column.

Kind of atom	Number on reactant side	Number on product side
Al	2	2
Cu	3	3
O	3	3

Convert the known mass of reactant to moles of reactant.

Use the molar mass of copper oxide as the conversion factor.

$$10.5 \cancel{\text{ g copper oxide}} \times \frac{1 \text{ mol copper oxide}}{79.55 \cancel{\text{ g copper oxide}}} = 0.132 \text{ mol copper oxide}$$

mass of copper oxide molar mass of copper oxide

Convert moles of copper oxide to moles of copper.

Use the coefficients in the balanced equation to set up a conversion factor between mol copper oxide and mol copper.

$$0.132 \cancel{\text{ mol copper oxide}} \times \frac{3 \text{ mol copper}}{3 \cancel{\text{ mol copper oxide}}} = 0.132 \text{ mol copper}$$

Convert moles of copper to grams of copper.

Use the molar mass of copper as a conversion factor.

$$0.132 \text{ \sout{mol copper}} \times \frac{63.55 \text{ g copper}}{1 \text{ \sout{mol copper}}} = 8.39 \text{ g copper}$$

mol of copper   molar mass of copper

Therefore, 10.5 g of copper oxide produces 8.39 g of copper.

99 Bioenergetics is the study of energy transfer in biological cells.

101 The difference in potential energy between reactants and products can be measured and is known as the change in enthalpy of the reaction, ΔH.

103 In an exothermic reaction, the products are lower in energy than the reactants. If the reaction is reversed, the products become the reactants and the reactants the products. In the reverse reaction, the products are higher in energy than the reactants; therefore, it is an endothermic reaction.

105
a. Exothermic. Heat is a product of the reaction.
b. Exothermic. Heat is a product of the reaction.
c. Endothermic. Heat needs to be added to the reaction.
d. Endothermic. Heat needs to be added to the reaction.

107 A calorimeter is used to measure the caloric content of substances.

109
a. Determine the Calories supplied by each type of biomolecule.

Calories provided by fat: $71 \text{ \sout{g}} \times \frac{9 \text{ Cal}}{\text{\sout{g}}} = 639 \text{ Cal}$

Calories provided by carbohydrate: $28 \text{ \sout{g}} \times \frac{4 \text{ Cal}}{\text{\sout{g}}} = 112 \text{ Cal}$

Calories provided by protein: $27 \text{ \sout{g}} \times \frac{4 \text{ Cal}}{\text{\sout{g}}} = 108 \text{ Cal}$

Sum these values to obtain the total Calories supplied by almonds:

639 Cal + 112 Cal + 108 Cal = 859 Cal or 860 Cal

b. Determine the Calories supplied by each type of biomolecule.

Calories provided by fat: $1 \text{ \sout{g}} \times \frac{9 \text{ Cal}}{\text{\sout{g}}} = 9 \text{ Cal}$

Calories provided by carbohydrate: $27 \text{ \sout{g}} \times \frac{4 \text{ Cal}}{\text{\sout{g}}} = 108 \text{ Cal}$

Calories provided by protein: $1\ \cancel{g} \times \frac{4\ \text{Cal}}{\cancel{g}} = 4\ \text{Cal}$

Sum these values to obtain the total Calories supplied by the banana:

9 Cal + 108 Cal + 4 Cal = 121 Cal or 120 Cal

c. Determine the Calories supplied by each type of biomolecule.

Calories provided by fat: $9\ \cancel{g} \times \frac{9\ \text{Cal}}{\cancel{g}} = 81\ \text{Cal}$

Calories provided by carbohydrate: $0\ \cancel{g} \times \frac{4\ \text{Cal}}{\cancel{g}} = 0\ \text{Cal}$

Calories provided by protein: $7\ \cancel{g} \times \frac{4\ \text{Cal}}{\cancel{g}} = 28\ \text{Cal}$

Sum these values to obtain the total Calories supplied by the cheddar cheese:

81 Cal + 0 Cal + 28 Cal = 109 Cal or 110 Cal

d. Determine the Calories supplied by each type of biomolecule.

Calories provided by fat: $10\ \cancel{g} \times \frac{9\ \text{Cal}}{\cancel{g}} = 90\ \text{Cal}$

Calories provided by carbohydrate: $23\ \cancel{g} \times \frac{4\ \text{Cal}}{\cancel{g}} = 92\ \text{Cal}$

Calories provided by protein: $2\ \cancel{g} \times \frac{4\ \text{Cal}}{\cancel{g}} = 8\ \text{Cal}$

Sum these values to obtain the total Calories supplied by the glazed donut:

90 Cal + 92 Cal + 8 Cal = 190 Cal

e. Determine the Calories supplied by each type of biomolecule.

Calories provided by fat: $4\ \cancel{g} \times \frac{9\ \text{Cal}}{\cancel{g}} = 36\ \text{Cal}$

Calories provided by carbohydrate: $0\ \cancel{g} \times \frac{4\ \text{Cal}}{\cancel{g}} = 0\ \text{Cal}$

Calories provided by protein: $22\ \cancel{g} \times \frac{4\ \text{Cal}}{\cancel{g}} = 88\ \text{Cal}$

Sum these values to obtain the total Calories supplied by the swordfish:

36 Cal + 0 Cal + 88 Cal = 124 Cal or 120 Cal

111 $\frac{2\ \cancel{\text{miles}}}{1\ \cancel{\text{trip}}} \times \frac{2\ \cancel{\text{trips}}}{\cancel{\text{day}}} \times 5\ \cancel{\text{days}} \times \frac{40\ \text{Cal}}{1\ \cancel{\text{mile}}} = 800\ \text{Cal}$

113 100% − 15% = percent of fat = 85%

$$0.85 \times 2.0 \text{ lbs} \times \frac{454 \text{ g}}{1 \text{ lb}} \times \frac{9 \text{ Cal}}{1 \text{ g}} = 7.0 \times 10^3 \text{ Cal}$$

115 The two types of metabolic reactions are anabolic reactions and catabolic reactions.

117 a and d Catabolic reactions release energy and break down larger molecules into smaller molecules.

119 $C_5H_{12} + 8\ O_2 \rightarrow 5\ CO_2 + 6\ H_2O$ + heat

121 Calorimetry is used to determine the number of calories (amount of energy) in foods. The amount of energy in a food is measured as heat energy. This heat energy is what fuels our bodies so we can walk, talk, move, etc.

123 The activation energy is the amount of energy that must be attained by the reactants in order for the reaction to occur. If the reactants do not have the required activation energy, they will only bounce off each other without reacting.

125 a. It illustrates an endothermic reaction. The products are higher in energy than the reactants.

b.

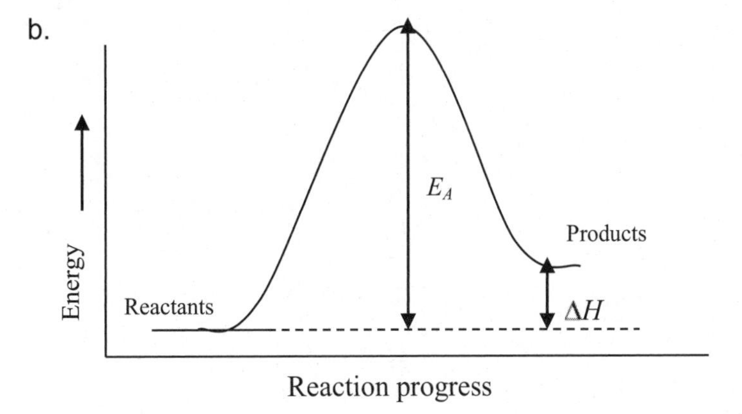

c. In the presence of a catalyst, the reaction curve will look like this.

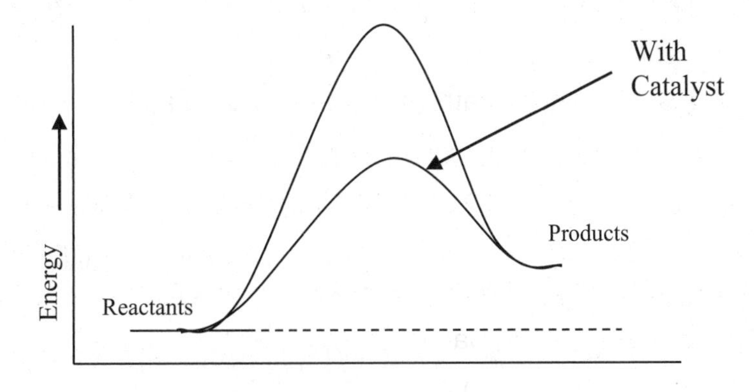

127 A catalyst does not affect the value of ΔH. A catalyst affects the value of E_A by making it smaller.

129 An enzyme contains an active site, where the reactant molecule(s) bind(s). The enzyme reduces the freedom of motion available to the reactant(s), thereby forcing the reactants into a spatial orientation conducive to reaction and lowering the activation energy for the reaction.

131 In the cell, chemical reactions occur at normal body temperature and at a relatively constant concentration. Therefore, to increase the rate of a biochemical reaction, enzymes are used. Enzymes reduce the freedom of motion available to reactants; they lower the activation energy by forcing reactants into a spatial orientation conducive for reaction.

133

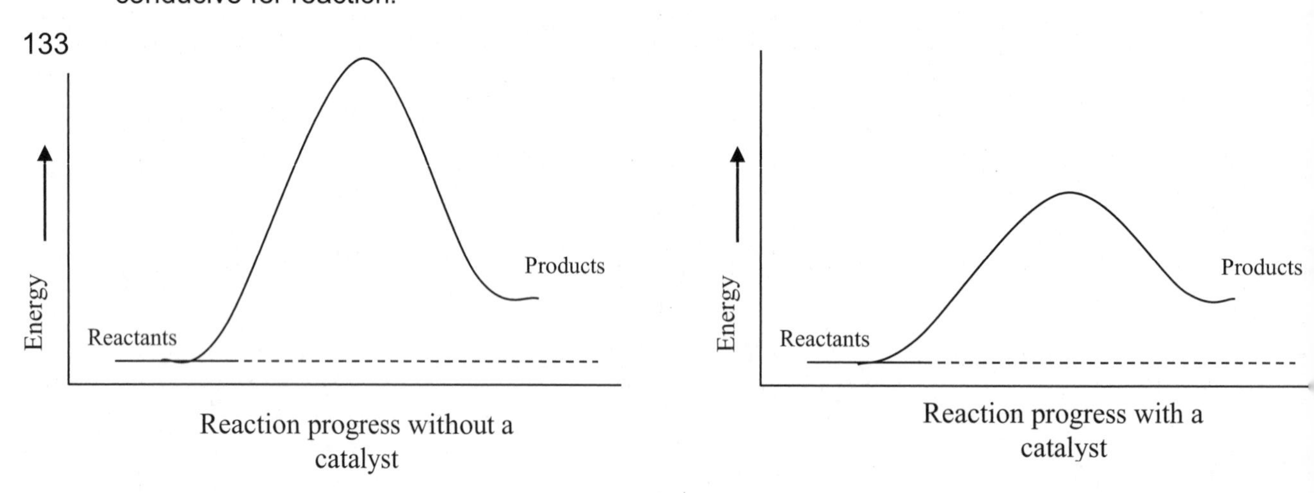

The reaction without a catalyst has the higher activation energy. The catalyst makes it easier for the reaction to occur by lowering the activation energy.

135 In a reversible reaction, the products combine to form the reactants. At the end of the reaction, both reactants and products are present. A reversible reaction is indicated by two half arrows pointing in opposite directions.

137 a. At equilibrium, the concentration of the reactants and products is constant.

139 a. Acetic acid, water, acetate, and hydronium ions are all present at equilibrium.

b. The concentrations of acetic acid and acetate are constant at equilibrium.

c. If more acetate is added to the reaction, the reaction will shift to the left in order to consume the excess acetate present.

d. If more acetic acid is added, the reaction will shift to the right in order to consume the excess acetic acid present.

141 a. Adding more reactant (H_2CO_3 or H_2O) or removing product (HCO_3^- or H_3O^+) will shift the reaction to the right.

b. Removing reactant (H_2CO_3 or H_2O) or adding more product (HCO_3^- or H_3O^+) will shift the reaction to the left.

143 A person on a mechanical respirator cannot be overfed or underfed. If they are underfed, they will be malnourished which can lead to coma and death. If they are overfed, oxygen consumption increases and the ventilator and lungs must work harder.

145 Direct calorimetry places an individual in a human calorimeter and measures the heat radiated from them. Indirect calorimetry measures a patient's oxygen uptake. Direct calorimetry is the gold standard. Indirect calorimetry is the more practical method since the patient only needs use a spirometer to measure the oxygen uptake, rather than a human calorimeter that requires considerable cost, time, and engineering skills.

Chapter 5

Changes of State and the Gas Laws

Chapter Summary

In this chapter, you have examined the factors that determine the physical states of matter and were introduced to the central role that energy plays in determining the physical state. You studied gases in more detail. Having a basic knowledge of the special behavior of gases helps you understand important medical issues such as breathing, decompression sickness, and anesthesia.

5.1 Changes of State

A change in state is the physical process of going from one state to another. Changes of state are physical changes, not chemical changes, because covalent bonds in the molecules are not formed or broken. Although covalent bonds remain intact during a change of state, intermolecular forces of attraction between molecules do change.

In a Nutshell: Energy and Changes of State

A change of state from the solid to the liquid phase is known as melting; the reverse change from liquid to solid is known as freezing. A change of state from the liquid to the gas phase is known as vaporization or evaporation, while the reverse change from gas to liquid is known as condensation. A solid can go directly to the gas phase without becoming a liquid first; a process known as sublimation. The reverse process changing from the gas phase to the solid phase is known as deposition.

Energy is an important factor in a change of state. During melting, vaporization, and sublimation, intermolecular forces of attraction are disrupted as the kinetic energy of the particle increases (temperature therefore increases as well). Conversely, during condensation, freezing, and deposition, intermolecular forces of attraction are created as the kinetic energy of the particles decreases (temperature therefore decreases as well). Energy must be added to achieve melting, vaporization, or sublimation, while energy must be removed to achieve freezing, condensation, or deposition.

In a Nutshell: Melting and Boiling Points

A heating curve is a graphical way to show how the temperature of a substance changes as energy is added at a constant rate. In the heating curve, the temperature increases in a linear fashion as you add heat, except in two places along the curve where the temperature remains constant even though heat continues to be added at the same rate. The lower temperature where the first plateau occurs is known as the freezing point or melting point of the substance. At this temperature, the energy added to the sample is used to disrupt the intermolecular forces of attraction rather than increasing the kinetic energy of the particles, which is why the temperature remains constant. The temperature at which a substance undergoes a change of state from the solid to the liquid is known as the melting point. The higher temperature where the second plateau occurs is known as the normal boiling point of the substance. At this temperature, the energy added to the sample is used to disrupt the remaining intermolecular forces of attraction rather than increasing the energy of the particles, and this is why the temperature remains constant.

The melting point and the boiling points of a substance are unique physical properties of a substance. Nonpolar covalent compounds tend to have lower melting points and boiling points because they have the weakest intermolecular forces of attraction, requiring the least amount of energy to disrupt. In contrast, compounds with O—H, and N—H bonds, which have the ability to hydrogen bond, have the strongest intermolecular forces of attractions. These compounds have higher melting and boiling points because it takes more energy to disrupt these stronger intermolecular forces of attraction.

The amount of energy (calories) required to melt a solid at its melting point is known as the heat of fusion or enthalpy of fusion: ΔH_{fus}. Similarly, the amount of energy required to vaporize a liquid at its boiling point is known as its heat of vaporization, or enthalpy of vaporization: ΔH_{vap}. Generally, heats of vaporization are greater than heats of fusion because the former requires all the intermolecular forces of attraction be disrupted.

Worked Example #1

Identify the change of state and the term that describes the change of state.

a. formation of ice cubes in the freezer
b. seeing your breath on a cold winter day
c. a puddle of water disappearing on a hot summer day

a. The formation of ice cubes is water changing from the liquid state to the solid state. This process is known as freezing.

b. Seeing your breath on a cold winter day is seeing water vapor (a gas) change into liquid water. This process is known as condensation.

c. When a puddle of water disappears, the liquid is changing into a gas. This process is known as evaporation.

Try It Yourself #1

Identify the change of state and the term that describes the change of state.

a. a solid block of metal heated so that it turns into a pool of molten metal
b. the mist coming from a humidifier
c. snow forming from water vapor in clouds

a. Change in state: ______________________________

Term to describe change in state: __________________________

b. Change in state: ______________________________

Term to describe change in state: __________________________

c. Change in state: ______________________________

Term to describe change in state: __________________________

In a Nutshell: Vapor Pressure

At the surface of a liquid, some of the liquid molecules gain sufficient kinetic energy to enter the gas phase, a process known as evaporation. It is called evaporation instead of vaporization because the phase change is occurring below the boiling point of the liquid. Simultaneously, some of the gas molecules directly above the liquid lose kinetic energy and return to the liquid phase, creating an equilibrium between the liquid and gas phases. This change is state is a result of vapor pressure. Vapor pressure is the pressure exerted by a gas in the liquid-gas equilibrium. Pressure (P) is defined as the force per unit area exerted by gas particles colliding against the walls of the container.

Vapor pressure is a physical property of a substance that depends on the chemical structure and on the kinetic energy of the substance as measured by its temperature. A liquid is volatile if it has a high vapor pressure, which means its molecules enter the gas phase readily. Volatile compounds have high vapor pressures because they have weak intermolecular forces of attraction, making it easier for molecules to enter the gas phase.

The vapor pressure of a liquid increases with temperature. At higher temperatures, molecules in the liquid phase have greater kinetic energy and therefore their vapor pressure is higher, allowing them to enter the gas phase more readily.

When the temperature is high enough that the vapor pressure of the substance equals the atmospheric pressure, the liquid boils. The normal boiling point of a liquid is defined as the temperature at which the vapor pressure of the liquid equals 1 atm, the atmospheric pressure at sea level. Substances with a higher vapor pressure (more volatile substances) will have lower boiling points because their vapor pressure curves cross the 760-mmHg (1-atm) line at a lower temperature.

Worked Example #2

Diethyl ether was has been used as an anesthetic. It has a vapor pressure of 400 mmHg at 17.9 °C. Is this compound volatile? Does it have a high or a low boiling point?

The vapor pressure is high; therefore, the compound is volatile and the boiling point should be low.

Try It Yourself #2

The vapor pressure of ethyl acetate (commonly used in nail polish remover) is 76 mmHg at 20 °C. The vapor pressure of phenol is 0.36 mmHg at 20 °C. Which would you expect to have the lower boiling point?

Compound with higher vapor pressure: ________________

Compound with lower boiling point: ____________________

In a Nutshell: Atmospheric Pressure

Pressure can be measured and reported in a variety of units. The most common unit used in medicine is mmHg. Atmospheres are the unit used to measure atmospheric pressure. The other common units for pressure are pounds per square inch (psi), torr, and Pascal.

Atmospheric pressure is the pressure created by the force of the mass of air at any given place (area) on the Earth. At sea level the atmospheric pressure is 1 atmosphere (1 atm). As you move to higher altitudes, atmospheric pressure decreases because there is less air above you.

Worked Example #3

A tire on an automobile is filled to 35 psi. What is this pressure equivalent to in mmHg?

Use dimensional analysis to set up the calculation using the correct form of the conversion factor that will allow the supplied units, psi, to cancel. Use Table 5-3 to find the conversion between psi and mmHg: 14.70 psi = 760 mmHg.

$$35\ \cancel{\text{psi}} \times \frac{760\ \text{mmHg}}{14.70\ \cancel{\text{psi}}} = 1.8 \times 10^{3}\ \text{mmHg}$$

Try It Yourself #3

A patient has an intraocular eye pressure of 12 mmHg. What is this pressure in Pa?

Tools: Table 5-3

Use dimensional analysis to set up the calculation using the correct form of the conversion factor that will allow the supplied units, mmHg, to cancel.

Practice Problems for Changes of State

1. Which changes in state involve transfer of heat from the surroundings?

2. Which molecule would you expect to have a higher boiling point: $CHCl_3$ or CH_3OH?

3. Identify the change of state and the term that describes the change of state.
   a. steam coming from your hot cup of coffee
   b. frost forming from water vapor in the air
   c. rain forming from water vapor in clouds

4. The vapor pressure of cinnamaldehyde (it gives cinnamon its flavor and smell) is 5.8 $\times 10^{-4}$ psi at 25 °C. What is the vapor pressure in mmHg? Do you expect cinnamaldyde to have a high or low boiling point?

5. A patient has an intraocular eye pressure of 2.0×10^3 Pa. If the intraocular eye pressure is greater than 21.5 mmHg, then the patient has a greater risk for developing glaucoma. What is the patient's intraocular eye pressure in mmHg? Does this patient have a risk for developing glaucoma?

5.2 The Gas Laws

Gases are compressible, which gives them unique physical properties. To understand the macroscopic properties of a gas, it is useful to consider the kinetic molecular view of gases:

The particles of a gas are in constant, random motion.
The total volume of all the gas particles in a container is negligible compared to the volume of the container.
The attractive forces among the particles of a gas are negligible.
The temperature of a gas depends on the average kinetic energy of the gas particles.

The macroscopic properties of a gas can be described by the following four interrelated variables: pressure (P), volume (V), temperature (T), and number of moles (n). These variables are interrelated, so changing one affects the others in a predictable way. The relationship between any two variables while the other two are constant is expressed in the gas laws: Boyle's law, Charles's law, and Avogadro's law.

Worked Example #4

Use kinetic molecular theory of gases to explain the following:

a. Dispersion forces are not important in the gas phase.
b. Perfume dropped in one part a room is quickly detected at the other end of the room.

Solutions

a. Molecules in the gas phase do not have intermolecular forces of attraction; therefore, they do not exert dispersion forces.
b. The particles of a gas do not have intermolecular forces of attraction; therefore, gas particles move at high speeds in all directions, filling the volume of the container they are in.

Try It Yourself #4

Use kinetic molecular theory of gases to explain the following:

a. A balloon filled with helium on a winter day would be smaller than a balloon filled with the same amount of helium on a summer day.
b. Gas molecules can be compressed to fit into a smaller volume.

Solutions

a.

b.

In a Nutshell: Pressure-Volume Relationship of Gases: Boyle's Law

As the volume of gas decreases, the pressure of the gas increases; and as the volume of a gas increases, the pressure decreases if the temperature, *T*, and the number of moles of gas, *n*, are constant. Pressure and volume are inversely related to each other, as one goes up, the other goes down. This relationship is known as Boyle's law. It follows from Boyle's law that, when a gas with an initial pressure, P_i, and an initial volume, V_i, undergoes a change to a final pressure, P_f, and final volume, V_f, these variables are related by the following equation:

$$P_iV_i = P_fV_f \quad (n \text{ and } T \text{ are constant})$$

The pressure volume relationship of gases is demonstrated every time you inhale and exhale. Upon inhalation, the pressure of the lungs decreases as the volume increases, and upon exhalation, the pressure of the lungs increases and the volume decreases.

Worked Example #5

The pressure gauge on a patient's full 13.9 L oxygen tank reads 10.2 atm. At constant temperature, how many liters of oxygen can the patient's entire tank hold at a pressure of 25.1 atm?

Since both P *and* V *are changing, while* n *and* T *are constant, we can use Boyle's law:* $P_iV_i = P_fV_f$.

Define the variables and select the variable to solve for.

The problem indicates that P_i *= 10.2 atm,* V_i *= 13.9 L, and* P_f *= 25.1 atm; therefore, you need to solve for* V_f *the final volume.*

Algebraically isolate the unknown variable on one side of the equation.

Use algebra to manipulate the equation so that V_f is isolated:

$$V_f = P_i \times \frac{V_i}{P_f}$$

Substitute the known values into the equation and solve for the unknown variable:

$$V_f = 10.2\ \cancel{atm} \times \frac{13.9\ L}{25.1\ \cancel{atm}} = 5.65\ L$$

Try It Yourself #5

An air bubble forms at the bottom of a lake where the total pressure is 4.35 atm. At this pressure, the bubble has a volume of 7.32 mL. What is the volume of the bubble when the bubble rises to the surface, where the pressure is 1.00 atm?

The variables that are changing are: __________

The equation to use is: ________________

Define the variables and select the variable to solve for.

Algebraically isolate the unknown variable on one side of the equation.

Substitute the known variables into the equation and solve for the unknown variable.

In a Nutshell: Volume-Temperature Relationship of Gases: Charles's Law

As the temperature of a gas increases, the kinetic energy of the particles of a gas increase, causing the volume of occupied by the gas increases if the number of moles of gas and the pressure of the gas remain constant (no gas escapes). The volume of a gas is directly proportional to the temperature of a gas: As one goes up, the other one goes up. The relationship between temperature and volume is known as Charles's law. In general, when a gas with and initial temperature, T_i, and an initial volume, V_i, undergoes a change to a final temperature, T_f, and final volume, V_f, these variables are related by the following equation, which follows from Charles's law:

$\frac{V_i}{T_i} = \frac{V_f}{T_f}$ (*n* and *P* are constant, *T* must be in kelvins)

When performing calculations using this or any other gas law equation, the Kelvin scale is used because it is an absolute scale.

Worked Example #6

Oxygen is warmed from 27 °C to 58 °C. The original volume of the gas was 19.8 L. What is the final volume of the gas assuming *P* and *n* have remained constant?

Since V *and* T *are changing and* P *and* n *are constant, we can use Charles's law:*

$$\frac{V_i}{T_i} = \frac{V_f}{T_f}$$

Define the variables and select the variable to solve for.

The problem indicates T_i = *27 °C,* V_i = *19.8 L, and* T_f = *58 °C; therefore, you need to solve for* V_f. *However, the initial and final temperatures must first be converted into kelvins:*

T_i = 27 °C + 273.15 = 300 K

T_2 = 58 °C + 273.15 = 331 K

Algebraically isolate the unknown variable on one side of the equation.

$$V_f = T_f \times \frac{V_i}{T_i}$$

Substitute the known variables into the equation and solve for the unknown variable.

$$V_f = 331\ \cancel{K} \times \frac{19.8\ L}{300\ \cancel{K}} = 21.8\ L$$

Try It Yourself #6

A balloon taken from the freezer has a temperature of 2 °C and a volume of 1.63 L. The balloon is warmed up and the final volume is 1.80 L. Assume that the pressure and the number of moles of gas in the balloon are constant. What is the final temperature of the gas in the balloon?

The variables that are changing are: __________

The equation to use is: _________________

Define the variables and select the variable to solve for. Remember, you must first convert temperatures to kelvins!

Algebraically isolate the unknown variable on one side of the equation.

Substitute the known variables into the equation and solve for the unknown variable.

In a Nutshell: Volume-Mole Relationship of Gases: Avogadro's Law

There is a direct relationship between the number of moles of a gas and its volume if the pressure and temperature are constant. The direct proportionality between moles of a gas and the volume is known as Avogadro's law. It follows from Avogadro's law that when a gas with an initial number of moles, n_i, at an initial volume, V_i, undergoes a change to a final number of moles, n_f, and final volume, V_f, these variables are related by the equation:

$$\frac{V_i}{n_i} = \frac{V_f}{n_f}$$

Worked Example #7

A boy with cystic fibrosis has lungs with a volume of 2.6 L that fill with 0.12 moles of air when he inhales. When he exhales, his lung volume decreases to 2.1 L. How many moles of gas remain in his lungs after he exhales? Assume constant temperature and pressure.

Since V *and* n *are changing and* P *and* T *are constant, we can use Avogadro's law:*

$$\frac{V_i}{n_i} = \frac{V_f}{n_f}$$

Define the variables and select the variable to solve for.

The problem indicates n_i *= 0.12 mol,* V_i *= 2.6 L, and* V_f *= 2.1 L; therefore, you need to solve for* n_f*.*

Algebraically isolate the unknown variable on one side of the equation.

$$n_f = V_f \times \frac{n_i}{V_i}$$

Substitute the known variables into the equation and solve for the unknown variable.

$$n_f = 2.1\ \cancel{L} \times \frac{0.12\ \text{mol}}{2.6\ \cancel{L}} = 0.10\ \text{mol}$$

Try It Yourself #7

A healthy female has lungs with a volume of 4.2 L that fill with 0.21 moles of air when she inhales. After she exhales, the number of moles of air in her lungs decreases to 0.14 moles. What is the volume of air in her lungs after she exhales, assuming constant temperature and pressure?

The variables that are changing are: __________

The equation to use is: ________________

Define the variables and select the variable to solve for.

Algebraically isolate the unknown variable on one side of the equation.

Substitute the known variables into the equation and solve for the unknown variable.

In a Nutshell: The Ideal Gas Law

By combining Boyle's law, Charles's law, and Avogadro's law, a single mathematical relationship, known as the ideal gas law, can be written. The mathematical equation that represents the ideal gas law contains the universal gas constant, *R*.

$$R = 0.0826\ \frac{\mathrm{L \cdot atm}}{\mathrm{mol \cdot K}}$$

The common form of the ideal gas law is:

$$PV = nRT$$

The ideal gas law can be used to determine one variable (*P*, *V*, *T*, or *n*) when the other three are known. When using the universal gas constant, *R*, units for the variables must match those in the universal gas constant: volume in liters, pressure in atm, and temperature in kelvin.

Worked Example #8

A piece of dry ice with a mass of 2.75 g is placed inside a balloon and sealed. After all the carbon dioxide has sublimed, what will the volume of the balloon be at 18 °C and 0.93 atm? You will need to convert the grams of carbon dioxide to moles of carbon dioxide using the molar mass.

Since n, P, *and* T *are given and* R *is a known constant, we can use the ideal gas law to solve for* V*:*

$$PV = nRT$$

Define the variables and identify the requested variable.

The problem indicates T *= 18 °C,* P *= 0.93 atm, the mass of* CO_2 *= 2.75 g (which will give the number of moles), and* $\mathrm{R} = 0.0826\ \frac{L \cdot atm}{mol \cdot K}$ *and requests the volume,* V*. You must also convert 18 °C into kelvin and 2.75 g into moles.*

18 °C + 273.15 = 291 K

In order to convert grams to moles, we need to first calculate the molar mass of CO_2 and then use dimensional analysis to set up the calculation using the correct form of the conversion factor that will allow the supplied units, grams, to cancel.

molar mass of CO_2 = 1 C + 2 O = 12.01 + (2 × 16.00) = 44.01 g/mol

$$2.75\ \text{g} \times \frac{1\ \text{mol}}{44.01\ \text{g}} = 0.0625\ \text{mol}$$

Algebraically isolate the unknown variable on one side of the equation.

$$V = \frac{nRT}{P}$$

Substitute the known variables into the equation and solve for the unknown variable.

$$V = \frac{0.0625\ \cancel{\text{mol}} \times 0.0826 \frac{\text{L}\cdot\cancel{\text{atm}}}{\cancel{\text{mol}\cdot\text{K}}} \times 291\ \cancel{\text{K}}}{0.93\ \cancel{\text{atm}}} = 1.6\ \text{L}$$

Try It Yourself #8

Calculate the pressure of 5.37 moles of hydrogen gas with a volume of 2.64 L at a pressure of 720 mmHg and a temperature of 25°C.

The variables that are changing are: __________

The equation to use is: ________________

Define the variables and identify the requested variable.

Remember, that you must convert units into L, K, and atm if necessary.

Algebraically isolate the unknown variable on one side of the equation.

Substitute the known variables into the equation and solve for the unknown variable.

The properties of a gas are described under a standard set of reference conditions to make comparisons between different gases. These conditions are known as STP. Standard temperature and pressure is 0 °C (273.15 K) and 1 atm. Under the conditions of STP, one mole of any gas occupies a volume of 22.4 L, the molar volume of a gas. The molar volume of a gas does not depend on the identity or molar mass of the gas.

Worked Example #9

What volume will be occupied by 10.2 moles of helium at STP? Would the volume be the same for 10.2 moles of xenon?

Use dimensional analysis to set up the calculation using the correct form of the conversion factor (molar volume) that will allow the supplied units, moles, to cancel.

$$10.2\ \cancel{\text{mol}} \times \frac{22.4\ \text{L}}{1\ \cancel{\text{mol}}} = 228\ \text{L}$$

Therefore, 228 L would be occupied by 10.2 moles of helium. Yes, the volume would be the same for 10.2 moles of xenon because the identity of the gas is not important.

Try It Yourself #9

What volume will be occupied by 2.7 moles of krypton at STP?

Use dimensional analysis to set up the calculation using the correct form of the conversion factor (molar volume) that will allow the supplied units, moles, to cancel.

The density of a gas at STP can be determined from the molar volume. Density is the molar mass divided by the molar volume. The density a gas is proportional to its molar mass; the greater the molar mass, the greater the density of the gas.

Worked Example #10

Calculate the density of nitrous oxide gas, N_2O, at STP. Nitrous oxide is also known as laughing gas and is used as an anesthetic and analgesic in dentistry.

Use the equation for density:

$$\text{Density} = \frac{\text{molar mass}}{\text{molar volume}}$$

In order to calculate the density, we need to calculate the molar mass of nitrous oxide.

Molar mass N_2O = 2 N + 1 O = (2 × 14.01) + (1 × 16.00) = 44.02 g/mol

Substitute the known quantities:

$$\text{Density} = \frac{44.02\ \frac{\text{g}}{\cancel{\text{mol}}}}{22.4\ \frac{\text{L}}{\cancel{\text{mol}}}} = 1.97\ \frac{\text{g}}{\text{L}}$$

Try It Yourself #10

Calculate the density of oxygen gas, O_2, at STP.

Use the equation for density:

Substitute the known quantities

Practice Problems for The Gas Laws

1. A sample of xenon has a volume of 12.3 L at a pressure of 672 mmHg and a temperature of 15 °C. Calculate the number of moles present in the sample.

2. A balloon is filled with helium and it has a volume of 17.8 L at 20 °C. What is the volume of this balloon when the temperature changes to 56 °C (and the pressure and the number of moles of gas remain constant)?

3. A sample of oxygen has a pressure of 746 Torr and a volume of 30.8 L. What is the pressure of the gas if the volume of the gas changes to 50.3 L (assume that the number of moles and the temperature remain constant)?

4. A healthy boy has lungs with a volume of 1.9 L that fill with 0.82 mole of air when he inhales. After he exhales, the number of moles of air in his lungs decreases to 0.21 mole. What is the volume of air in his lungs after he exhales, assuming the temperature and pressure in his lungs is constant?

5. Calculate the density of neon gas at STP.

5.3 Gas Mixtures and Partial Pressures

Dalton's law states that, for a mixture of gases, each gas exerts a pressure independent of the other gases, and each gas will behave as it alone occupied the total volume. The pressure exerted by a gas in a mixture is known as the partial pressure, P_n. Dalton's law states that the sum of the partial pressures of each gas present in the mixture equals the total pressure (P_{total}) exerted by the mixtures of gases:

$$P_{total} = P_1 + P_2 + P_3 + \ldots\ldots.P_n$$

where P_1, P_2, P_3,.....P_n, represent the partial pressure of each gas in the mixture.

Worked Example #11

SCUBA divers often use a mixture of nitrogen and gas in their air tanks. An air tank has a total pressure of 4.67 atm and contains a mixture of oxygen and nitrogen. If the partial pressure of nitrogen is 3.18 atm, what is the partial pressure of oxygen?

Since partial pressures are provided, use the following equation:

$$P_{\text{total}} = P_{\text{oxygen}} + P_{\text{nitrogen}}$$

Define the variables and select the variable to solve for.

The problem indicates that P_{total} *= 4.67 atm and* $P_{nitrogen}$ *= 3.18 atm; therefore, we need find* P_{oxygen}.

Algebraically isolate the unknown variable on one side of the equation:

$$P_{\text{oxygen}} = P_{\text{total}} - P_{\text{nitrogen}}$$

Substitute the known variables into the equation and solve for the unknown variable:

$$P_{\text{oxygen}} = P_{\text{total}} - P_{\text{nitrogen}} = 4.67 \text{ atm} - 3.18 \text{ atm} = 1.49 \text{ atm}$$

Try It Yourself #11

A SCUBA diver is going to dive to a depth of 330 ft. He will be using an air tank filled with a mixture of helium, nitrogen, and oxygen. The total pressure of the gas mixture is 8.45 atm. If the partial pressure of helium is 5.91 atm and the partial pressure of nitrogen is 1.69 atm, what is the partial pressure of oxygen in the tank?

The variables that are changing are: __________

The equation to use is: ________________

Define the variables and identify the requested variable.

Algebraically isolate the unknown variable on one side of the equation.

Substitute the known variables into the equation and solve for the unknown variable.

In a Nutshell: Henry's Law

Henry's law states that the number of gas molecules dissolved in a liquid is directly proportional to the partial pressure of the gas. Thus, as the partial pressure of the gas increases, the number of gas molecules dissolved in the solution increases. At constant temperature, Henry's law can be expressed by the following equation:

$$P = kC$$

where P is the partial pressure of the gas, k is Henry's constant, which is unique for each gas, and C is the concentration of the dissolved gas.

Henry's law is useful in the field of anesthesiology when gaseous anesthetics are inhaled. Anesthetics have different Henry's constants, k, and therefore are present in different concentrations when dissolved in blood at given pressure. The smaller the Henry's constant of an anesthetic, the higher the concentration of anesthetic dissolved in blood.

Worked Example #12

Where would you expect a glass of soda to have more bubbles: at the beach or in the mountains?

Henry's law shows that there is a direct relationship between the pressure above a solution and the concentration of gas in a solution. Therefore, at lower altitudes and higher atmospheric pressure, you would have a corresponding higher concentration of carbon dioxide (the bubbles) in the solution. Thus, there would be more bubbles in the glass of soda on the beach. In the mountains, the CO_2 would escape more quickly once the bottle is opened and depressurized, causing the soda to go "flat" sooner.

Try It Yourself #12

Desflurane and sevoflurane have large Henry's constants compared to diethyl ether. Desflurane, sevoflurane, and diethyl ether are used as anesthetics. Which anesthetic would allow a patient to regain conscious more quickly? Explain.

Solution:

Practice Problems for Gas Mixtures and Partial Pressures

1. A mixture of two gases, argon and oxygen, is sometimes used in SCUBA diving. If the total pressure in air tank containing a mixture of argon and oxygen is 12.2 atm and the partial pressure of oxygen is 2.4 atm, what is the partial pressure of argon?

2. When a patient has pneumonia, a smaller volume of air enters the lungs, and therefore a lower oxygen partial pressure exists in the lungs and consequently the concentration of oxygen in the blood is lower. How is the partial pressure of oxygen in the blood affected when a patient has pneumonia? Explain using Henry's law.

3. As a climber stands on the top of Mt. Everest, would you expect the concentration of dissolved oxygen in the climber's blood to be higher or lower than the amount of dissolved oxygen in the climber's blood at the base of the mountain? Explain using Henry's law.

Chapter 5 Quiz

1. Sometimes, instead of melting, snow will turn directly into water vapor. What is the name of this change in state? Does this process release or absorb energy?

2. When frost forms, water vapor is turned directly into snowflakes without becoming liquid water. What is the name of this change in state? Does this process release or absorb energy?

3. If a compound has a low vapor pressure, do you expect it to have a low or high boiling point? Explain your answer

4. A sample of nitrogen has a pressure of 12.1 psi and a volume of 53.8 L. What is the volume of the gas if the pressure of the gas changes to 3.2 psi (assume that the number of moles and the temperature remain constant)?

5. A balloon is filled with helium and it has a volume of 345 mL at 18 °C. What is the temperature, in Celsius, of the helium in this balloon when the volume changes to 1098 mL (and the pressure and the number of moles of gas remain constant)?

6. If a balloon filled with 0.48 moles of helium has a volume of 8.3 L, what is the final number of moles if an additional 20 L of helium is added to the balloon at constant temperature and pressure?

7. A sample of argon has a volume of 5.3 L at a pressure of 270 torr and a temperature of 25 °C. Calculate the number of moles of argon in the sample.

8. What is the density of xenon gas at STP?

9. A tank that contains a mixture of carbon dioxide, nitrous oxide, and oxygen has a total pressure of 5.63 atm. The partial pressure of carbon dioxide is 0.56 atm, and the partial pressure of oxygen is 3.66 atm. What is the partial pressure of nitrous oxide?

10. Who has more dissolved oxygen in their blood: a scuba diver standing on the boat at the surface of the water or her companion who is 25 ft below the surface of the ocean?

Chapter 5

Answers to Additional Exercises

51 The mass of air and seawater contribute to the pressure you experience when diving below the surface of the sea.

53 Confusion, weakness, headache, itching, and joint pains are some of the symptoms of the bends. When the diver is ascending back to the surface, the pressure on the diver decreases and additional nitrogen molecules come out of solution. If a diver ascends too quickly, the dissolved nitrogen gas will diffuse out the blood too quickly, forming bubbles in the bloodstream.

55 a. Chemical reaction—the chemical composition of the metal has changed. b. Change of state—liquid is turning into water vapor. c. Change of state—liquid is turning into solid. d. Chemical change—the chemical composition of the wick has changed.

57 Energy must be removed for freezing, condensation, and deposition to occur.

59 a. melting (solid → liquid) b. sublimation (solid → gas) c. freezing (liquid → solid) d. vaporization (liquid → gas) e. condensation (gas → liquid) f. vaporization (liquid → gas) g. freezing (liquid →solid)

61 Vaporization going from a liquid to a gas makes mercury dangerous.

63 At the melting point or boiling point, the added energy is needed to disrupt all the intermolecular forces of attraction present.

65 Ethanol has the ability to hydrogen bond, which is the strongest intermolecular force of attraction. It takes more energy to disrupt these stronger intermolecular forces of attraction. Carbon dioxide does not have the ability to hydrogen bond.

67 Steam causes burns because of the change of state that occurs when steam comes in contact with your skin. Steam condenses when it comes in contact with the skin, which requires heat to be removed from the steam by an amount equivalent to the heat of vaporization of water. The heat is removed from the steam and transferred to your skin. Additional heat is removed as the liquid cools from 100 °C to 37 °C.

69 Pressure is defined as the force per unit area exerted by gas particles colliding against the walls of the container.

71 Phenol would have the lower boiling point. Phenol has the higher vapor pressure; it is more volatile than mercury.

73 The boiling point of water should be lower at the base camp for Mt. Everest because the atmospheric pressure is lower there.

75 At 14,000 ft above sea level, there are fewer air molecules around us than at sea level. At 14,000 feet below sea level, not only is the whole atmosphere pressing down on us, but so is 14,000 ft of water.

77 Use dimensional analysis to set up the calculation using the correct form of the conversion factor that will allow pascal to cancel:

$$4.7 \times 10^3 \text{ Pa} \times \frac{1 \text{ atm}}{1.013 \times 10^5 \text{ Pa}} \times \frac{760 \text{ mmHg}}{1 \text{ atm}} = 35 \text{ mmHg}$$

$$4.7 \times 10^3 \text{ Pa} \times \frac{1 \text{ atm}}{1.013 \times 10^5 \text{ Pa}} = 0.046 \text{ atm}$$

79 Use dimensional analysis to set up the calculation using the correct form of the conversion factor that will allow pascal to cancel:

$$5.69 \times 10^3 \text{ Pa} \times \frac{1 \text{ atm}}{1.013 \times 10^5 \text{ Pa}} \times \frac{760 \text{ mmHg}}{1 \text{ atm}} = 43 \text{ mmHg}$$

No, the number of breaths given by the mechanical ventilator does not need to be adjusted.

81 At the lower elevation in Denver, the atmospheric pressure is higher compared with the mountains. The pressure of a gas is inversely proportional to the volume of the gas. As the pressure increases as you come down from the mountains, the volume of gas in the water bottle decreases.

83 Since both *P* and *V* are changing, while *n* and *T* are constant, we can use Boyle's law: $P_iV_i = P_fV_f$.

Define the variables and select the variable to solve for.
The problem indicates that $P_i = 0.91$ atm, $V_i = 1.1$ L, and $V_f = 3.1$ L; therefore, you need to solve for P_f, the final pressure.

Algebraically isolate the unknown variable on one side of the equation.
Use algebra to manipulate the equation, so that P_f is isolated:

$$P_f = P_i \times \frac{V_i}{V_f}$$

Substitute the known values into the equation and solve for the unknown variable:

$$P_f = 0.91 \text{ atm} \times \frac{1.1 \text{ L}}{3.1 \text{ L}} = 0.32 \text{ atm}$$

There is less pressure; therefore, the child went up in altitude with the balloon.

85 Upon inhalation, the pressure of the lungs *decreases* as the volume of the lungs *increases*.

87 Boyle's law states the pressure of a gas is inversely proportional to the volume of a gas; thus, if one goes up, the other will go down. Upon inhalation, the pressure of the lungs decreases as the volume of the lungs increases. Upon exhalation, the pressure of the lungs increases as the volume of the lungs decreases.

89 Charles's law states that temperature and volume are directly proportional to each other. As you heat the cake in the oven, the carbon dioxide in the cake heats up; therefore, the volume of the carbon dioxide increases and the cake rises.

91 Since both n and V are changing, while P and T are constant, we can use Avogadro's law: $\frac{V_i}{n_i} = \frac{V_f}{n_f}$.

Define the variables and select the variable to solve for.

The problem indicates that n_i = 0.10 mol, V_i = 2.5 L, and V_f = 2.0 L; therefore, you need to solve for n_f.

Algebraically isolate the unknown variable on one side of the equation.

Use algebra to manipulate the equation, so that n_f is isolated:

$$n_f = n_i \times \frac{V_f}{V_i}$$

Substitute the known values into the equation and solve for the unknown variable:

$$n_f = 0.10 \text{ mol} \times \frac{2.0 \text{ L}}{2.5 \text{ L}} = 0.080 \text{ mol}$$

93 Since both n and V are changing, while P and T are constant, we can use Avogadro's law: $\frac{V_i}{n_i} = \frac{V_f}{n_f}$.

Define the variables and select the variable to solve for.

The problem indicates that n_i = 0.30 mol, V_i = 5.5 L, and V_f = 3.5 L; therefore, you need to solve for n_f.

Algebraically isolate the unknown variable on one side of the equation.

Use algebra to manipulate the equation, so that n_f is isolated:

$$n_f = n_i \times \frac{V_f}{V_i}$$

Substitute the known values into the equation and solve for the unknown variable:

$n_f = 0.30 \text{ mol} \times \frac{3.5 \text{ L}}{5.5 \text{ L}} = 0.19 \text{ mol}$

95 Since both V and T are changing, while n and P are constant, we can use Charles's law:

$\frac{V_i}{T_i} = \frac{V_f}{T_f}$.

Define the variables and select the variable to solve for.

The problem indicates that V_i = 8.7 L, T_i = 12 °C, and T_f = 34 °C; therefore, you need to solve for V_f, the final volume.

Algebraically isolate the unknown variable on one side of the equation.

Use algebra to manipulate the equation, so that V_f is isolated:

$$V_f = T_f \times \frac{V_i}{T_i}$$

Substitute the known values into the equation and solve for the unknown variable.

REMEMBER, YOU MUST FIRST CONVERT TEMPERATURES TO KELVIN!

T_i = 12 °C + 273.15 = 285 K

T_f = 34 °C + 273.15 = 307 K

$V_f = 307 \text{ K} \times \frac{8.7 \text{ L}}{285 \text{ K}} = 9.4 \text{ L}$

97 Since V, P, and T are given and R is a known constant, we can use the ideal gas law to solve for n: $PV = nRT$.

Define the variables and select the variable to solve for.

The problem indicates that V = 18 L, T = 501 K, and P = 4.3 atm.

Algebraically isolate the unknown variable on one side of the equation.

Use algebra to manipulate the equation so that n is isolated:

$$n = \frac{PV}{RT}$$

Substitute the known values and the universal gas constant into the equation and solve for the unknown variable.

$n = \frac{(4.3 \text{ atm})(18 \text{ L})}{(0.08206 \frac{\text{L}\cdot\text{atm}}{\text{mol}\cdot\text{K}})(501 \text{K})} = 1.9 \text{ mol}$

99 Since n, V, and P are given and R is a known constant, we can use the ideal gas law to solve for T: $PV = nRT$.

Define the variables and select the variable to solve for.

The problem indicates that $n = 7.00 \times 10^{-3}$ mol, $P = 712$ mmHg, and $V = 205$ mL and requests the temperature, T. You must also convert units into L, K, and atm if necessary.

$$P = 712\ \cancel{\text{mmHg}} \times \frac{1\ \text{atm}}{760\ \cancel{\text{mmHg}}} = 0.937\ atm$$

$$V = 205\ \cancel{\text{mL}} \times \frac{1\ \text{L}}{10^3\ \cancel{\text{mL}}} = 0.205\ \text{L}$$

Algebraically isolate the unknown variable on one side of the equation.

Use algebra to manipulate the equation, so that T is isolated:

$$T = \frac{PV}{Rn}$$

Substitute the known values and the universal gas constant into the equation and solve for the unknown variable.

$$T = \frac{(0.937\ \cancel{\text{atm}})(0.205\ \cancel{\text{L}})}{(0.08206\frac{\cancel{\text{L}\cdot\text{atm}}}{\cancel{\text{mol}}\cdot\text{K}})(7.00 \times 10^{-3}\ \cancel{\text{mol}})} = 334\ \text{K}$$

101 Use dimensional analysis to set up the calculation using the correct form of the conversion factor that will allow liters to cancel:

$$11.2\ \cancel{\text{L}} \times \frac{1\ \text{mol}}{22.4\ \cancel{\text{L}}} = 0.500\ \text{mol}$$

103 Use dimensional analysis to set up the calculation using the correct form of the conversion factor that will allow liters to cancel:

$$15.3\ \cancel{\text{L}} \times \frac{1\ \text{mol}}{22.4\ \cancel{\text{L}}} = 0.683\ \text{mol}$$

105 Use dimensional analysis to set up the calculation using the correct form of the conversion factor that will allow moles to cancel:

$$0.2\ \cancel{\text{mol}} \times \frac{22.4\ \text{L}}{1\ \cancel{\text{mol}}} = 4\ \text{L}$$

107 Since V, P, and T are given and R is a known constant, we can use the ideal gas law to solve for n: $PV = nRT$.

Define the variables and select the variable to solve for.

The problem indicates that $V = 5.58$ L, $T = 21$ °C, and $P = 25$ mmHg and requests moles. You must also convert units into L, K, and atm if necessary.

$$P = 25\ \cancel{\text{mmHg}} \times \frac{1\ \text{atm}}{760\ \cancel{\text{mmHg}}} = 0.033\ \text{atm}$$

$$T = 21\ °\text{C} + 273.15\text{K} = 294\text{K}$$

Algebraically isolate the unknown variable on one side of the equation.

Use algebra to manipulate the equation so that *n* is isolated:

$$n = \frac{PV}{RT}$$

Substitute the known values and the universal gas constant into the equation and solve for the unknown variable.

$$n = \frac{(0.033\ \cancel{\text{atm}})(5.58\ \cancel{\text{L}})}{(0.08206\frac{\cancel{\text{L}}\cdot\cancel{\text{atm}}}{\text{mol}\cdot\cancel{\text{K}}})(294\ \cancel{\text{K}})} = 7.6 \times 10^{-3}\ \text{mol}$$

109 Since partial pressures are provided, use the following equation:

$$P_{TOT} = P_{\text{oxygen}} + P_{\text{nitrogen}}$$

Define the variables and select the variable to solve for.

The problem indicates that P_{TOT} = 4.30 atm and P_{nitrogen} = 0.92 atm; therefore, we need find P_{oxygen}.

Algebraically isolate the unknown variable on one side of the equation:

$$P_{\text{oxygen}} = P_{TOT} - P_{\text{nitrogen}}$$

Substitute the known variables into the equation and solve for the unknown variable:

$$P_{\text{oxygen}} = P_{TOT} - P_{\text{nitrogen}} = 4.30\ \text{atm} - 0.92\ \text{atm} = 3.38\ \text{atm}$$

111 As the patient breathes out, the partial pressure of the carbon dioxide in the bag increases.

113 A glass of soda should have more bubbles in it on the beach since the atmospheric pressure is higher than in the mountains. Henry's law states that, the higher the pressure above a liquid, the higher the concentration of the gas in the liquid.

115 Since partial pressures are provided, use the following equation:

$$P_{TOT} = P_{\text{nitrogen}} + P_{\text{oxygen}}$$

Define the variables and select the variable to solve for.

The problem indicates that P_{TOT} = 3.83 atm and P_{oxygen} = 1.38 atm; therefore, we need find $P_{nitrogen}$.

Algebraically isolate the unknown variable on one side of the equation.

$$P_{\text{nitrogen}} = P_{\text{TOT}} - P_{\text{oxygen}}$$

Substitute the known variables into the equation and solve for the unknown variable:

$$P_{\text{nitrogen}} = P_{TOT} - P_{\text{oxygen}} = 3.83 \text{ atm} - 1.38 \text{ atm} = 2.45 \text{ atm}$$

The percentage of nitrogen in the tank is $\frac{2.45}{3.83} \times 100\% = 64\%$.

In order to calculate the percentage of nitrogen in the air, the total pressure of the air must be calculated.

$$P_{\text{TOT}} = P_{\text{nitrogen}} + P_{\text{oxygen}} + P_{\text{carbon dioxide}} = 0.77 + 0.18 + 0.05 = 1.00\ atm$$

The percentage of nitrogen in the air is $\frac{0.77}{1.00} \times 100\% = 77\%$.

The percentage of nitrogen in the tank is less than the percentage of nitrogen in the air.

117 A patient would need to use hyberbaric oxygen therapy if he had the bends, carbon monoxide poisoning, diabetic wounds, or an infection of necrotizing fasciitis.

119 Hemoglobin binds oxygen. Before oxygen can diffuse out of the blood, it first must be released from hemoglobin. Thus, oxygen diffuses out of the blood much more slowly than nitrogen.

121 HBOT significantly reduces the amount of time needed to drive out the CO present in the blood.

123 a. Since both *P* and *V* are changing, while *n* and *T* are constant, we can use Boyle's law: $P_iV_i = P_fV_f$.

Define the variables and select the variable to solve for.

The problem indicates that P_i = 2.81 atm, V_i = 0.021 mL, and P_f = 1.00 atm; therefore, you need to solve for V_f, the final volume.

Algebraically isolate the unknown variable on one side of the equation.

Use algebra to manipulate the equation, so that V_f is isolated:

$$V_{\text{f}} = P_{\text{i}} \times \frac{V_{\text{i}}}{P_{\text{f}}}$$

Substitute the known values into the equation and solve for the unknown variable:

$$V_f = 2.81\ \cancel{\text{atm}} \times \frac{0.021\ \text{mL}}{1.00\ \cancel{\text{atm}}} = 0.059\ \text{mL}$$

b. Since both *P* and *V* are changing, while *n* and *T* are constant, we can use Boyle's law: $P_iV_i = P_fV_f$.

Define the variables and select the variable to solve for.
The problem indicates that P_i = 1.00 atm, V_i = 0.059 mL, and P_f = 2.25 atm; therefore, you need to solve for V_f, the final volume.

Algebraically isolate the unknown variable on one side of the equation.
Use algebra to manipulate the equation, so that V_f is isolated:

$$V_f = P_i \times \frac{V_i}{P_f}$$

Substitute the known values into the equation and solve for the unknown variable:

$$V_f = 1.00\ \cancel{\text{atm}} \times \frac{0.059\ \text{mL}}{2.25\ \cancel{\text{atm}}} = 0.026\ \text{mL}$$

Chapter 6

Organic Chemistry: Hydrocarbons

Chapter Summary

In this chapter, you started your focus on the fundamentals of organic chemistry. You have learned about the four types of hydrocarbons: alkanes, alkenes, alkynes, and aromatic hydrocarbons. You were introduced to the basic rules for naming these hydrocarbons. You also studied the structural characteristics of these compounds, including the conformations that alkanes can adopt, and learned about structural isomers and geometric isomers. Most biological compounds are organic compounds; therefore, understanding organic chemistry will help you understand the chemistry of the human body.

6.1 Introduction to Hydrocarbons

In a Nutshell: Types of Hydrocarbons

Hydrocarbons are molecules that contain exclusively hydrogen and carbon. Hydrocarbons are divided into four categories: alkanes, alkenes, alkynes, and aromatic hydrocarbons. Alkanes contain only carbon-carbon single bonds and include cycloalkanes, which contain ring structures. Alkenes contain one or more carbon-carbon double bond, and alkynes contain one or more carbon-carbon triple bond. Aromatic hydrocarbons are distinguished by their unique six-carbon ring structure containing three carbon-carbon double bonds. Saturated hydrocarbons contain the maximum number of hydrogen atoms for a given number of carbon atoms. The general formula for a saturated hydrocarbon is C_nH_{2n+2}, where n is equal to the number of carbon atoms in the formula. Saturated hydrocarbons are alkanes but do not include cycloalkanes. Cycloalkanes (alkanes containing rings), alkenes, alkynes, and aromatic compounds are classified as unsaturated hydrocarbons because they contain fewer than the maximum number of hydrogen atoms per carbon atom.

As a result of carbon's unique ability to bond up to four other carbon atoms, we find both straight-chain and branched-chain hydrocarbons. Straight-chain hydrocarbons have a chemical structure composed of a chain of carbon atoms in which each carbon has a bond to two other carbon atoms (except the carbon atoms at the end of the chain). A branched-chain hydrocarbon has one or more carbon atoms in the chain with bonds to three or four carbon atoms instead of hydrogen atoms, creating a "branch" in the chain.

All hydrocarbons are nonpolar molecules. Nonpolar molecules interact through dispersion forces, the weakest of the intermolecular forces of attraction. Consequently, hydrocarbons have some of the lowest boiling points compared to other compounds with a comparable mass. Hydrocarbons with more carbon atoms will have higher boiling points than hydrocarbons with fewer carbon atoms. For hydrocarbons with the same chemical formula, straight-chain hydrocarbons have higher boiling points than branched-chain hydrocarbons because they have more surface area, resulting in more dispersion forces. The insolubility of hydrocarbons in water gives hydrocarbons hydrophobic properties—from the Latin "water fearing."

In a Nutshell: Naming Simple Organic Compounds

Many compounds have common names that were given to them when they were first discovered. Every organic compound has been assigned a unique name based on the IUPAC system. In addition to its IUPAC name, an active pharmaceutical ingredient is given a generic name, used to identify a drug and a brand name which is associated with the pharmaceutical company that makes the drug.

An IUPAC name is generally composed of three parts: the prefix, the root, and the ending. We start by assigning the root, which indicates the number of contiguous carbon atoms in the main hydrocarbon chain of a straight-chain or branched-chain hydrocarbon. We then change the ending of the root to indicate the type of compound: Alkanes end with *-ane*, alkenes end with *-ene*, and alkynes end with *-yne*. For branched-chain hydrocarbons, we add a prefix before the root to indicate what type of branches and where along the main chain they are. To assign an IUPAC name to a straight-chain alkane, use the following rules: 1) assign the root, 2) assign the ending, and 3) assign a locator number indicating the position of the first carbon of the carbon-carbon double or triple bond to the root if a multiple bond is present. To assign the root, count the number of carbons in the chain. The ending in an IUPAC name indicates the type of hydrocarbon it is: Alkanes end in *-ane*, alkenes end in *-ene*, and alkynes end in *-yne*. A locator number is used to indicate the location of a multiple bond in an alkene or alkyne. When the main chain has more than three carbons, number the contiguous chain of carbon atoms starting from the end closer to the carbon-carbon double or triple bond. Then place a locator number separated by hyphens (-) between the root and the ending, indicating where the first carbon atom of the double or triple bond appears in the chain.

Worked Example #1

From the chemical formula, determine which one of the hydrocarbons listed below is a saturated hydrocarbon.

a. C_6H_6
b. C_6H_{14}
c. C_6H_{10}

For six carbons, a saturated hydrocarbon should contain 14 hydrogen atoms. When 6 is substituted for n in the formula C_nH_{2n+2}, we see that there are 14 hydrogen atoms. Therefore, (b), C_6H_{14}, is a saturated hydrocarbon.

Try It Yourself #1

From the chemical formula, determine which one of the hydrocarbons listed below is a saturated hydrocarbon.

a. C_8H_{18}
b. C_8H_{14}
c. C_8H_{16}

For a saturated hydrocarbon with eight carbon atoms:

Number of hydrogen atoms: ____________________

Chemical formula: __________________________

Worked Example #2

For each of the Lewis structures shown below, identify whether the molecule is an alkane, alkene, or alkyne. Indicate whether each compound is saturated or unsaturated. Assign the IUPAC name to each compound.

a.
```
  H H H H H H H H H
  | | | | | | | | |
H-C-C-C-C-C-C-C-C-C-H
  | | | | | | | | |
  H H H H H H H H H
```

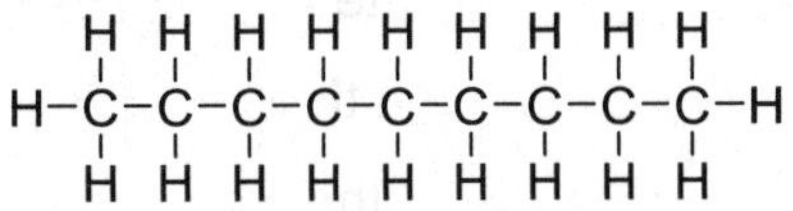

b.
```
     H
     |
  H-C-C≡C-H
     |
     H
```

c.
```
        H H H
        H\| |
   H     C-C-H
    \   /  |
     C=C   H
    /   \
   H     H
```

Solutions

a. *The compound is an alkane because it contains only carbon-carbon single bonds and (C—H) bonds. It is saturated because there are only single bonds and it has the maximum number of hydrogen atoms (twenty) for a hydrocarbon with nine carbons. The IUPAC name is nonane because it contains a chain of nine carbons, which has the root nonane. The ending remains* -ane *because it is an alkane.*

b. *The compound is an alkyne because it contains a carbon-carbon triple bond. It is unsaturated because it contains a carbon-carbon triple bond and therefore fewer hydrogens (four) than the maximum number of hydrogen atoms (eight) for a hydrocarbon with three carbons. The IUPAC name is propyne because it contains a chain of three carbons (root = propane) with a triple bond, so the ending is changed from* -ane *to* -yne*. There is no locator number because there is only one place the triple bond can be.*

c. *The compound is an alkene because it contains a carbon-carbon double bond. It is unsaturated because it contains a carbon-carbon double bond and therefore fewer hydrogens (eight) than the maximum number of hydrogen atoms (ten) for a hydrocarbon with four carbons. The IUPAC name is but-1-ene because it contains a chain of four carbons (root = butane) with a double bond, so the ending is changed from -ane to -ene. We then add a locator number to indicate where the double bond first appears when numbering the chain of carbons starting from the left side because that side is closer to the double bond. Since the carbon-carbon double bond is between C(1) and C(2), the locator number 1- is inserted between the root and the ending: but-1-ene.*

Try It Yourself #2

For each of the Lewis structure shown below, identify whether the molecule is an alkane, alkene, or alkyne. Indicate whether each compound is saturated or unsaturated. Assign the IUPAC name to each compound.

a.
```
  H H       H
  | |       |
H-C-C-C≡C-C-H
  | |       |
  H H       H
```

b.
```
          H
          |
  H      C-H
   \    /   \
H   C=C     H
 \ /     \
  C       H
 / \
H   H
```

c.
```
  H H H H H H H H
  | | | | | | | |
H-C-C-C-C-C-C-C-C-H
  | | | | | | | |
  H H H H H H H H
```

Solutions

a. *Type of carbon-carbon bonds present:* ____________

Type of hydrocarbon: ____________________________

Number of carbon atoms: ____________________

Number of hydrogen atoms: ____________________

General formula for a saturated hydrocarbon: _________________________

The hydrocarbon is (saturated or unsaturated).

Number of carbon atoms in the main chain: ______

Root name: __________________

Ending for the carbon-carbon multiple bond: __________

Locator number for the double bond: __________

IUPAC name: ______________________

b. *Type of carbon-carbon bonds present:* ____________

Type of hydrocarbon: ____________________________

Number of carbon atoms: ____________________

Number of hydrogen atoms: ____________________

General formula for a saturated hydrocarbon: _________________________

The hydrocarbon is (saturated or unsaturated).

Number of carbon atoms in the main chain: ______

Root name: ________________

Ending for the carbon-carbon multiple bond: __________

Locator number for the double bond: __________

IUPAC name: ______________________

c. *Type of carbon-carbon bonds present:* ____________

Type of hydrocarbon: __________________________

Number of carbon atoms: ____________________

Number of hydrogen atoms: ____________________

General formula for a saturated hydrocarbon: ________________________

The hydrocarbon is (saturated or unsaturated).

Number of carbon atoms in the main chain: ______

Root name: ________________

IUPAC name: ______________________

Practice Problems for Hydrocarbons

1. Which of the following structures represent a saturated hydrocarbon?
   a. C_3H_6

   b. C_4H_{10}

   c. $C_{12}H_{24}$

   d. $C_{11}H_{22}$

   e. $C_{13}H_{28}$

2. For each of the structural formulas shown below, identify whether it is an alkane, alkene, or alkyne.

   a.

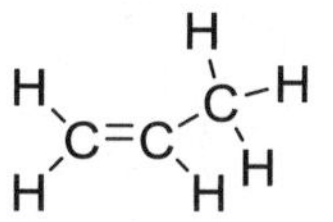

```
          H
 H        |
  \      C-H
   C=C /  \
  /    \   H
 H      H
```

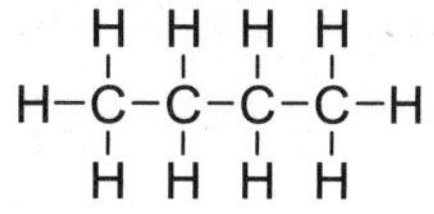

   b.

```
   H H H H
   | | | |
 H-C-C-C-C-H
   | | | |
   H H H H
```

   c.

```
   H       H
   |       |
 H-C-C≡C-C-H
   |       |
   H       H
```

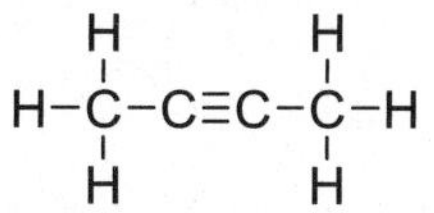

3. Classify the hydrocarbons in Question 2 as saturated or unsaturated. Assign an IUPAC name to each hydrocarbon in Question 2.

6.2 Writing Alkane and Cycloalkane Structures

Alkanes contain only carbon-carbon single bonds as well as carbon-hydrogen bonds; they contain no multiple bonds. Every carbon atom in an alkane has a tetrahedral shape. The tetrahedral shape of the carbon atoms gives these molecules an overall zigzag shape when there are three or more carbon atoms present.

Lewis structures show every bond and atomic symbol; writing Lewis structures for large organic molecules is tedious and time consuming. Two simpler notations for writing chemical structures are in common use: condensed structures and skeletal line structures.

In a Nutshell: Condensed Structures

In writing a condensed structure, we start at one end of the molecule and work our way to the other end of the molecule writing each carbon atom and its attached hydrogen atom(s) as a group: C, CH, CH_2, or CH_3. Bonds are omitted, except in the case of branch points. In writing the condensed structure for long, straight-chain alkanes, the repeating CH_2 groups are often indicated by writing a single CH_2 enclosed in parentheses followed by a subscript that indicates the number of repeating CH_2 groups.

The condensed structure for a branched-chain hydrocarbon is written the same way as that of a straight-chain hydrocarbon, except branch points along the chain are shown by including the branching bond(s) as a line pointing up or down from the carbon chain at its branch point. The branch itself is then written as a sequence of carbon atoms and its attached hydrogen atoms.

Worked Example #3

For each condensed structures below, determine whether it represents a straight-chain alkane or a branched-chain alkane, then write the corresponding Lewis structure.

a. $CH_3CH_2CH_2CH_2CH_2CH_2CH_2CH_3$

b.
$$\begin{array}{l} CH_3 \\ | \\ CH_3CHCH_2CH_2CH_3 \end{array}$$

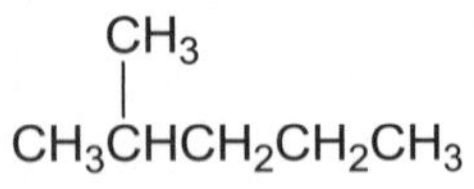

c.
$$\begin{array}{l} CH_3 \\ | \\ CH_3CH_2CHCH_2CHCH_2CH_3 \\ | \\ CH_2CH_3 \end{array}$$

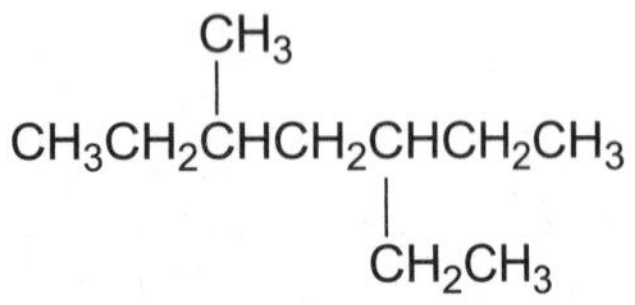

a. *Straight-chain alkane because there are no bonds indicating branch points along the chain.*

$$\begin{array}{c} HHHHHHHH \\ H-C-C-C-C-C-C-C-C-H \\ HHHHHHHH \end{array}$$

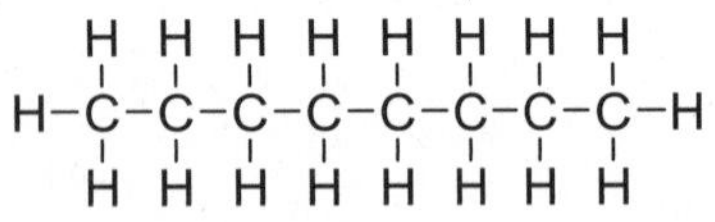

b. *Branched-chain alkane. There is one branching group off the second carbon in the chain, indicated by the bond shown projecting from the carbon atom.*

```
       H
       |
     H-C-H
   H   |   H H H
   |   |   | | |
 H-C---C---C-C-C-H
   |   |   | | |
   H   H   H H H
```

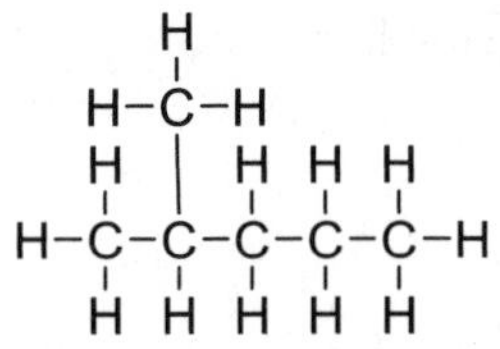

c. *Branched-chain alkane. There are two branching groups off the third and fifth carbons in the chain, indicated by the bonds shown projecting from the carbon atom.*

```
         H
         |
       H-C-H
   H H   |   H H H H
   | |   |   | | | |
 H-C-C---C---C-C-C-C-H
   | |   |   |   | |
   H H   H   H   H H
             |   H
             |   |
           H-C---C-H
             |   |
             H   H
```

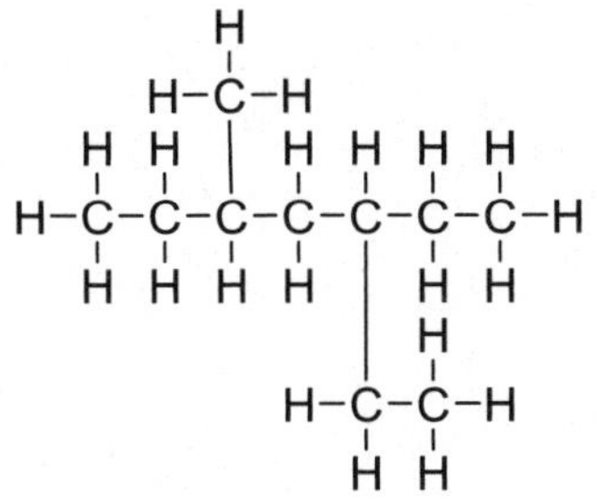

Try It Yourself #3

For each condensed structure below, determine whether it represents a straight-chain alkane or a branched-chain alkane, then write the corresponding Lewis structure.

a.
```
   CH2CH3
   |
CH3CHCH2CHCH3
        |
        CH3
```

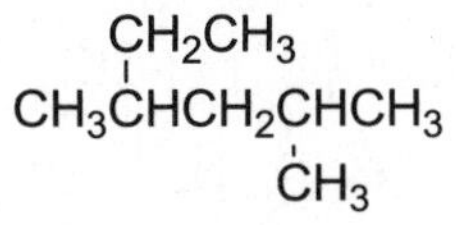

b. $CH_3(CH_2)_5CH_3$

c.
```
   CH3
   |
CH3CH
   |
   CH3
```

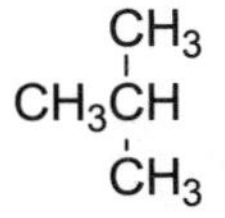

a. *Are there branching points in the chain?* ________

It is a ______________________ *-chain alkane.*

Lewis structure:

b. Are there branching points in the chain? ________

It is a ______________________ -chain alkane.

Lewis structure:

c. Are there branching points in the chain? ________

It is a ______________________ -chain alkane.

Lewis structure:

In a Nutshell: Skeletal Line Structures

Skeletal-line structures are an even more efficient shorthand for writing large molecular structures and have a clean appearance. They only show the carbon atom linkages uncluttered by hydrogen atoms. There are general rules for writing skeletal line structures:

1. Carbon-carbon bonds in a contiguous chain of carbon atoms are written in a zigzag fashion, with the symbol for carbon, C, omitted as well as all hydrogen atoms and C—H bonds. A carbon atom is implied wherever two lines (representing carbon-carbon bonds) come together as a point and at the terminal ends of a line.
2. Double bonds are shown as two parallel lines, ═, and triple bonds are shown as three parallel lines, ≡.
3. Carbon branches are drawn above the zigzag chain when the branching point is up and below the zigzag chain when the branching point is down.
4. A heteroatom (an atom other than carbon and hydrogen) must be written in at a point in order to distinguish them from carbon atoms. All hydrogen atoms bonded to heteroatoms (O—H, NH_2, etc.) must also be written in.
5. To determine the number of hydrogen atoms on a particular carbon atom from a skeletal-line structure, count the number of bonds shown and subtract this value from 4 because the octet rule tells us a carbon atom must always have four bonds.

Worked Example #4

Write the skeletal line structure that corresponds to each of the condensed structures shown below:

a. $CH_3(CH_2)_5CH_3$

b.

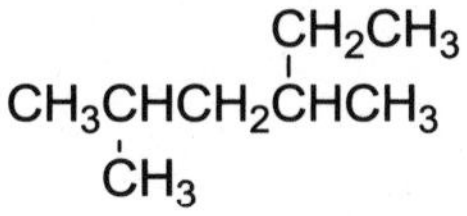

$$\begin{array}{c} \qquad\qquad CH_2CH_3 \\ CH_3CHCH_2CHCH_3 \\ CH_3 \qquad\qquad \end{array}$$

Solutions

a.  *Draw a line for every carbon-carbon bond.*

b. 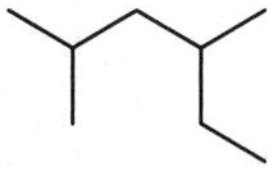 *Draw a line for every carbon-carbon bond.*

Try It Yourself #4

Write the skeletal line structure that corresponds to each of the condensed structures shown below:

a.

$$\begin{array}{c} CH_2CH_3 \qquad\qquad \\ CH_3CHCHCH_2CH_3 \\ CH_2CH_3 \qquad \end{array}$$

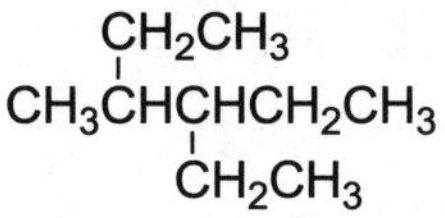

b. $CH_3(CH_2)_9CH_3$

Solutions

a. *Skeletal line structure:*

b. *Skeletal line structure:*

Worked Example #5

Write the Lewis structure and the condensed structure for the skeletal line structures shown below.

a.

b.

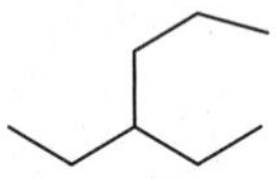

Solutions

a. *The end of each line and the intersection of two lines represent a carbon atom. The condensed structure is $CH_3CH_2CH_2CH_3$. The Lewis structure is:*

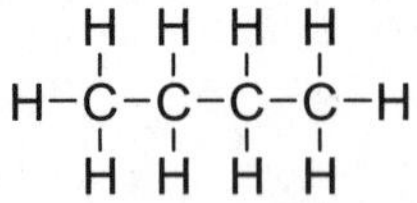

```
  H H H H
  | | | |
H-C-C-C-C-H
  | | | |
  H H H H
```

b. *The end of each line and the intersection of two lines represent a carbon atom. The condensed structure is* $\begin{matrix} & CH_2CH_2CH_3 \\ CH_3CH_2 & CHCH_2CH_3 \end{matrix}$ *. The Lewis structure is:*

```
      H
      |
    H-C-H
      |
    H-C-H
      |
    H-C-H
      |
  H H | H H
  | | | | |
H-C-C-C-C-C-H
  | | | | |
  H H H H H
```

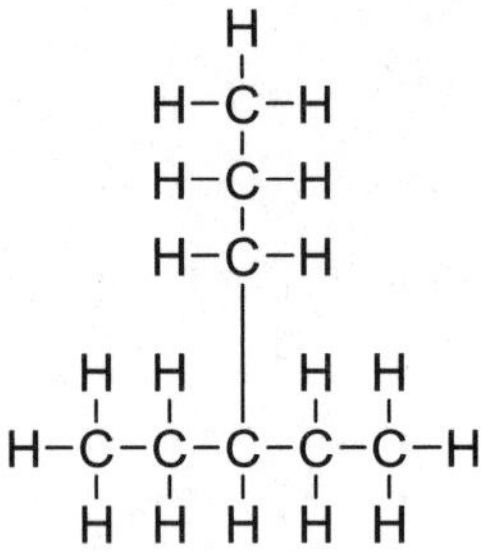

Try It Yourself #5

Write the Lewis structure and the condensed structure for the skeletal line structures shown below.

a.

b.

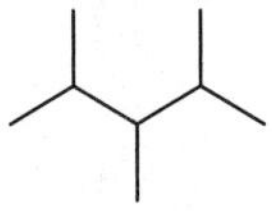

a. Lewis structure:

Condensed structure:

b. Lewis structure:

Condensed structure:

Worked Example #6

Indicate the number of hydrogen atoms on each of the carbon atoms in the skeletal line structures below:

a.

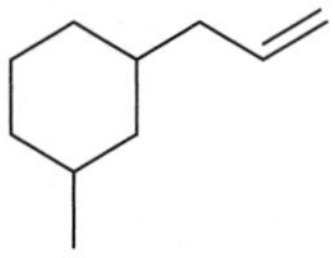

b.

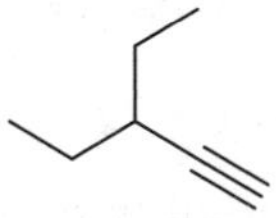

Solutions

a.

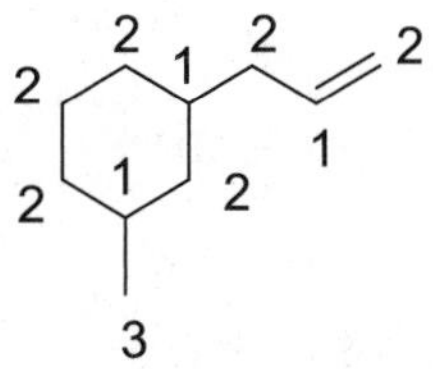

b.

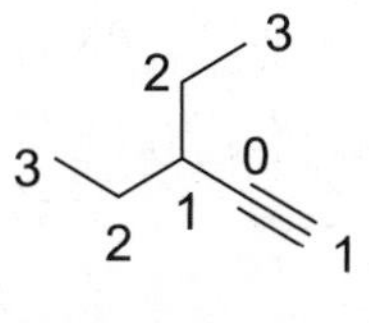

Try It Yourself #6

Indicate the number of hydrogen atoms on each of the carbon atoms in the skeletal line structures below:

a.

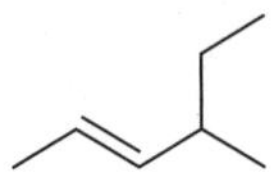

b.

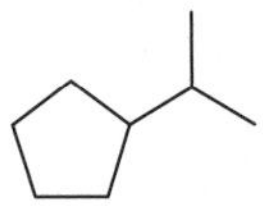

Solutions:

a.

b.

In a Nutshell: Cycloalkanes

A cycloalkane is an alkane whose chain of carbon atoms is joined in a way that forms a ring structure. Cycloalkanes are unsaturated hydrocarbons because they contain fewer than the maximum number of hydrogen atoms for a given number of carbon atoms, *n*. The structure of a cycloalkane is written as a skeletal line structure. Cycloalkanes are written as polygons with three or more sides. A carbon atom is present at each corner of the polygon with two hydrogen atoms attached.

The most common ring sizes encountered in nature are five-and six-membered rings. Three- and four-membered rings are less common because the rings are strained, which makes them less stable. Ring strain exists in cycloalkanes that have geometries where they can't achieve the required 109.5° bond angle that is necessary for a tetrahedral molecular shape. Cycloalkanes, with the exception of cyclopropane, are not flat planar structures.

Worked Example #7

Name the cycloalkanes show below.

a. 

b. 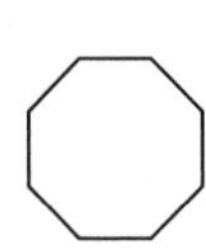

a. The IUPAC name for this compound is cycloheptane because it is a ring (cyclo) containing seven (heptane) carbon atoms with only carbon-carbon single bonds (-ane).

b. The IUPAC name for this compound is cyclooctane because it is a ring (cyclo) containing eight (octane) carbon atoms with only carbon-carbon single bonds (-ane).

Try It Yourself #7

Write the IUPAC names for the cycloalkanes shown below.

a. 

b. 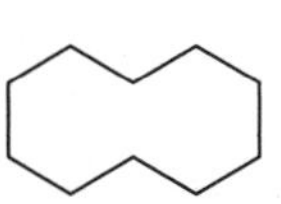

a. Number of carbon atoms in cycloalkane: _______

*Root name:*_________________

*IUPAC name:*________________

b. Number of carbon atoms in cycloalkane: _______

*Root name:*___________________

*IUPAC name:*_____________________

Practice Problems for Writing Alkane and Cycloalkane Structures

1. Write the Lewis structure and the condensed structure for the skeletal-line structures below.

   a.

   b.

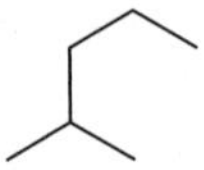

   c.

2. Write the skeletal line structure that corresponds to each of the condensed structures below.

   a.
   $$\begin{array}{c} \quad CH_2CH_3 \\ CH_3CH_2CHCH_2CHCH_3 \\ \qquad\quad CH_3 \end{array}$$

b. $CH_3(CH_2)_{10}CH_3$

c.

$$\begin{array}{c} \quad CH_3 \\ \quad | \\ CH_3CHCHCH_2CH_2CH_2CH_3 \\ \qquad | \\ \qquad CH_3 \end{array}$$

3. In Questions 1 and 2, which molecules are branched-chain alkanes and which molecules are straight-chain alkanes?

4. How many carbon atoms are there in the following cycloalkanes? Write the IUPAC name for the cycloalkane in part c.

a.

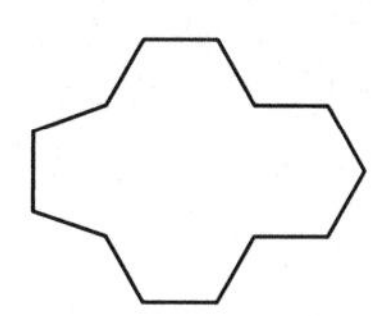

b.

c.

6.3 Alkane Conformations and Structural Isomers

In a Nutshell: Conformations

Alkanes are freely rotating about each of their carbon-carbon single bond (unless the bond is part of a ring or other structural constraint). An alkane has many different rotational forms, known as conformations. One conformation can be converted to another by rotation about one or more carbon-carbon single bonds. Whenever we write a three-dimensional representation of a molecule or build a model, we are choosing one of many possible conformations.

In a Nutshell: Structural Isomers

Molecules with the *same* chemical formula but a *different* chemical structure are known as structural isomers. Structural isomers are different chemical compounds with different chemical properties and different IUPAC names. The straight-chain isomer will have a higher boiling point than the most highly branched isomer because the straight-chain isomer has a greater surface area that creates more opportunity for dispersion forces. The number of structural isomers that exist for any chemical formula increases as the number of carbon atoms in the chemical formula increases.

Worked Example #8

Do the following pairs of molecules represent different conformations of the same molecule or different molecules? If they are different molecules, explain what makes them different.

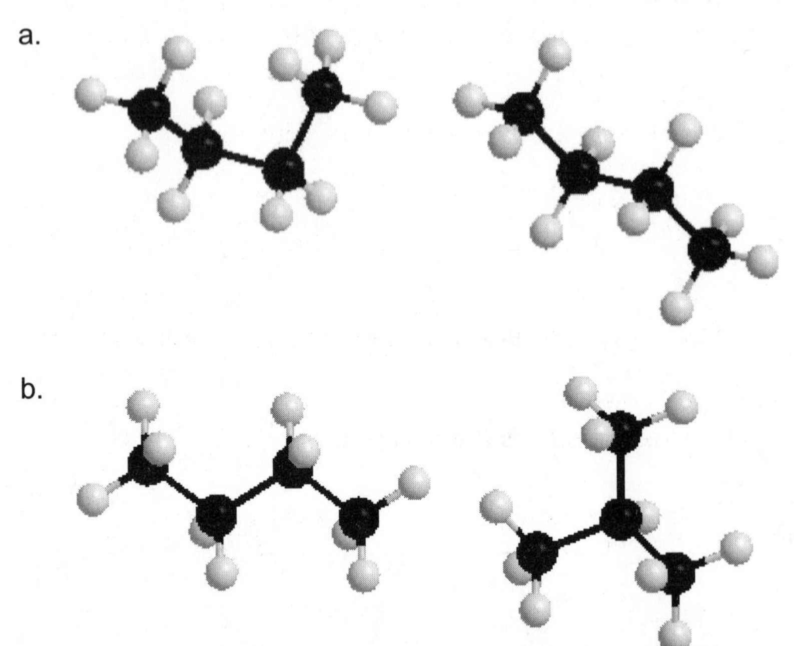

Solutions

a. *The two molecules are the same molecule; they are conformations of the same molecule. The carbon-carbon bond in the middle of the molecule has been rotated.*
b. *These two molecules are different. They have a different connectivity of the atoms. The first model has the four carbon atoms connected in a continuous chain. The second model has a central carbon atom bonded to three other carbon atoms.*

Try It Yourself #8

Do the following pairs of molecules represent different conformations of the same molecule or different molecules? If they are different molecules, explain what makes them different.

a.

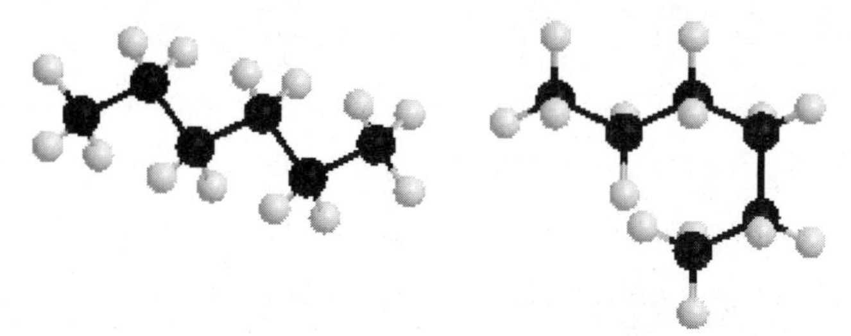

b.

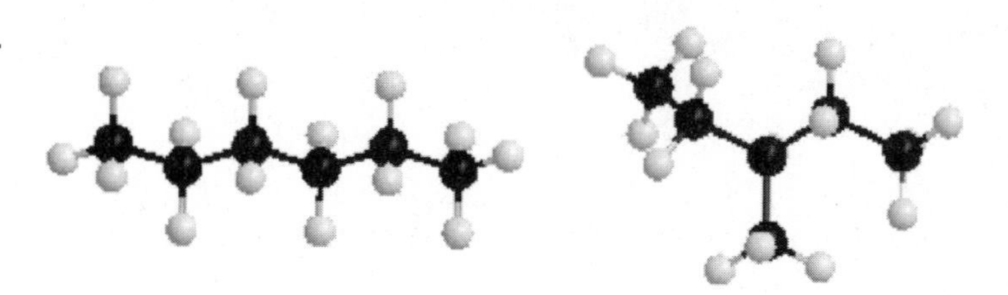

a. *The number of carbon and hydrogen atoms in the molecules is the (same or different) ________________________.*

The connectivity of the carbon atoms is the (same or different)

______________________.

They are the (same or different) ________________________ molecule(s).

b. *The number of carbon and hydrogen atoms in the molecules is the (same or different) ________________________.*

The connectivity of the carbon atoms is the (same or different)

______________________.

They are the (same or different) ________________________ molecule(s).

Worked Example #9

Which of the following pairs are not structural isomers?

a.

```
                              H
                              |
                            H-C-H
   H  H  H  H           H  H  |  H
   |  |  |  |           |  |  |  |
 H-C--C--C--C-H       H-C--C--C--C-H
   |  |  |  |           |  |  |  |
   H  |  H  H           H  H  H  H
      |
    H-C-H
      |
      H
```

b.

```
   H  H  H  H            H  H  H  H  H
   |  |  |  |            |  |  |  |  |
 H-C--C--C--C-H        H-C--C--C--C--C-H
   |  |  |  |            |  |  |  |  |
   H  |  H  H            H  H  H  H  H
      |
    H-C-H
      |
      H
```

c.

Solutions

a. *The two molecules are not structural isomers. They both have a four-carbon chain with a CH_3 group bonded to the second carbon from the end of the chain. Note: It doesn't matter which end you start counting.*

b. *The two molecules are structural isomers. They have the same number of carbon atoms, but the connectivity of the carbon atoms is different.*

c. *The two molecules are not structural isomers. They have the same number of carbon atoms, but the atoms have been rotated around a bond.*

Try It Yourself #9

Which of the following pairs are not structural isomers?

a.

```
                              H
                              |
                            H-C-H
                              |
  H H H H                   H | H
  | | | |                   | | |
H-C-C-C-C-H               H-C-C-C-H
  | | | |                   | | |
  H H H H                   H H H
```

b.

```
  H H H H                   H H H H H H
  | | | |                   | | | | | |
H-C-C-C-C-H               H-C-C-C-C-C-C-H
  | | | |                   | | | | | |
  H | H H                   H H H H H H
    |
  H-C-H
    |
  H-C-H
    |
    H
```

c.

```
  H H H                       H
  | | |                       |
H-C-C-C-H                   H-C-H
  | | |                       |
  H | H                     H | H
    |                       | | |
  H-C-H                   H-C-C-C-H
    |                       | | |
    H                       H H H
```

a. *The molecules (are or are not) ________________ structural isomers.*

b. *The molecules (are or are not) ________________ structural isomers.*

c. *The molecules (are or are not) ________________ structural isomers.*

Worked Example #10

Write the Lewis structure for two of the nine structural isomers for C_7H_{16}.

Start with the straight chain isomer.

```
  H H H H H H H
  | | | | | | |
H-C-C-C-C-C-C-C-H
  | | | | | | |
  H H H H H H H
```

For the next isomer, move a CH_3 group from the end carbon to the next carbon over.

```
   H  H  H  H  H  H
   |  |  |  |  |  |
H--C--C--C--C--C--C--H
   |  |  |  |  |  |
   H  |  H  H  H  H
      |
   H--C--H
      |
      H
```

To find the next isomer, keep moving the CH_3 to the next carbon on the backbone.

Try It Yourself #10

Write the Lewis structure for two more of the nine structural isomers for C_7H_{16} (not the ones listed in Worked Example #10).

Structure of first isomer:

Structure of second isomer:

Practice Problems for Alkane Conformations and Structural Isomers

1. Do the following pairs of molecules represent different conformations of the same molecule or different molecules? If they are different molecules, explain what makes them different.

   a.

```
                                     H
                                     |
                                  H--C--H
                                     |
                                  H--C--H
   H  H  H  H  H               H  H  |
   |  |  |  |  |               |  |  |
H--C--C--C--C--C--H         H--C--C--C--H
   |  |  |  |  |               |  |  |
   H  H  H  H  H               H  H  H
```

b.

```
                                  H
                                  |
                                H-C-H
                                  |
                                H-C-H
    H H H H H               H H   |   H
    | | | | |               | |   |   |
  H-C-C-C-C-C-H           H-C-C-C-C-H
    | | | | |               | | | |
    H H H H H               H H H H
```

2. Which of the following pairs are structural isomers?

a.

```
    H H H H             H H H H
    | | | |             | | | |
  H-C-C-C-C-H         H-C-C-C-C-H
    | | | |             | | | |
    H H H H             H | H H
                          |
                        H-C-H
                          |
                          H
```

b.

```
    H H H H H           H H H H H H
    | | | | |           | | | | | |
  H-C-C-C-C-C-H       H-C-C-C-C-C-C-H
    | | | | |           | | | | | |
    H | H H H           H H H H H H
      |
    H-C-H
      |
      H
```

c.

6.4 Alkenes and Alkynes

Alkenes, alkynes, and aromatic hydrocarbons are unsaturated hydrocarbons because they contain one or more carbon-carbon double or triple bonds; therefore. they have fewer than the maximum number of hydrogen atoms per carbon.

In a Nutshell: Alkenes

An alkene is a hydrocarbon that contains one or more carbon-carbon double bonds. The bond angles around each of the carbon atoms that form a double bond are 120°. In a condensed structure for an alkene, the double bond may be drawn in or inferred. In the skeletal-line structure, an alkene is represented by two parallel lines. An alkene containing two carbon-carbon double bonds is referred to as a diene. An alkene containing several carbon-carbon double bonds is called a polyene. Polyenes with alternating double and single bonds are known as conjugated polyenes.

Worked Example #11

Identify the skeletal line structure below as a simple alkene, diene, or polyene.

a.

b.

c.

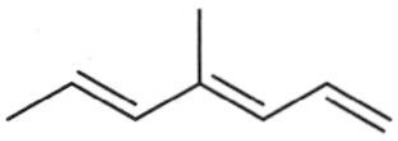

Solutions

a. *There are two double bonds in the structure; therefore, it is a diene.*
b. *There is one double bond in the structure; therefore, it is a simple alkene.*
c. *There are more than two double bonds in the structure; therefore, it is a polyene.*

Try It Yourself #12

Identify the skeletal line structure below as a simple alkene, diene, or polyene.

a.

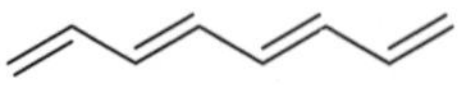

b.

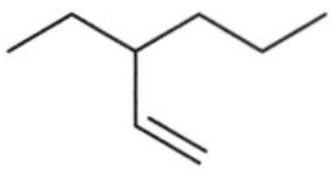

c.

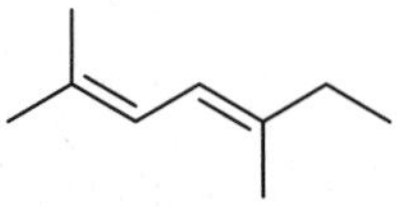

Solutions

a. *Number of double bonds:* _______________

Type of alkene: ____________________

b. *Number of double bonds:* ______________

Type of alkene: ____________________

c. *Number of double bonds:* ______________

Type of alkene: ____________________

In a Nutshell: Geometric Isomers

The most significant difference between carbon-carbon single bonds and carbon-carobn double bonds is that double bonds cannot rotate freely about the double bond; therefore, geometric isomers are possible. Geometric isomers have the same chemical formula *and* same connectivity of atoms, but a different three-dimensional orientation of the atoms as the result of the restricted rotation about the carbon-carbon double bond. When two larger groups are on the same side of the double bond, the geometric isomer is called the cis isomer. When two larger groups are on opposite sides of the double bond, the geometric isomer is called the trans isomer. Geometric isomers are named according to the IUPAC rules for naming alkenes with the added *cis*- or *trans*- included as a prefix in front of the root name.

Worked Example #13

Draw the geometric isomer for the following compounds. Indicate if the isomer shown is cis or trans.

a.

b.

a. *To draw the geometric isomer, exchange the two groups attached to one carbon of the double bond.*

The isomer shown in the question is the trans isomer; the two groups (CH_3 and CH_2CH_3) are on opposite sides of the double bond.

b. *To draw the geometric isomer, exchange the two groups attached to one carbon of the double bond.*

The isomer shown in the question is the cis isomer; the two groups (CH_3CH_2 and CH_3CH_2) are on the same side of the double bond.

Try It Yourself #13

Draw the geometric isomer for the following compounds. Indicate if the isomer shown is cis or trans.

a.

b.

a. *Are the two groups on the same side of the double bond?* _______

The geometric isomer shown is: ___________

The other geometric isomer: _____________

The structure of the other geometric isomer:

b. Are the two groups on the same side of the double bond? _______

The geometric isomer shown is: ___________

The other geometric isomer: ______________

The structure of the other geometric isomer:

Worked Example #14

Draw the skeletal line structure of the following compound, *trans*-3-hexene.

Solution

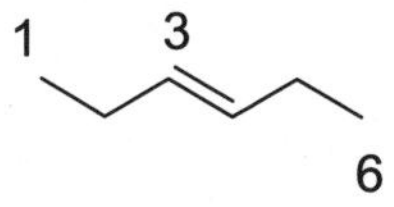

1 3

6

Write a zigzag structure showing six points and place a double bond between the third and fourth carbons. To indicate the trans isomer, place the two large groups on the opposite sides of the double bond.

Try It Yourself #14

Draw the skeletal line structure of the *cis*-3-octene.

Skeletal line structure:

In a Nutshell: Alkynes

An alkyne contains one or more carbon-carbon triple bond. Alkynes have a linear shape around the carbon atoms that form the carbon-carbon triple bond. When writing the skeletal line structure of a triple bond use three parallel lines and form a bond with 180° bond angles around the triple bond carbon(s). Alkynes are not common in nature, although they are found in some compounds.

Worked Example #15

Write the skeletal line structure and the Lewis structure for the following alkynes:

a. hept-2-yne
b. hex-3-yne

a. *The root for the alkyne is heptane, which indicates seven carbon atoms in the chain. The -yne ending indicates that a triple bond is present. The -2- in hept-2-yne indicates that the triple bond is located between C(2) and C(3). Generally, bonds attached to the triple bond carbon are shown in a linear arrangement rather than a zigzag.*

```
  H H H H       H
  | | | |       |
H-C-C-C-C-C≡C-C-H
  | | | |       |
  H H H H       H
```

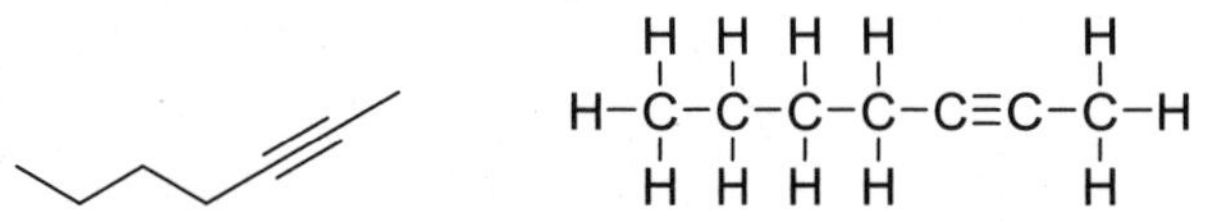

b. *The root for the alkyne is hexane, which indicates six carbon atoms in the chain. The -yne ending indicates that a triple bond is present. The -3- in hex-3-yne indicates that the triple bond is located between C(3) and C(4). Generally, bonds attached to the triple bond carbon are shown in a linear arrangement rather than a zigzag.*

```
  H H     H H
  | |     | |
H-C-C-C≡C-C-C-H
  | |     | |
  H H     H H
```

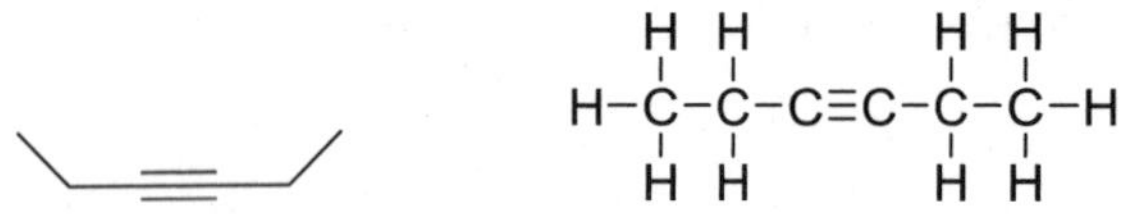

Try It Yourself #15

Write the skeletal line structure and the Lewis structure for the following alkynes:

a. oct-1-yne
b. hept-3-yne

Solutions

a. *Root name:*________________

*Number of carbons in chain:*________________

-yne *indicates:*________________

Locator number:______________

The triple bond is between C___ and C___.

Structural line structure:

Lewis structure:

b. *Root name:______________*

Number of carbons in chain:________________

-yne *indicates:______________*

Locator number:______________

The triple bond is between C___ and C___.

Structural line structure:

Lewis structure:

Practice Problems for Unsaturated Hydrocarbons: Alkenes and Alkynes

1. Identify the structure below as a simple alkene, diene, or polyene.

   a.

   b.

c.

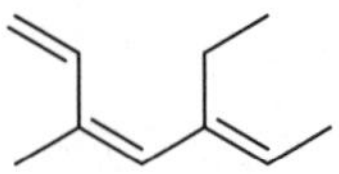

2. In the following molecule, identify the alkene and the alkyne.

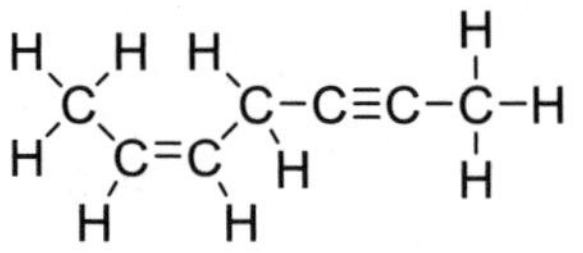

```
H   H H           H
 \ /   \          |
  C     C-C≡C-C-H
 / \   /  \       |
H   C=C    H      H
    |  \
    H   H
```

3. Write the geometric isomer for the compounds shown. Indicate whether the isomer that you have drawn is the cis or trans isomer.

a.

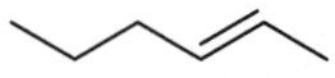

b.

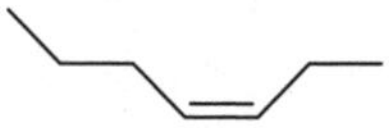

6.5 Naming Branched-Chain Hydrocarbons

In branched-chain hydrocarbons, one or more hydrogen atoms along the main carbon chain are *substituted* by one or more carbon branches containing one or more carbons, known as substituents. The root is assigned based on the number of carbons in the longest contiguous chain—known as the main chain—or parent chain. A prefix is then inserted in front of the root that indicates where and what substituents are found in the main chain. An alkyl substituent consists only of carbon and hydrogen atoms, using the same guidelines to name the main chain, except the ending is changed from *-ane* to *-yl.* To assign an IUPAC name to a branched-chain alkane, use the following rules: 1a) Locate and assign the root to the main chain; 1b) if there is a carbon-carbon double or triple bond, select the longest contiguous chain that contains both carbon atoms of the double or triple bond as the main chain; 1c) if a molecule has two possible main chains with the same number of carbon atoms, choose the chain with the greater number of substituents as the main chain; 2) name each substituent; 3) assign a locator number to each substituent; 4a) assemble the prefix; 4b) use a multiplier if a substituent appears more than once; and 5) assemble the IUPAC name.

The root name is determined by finding the longest continuous carbon chain containing the carbon atoms in the carbon-carbon double or triple bond if one exists, even if there is a longer chain in the molecule. Note: The main chain is not always the one that lies horizontally on the page. The ending is changed to reflect whether the molecule is an alkane, alkene, or alkyne. If there is a multiple bond, a locator number is placed immediately before the root name to indicate the location of the first carbon of a multiple bond. The substituents are assigned a name based on the number of carbon atoms it contains. Change the ending from *-ane* to *-yl* to signify that it is a substituent and not the main chain. Assign a locator number to each substituent. Number the main chain starting from the end closer to the first substituent (first branch point), unless there is a multiple bond, in which case the numbering must begin from the end closer to the multiple bond. Assign a locator number to each substituent based on where along the main chain it is located. If there are *two* substituents on the same carbon atom, cite the locator number twice, once for each substituent. Place each locator number before its associated substituent name separated by a hyphen. Assemble the prefix. List the substituent names in alphabetical order preceded by each of their locator numbers. Place the prefix before the root. Letters should be separated from numbers by a hyphen and numbers separated from numbers by commas. Use a multiplier if a substituent appears more than once. If the same substituent appears more than once along the main chain, insert the multiplier *di-* (for 2), *tri-* (for 3), or *tetra-* (for 4) in front of the repeating name to indicate how many times the substituent appears along the main chain. Place locator numbers in front of the multiplier, separated by commas, corresponding to each of the repeated substituents. If there are two substituents on the same carbon atom of the main chain, the locator number must be written twice, once for each substituent. Assemble the full IUPAC name by writing the prefix, the root and the ending in that order. Separate number from letters with a hyphen, and separate numbers from numbers with commas.

Worked Example #16

Provide the IUPAC name for the structure shown below:

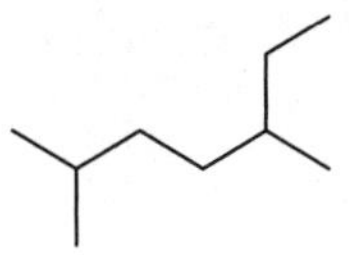

The IUPAC name is 2,5-dimethylheptane. The longest continuous chain of carbon atoms contains seven carbon atoms, so according to Table 6-4 the root name is heptane. The heptane main chain is numbered from the end closer to the substituents, so numbering starts from the left side as drawn.

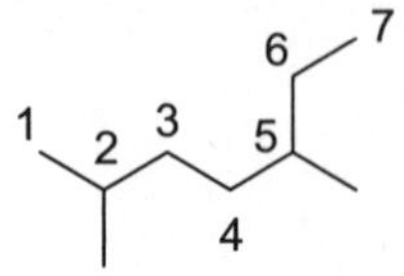

The prefix is 2,5-dimethyl because there are two –CH_3 groups: one on carbon 2 and one on carbon 5. Assembling the prefix, main chain, and ending gives 2,5-dmethylheptane. Write the prefix followed by the root and suffix. The name of the compound is 2,5-dimethylheptane.

Try It Yourself #16

Write the IUPAC name for the following branched chain alkane.

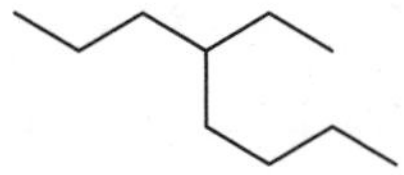

Number of carbon atoms in the longest contiguous chain: ________________

Root name: __________________________

Ending: ______________

Name each substituent.

Number of carbon atoms in substituent: ________________

Root name for substituent: ______________________

Substituent name: ___________________________

Assign a locator number to the substituent.

Locator number: _________

Prefix: ____________________________________

Assemble the IUPAC name.

IUPAC name: ______________________________

Worked Example #17

Write the skeletal line structure of 2-methyl-3-ethylpent-2-ene.

Solution

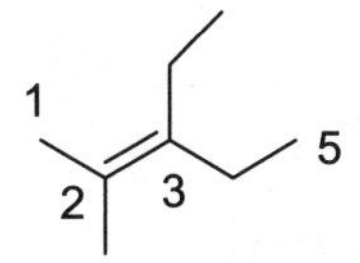

1
5
2 3

The root name "pentene" indicates a five-carbon chain. The ending -ene *indicates that there is a double bond, and the locator number 2 indicates that the double bond is between C(1) and C(2). A methyl group, CH_3-, is placed on C(2) and an ethyl group, CH_3CH_2-, is placed on C(3).*

Try It Yourself #17

Write the skeletal line structure of 4-ethyl-3-methyloct-2-ene.

Solution

Number of carbon atoms in the main chain: _____________

Any multiple bonds present? _________________

*What type of multiple bond and where is it located?*______________________

Substituents on carbon number(s): ____________________

Type of substituent: _____________________________

Skeletal line structure:

Practice Problems for Naming Branched-Chain Hydrocarbons

1. Write the IUPAC name of the following compounds. (There are no cis-trans isomers).

   a.

   b.

   c.

   d.

2. Write the Lewis structure and skeletal line structure for the following compounds.

   a. 3-methylhex-3-ene

b. methylcyclohexane

c. 2-methylpentane

d. 2,3,4,6-tetramethylheptane

6.6 Aromatic Hydrocarbons

In a Nutshell: Benzene

The simplest aromatic hydrocarbon is benzene. From the skeletal line structure, each carbon atom in benzene appears to have one carbon-carbon single bond and one carbon-carbon double bond. However, since it is a conjugated ring system the six electrons from the second pair of electrons in double bond are distributed evenly across all six carbon atoms, which we refer to as delocalized electrons. Hence, all the carbon-carbon bonds in benzene are the same length and strength, somewhere longer and weaker than a double bond but shorter and stronger than a single bond. The delocalization of electrons in an aromatic ring makes the benzene ring a particularly stable molecule. The stability of an aromatic ring makes it much less likely to undergo chemical reactions that involve breaking the aromatic ring carbon-carbon bonds. Each carbon atom in benzene has a trigonal planar geometry and every C—C—C bond angle is 120°, giving the overall molecule a flat two-dimensional shape. The condensed structure used to represent a benzene ring is C_6H_5- or Ph- (phenyl).

In a Nutshell: Naming Substituted Benzenes

Benzene is the IUPAC name for C_6H_6 and is the root name for hydrocarbons containing substituted benzene rings. The substituent is named according to the number of carbon atoms that it contains, with the ending changed to *-yl* to indicate that it is a substituent and not the main chain. No locator name is needed for one substituent because every position is identical. Locator numbers are needed when there is more than one substituent to indicate the relative position of the substituents. The benzene ring is numbered beginning at one substituent and working clockwise or counterclockwise, whichever gives a lower number to the second substituent. If there are two equal choices, numbering begins on the carbon containing the substituent that appears first alphabetically. When two identical substituents are present, the multiplier *di-* is inserted in front of the substituent name.

Worked Example #18

Provide the IUPAC name for the following compound.

CH_3

CH_2CH_3

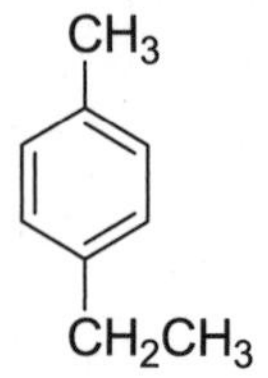

The IUPAC name is 1-ethyl-4-methylbenzene. The root is benzene because the main chain is the aromatic ring. The ethyl group, CH_3CH_2-, comes first alphabetically, so the carbon containing the ethyl group is given the locator number 1. Counting around to the methyl group, CH_3, gives the locator number 4.

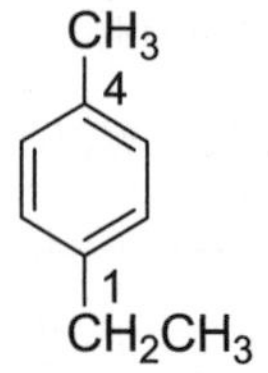

CH_3
4

1
CH_2CH_3

Try It Yourself #18

Provide the IUPAC name for the following compound.

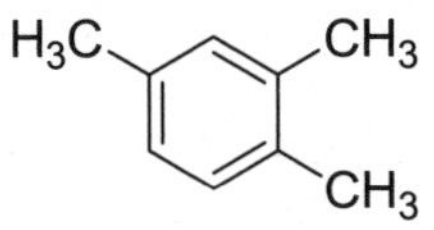

H_3C CH_3 CH_3

Root name: ____________________________

Name each substituent.

Number of carbon atoms in substituent: __________________

Root name for substituent: ________________________

Substituent name: ______________________________

Assign a locator number to the substituent.

Locator number: _________

Prefix: ____

Assemble the IUPAC name.

IUPAC name: __________________________________

Worked Example #19

Provide a skeletal line structure for 1-methyl-3-propylbenzene.

Solution

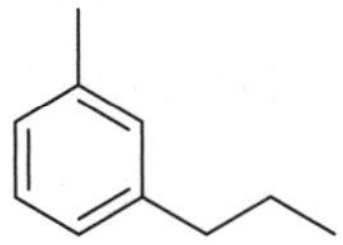

A methyl group is attached to C(1) and a propyl (three carbon atoms) group is attached to C(3).

Try It Yourself #19

Provide a skeletal line structure for 1,2,4-triethylbenzene.

Structure:

Practice Problems for Aromatic Hydrocarbons

1. Write the IUPAC name of the following compounds:

   a. CH_3 / $CH_2CH_2CH_3$

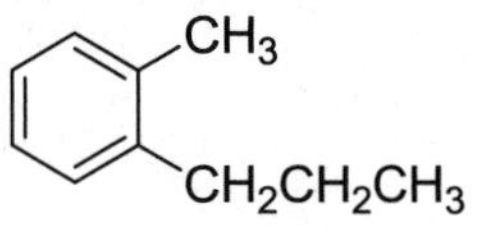

   b. $H_3CH_2CH_2C$ / $CH_2CH_2CH_3$ / $CH_2CH_2CH_3$

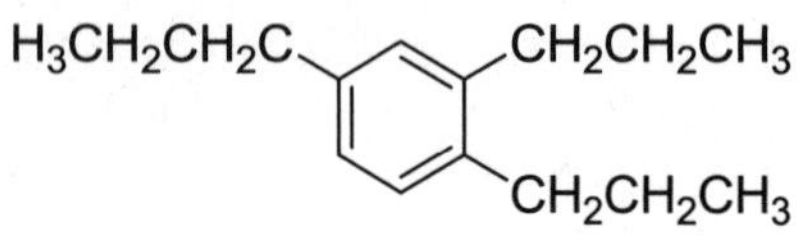

   c. CH_3 / H_3CH_2C / CH_3

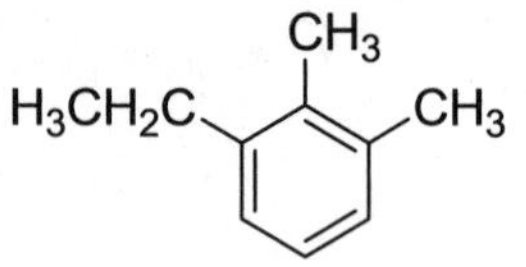

2. Provide a structure for the following compounds:

   a. 5-ethyl-1,3-dimethylbenzene

   b. 1-butyl-3-methylbenzene

c, 1,2-diethylbenzene

Chapter 6 Quiz

1. Which of the following compounds represent a saturated hydrocarbon?
    a. $C_{12}H_{26}$

    b. $C_{10}H_{20}$

    c. $C_{15}H_{32}$

    d. C_6H_6

    e. C_9H_{14}

2. What type of intermolecular force of attraction exists between hydrocarbons? How do the boiling points of hydrocarbons compare to other molecules with roughly the same size?

3. What is the C—C—C bond angle in the following compounds?
    a. propane

    b. propene

    c. propyne

d. benzene

4. Identify the structures below as a simple alkene, diene, polyene, alkyne, or aromatic hydrocarbon.

a.

b.

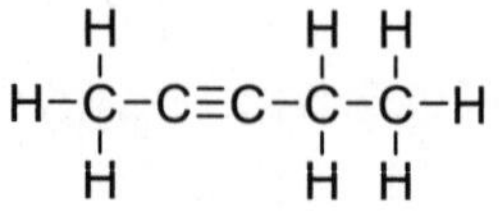

```
     H          H  H
     |          |  |
   H-C-C≡C-C-C-H
     |          |  |
     H          H  H
```

c.

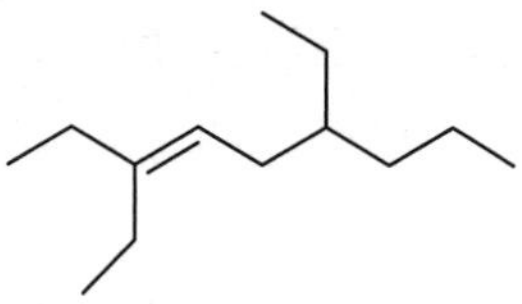

d.

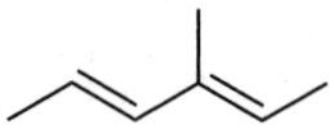

e.

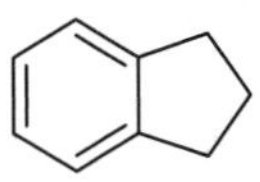

5. Identify the following pairs of molecules as structural isomers, conformational isomers, or geometric isomers.

a.

b.

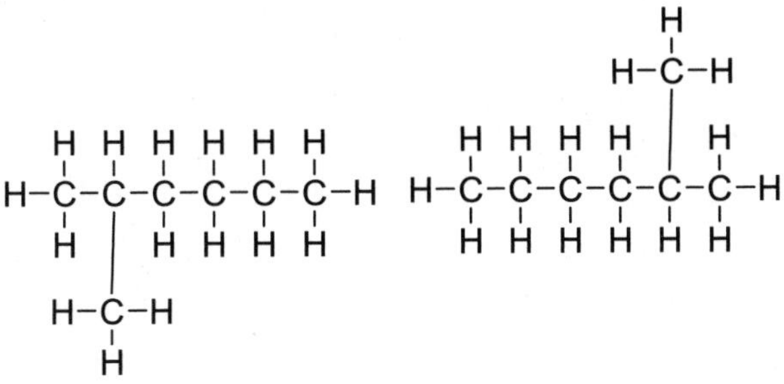

```
                                          H
                                          |
                                        H-C-H
                                          |
  H H H H H H            H H H H   |  H
  | | | | | |            | | | |   |  |
H-C-C-C-C-C-C-H        H-C-C-C-C-C-C-H
  |   | | | |            | | | | | |
  H | H H H H            H H H H H H
    |
  H-C-H
    |
    H
```

c.

```
                            H
                            |
                          H-C-H
                            |
  H H H H H               H | H H H
  | | | | |               | | | | |
H-C-C-C-C-C-H           H-C-C-C-C-C-H
  | |   | |               | | | | |
  H H | H H               H H H H H
      |
    H-C-H
      |
      H
```

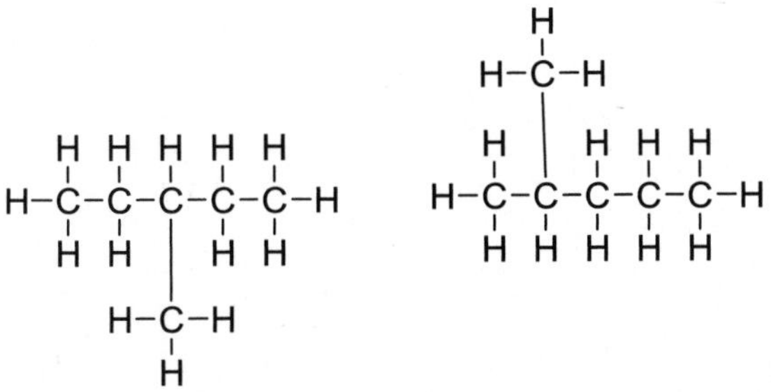

6. Provide a skeletal line structure for the following compounds:
   a. 2-ethyl-3-methyl-1-propylbenzene

   b. *cis*-5-ethyl-2,2-dimethylnon-3-ene

   c. *trans*-4-methylpent-2-ene

   d. 2-methylhex-3-yne

   e. 2,4-dimethyl-3-ethylhexane

   f. cyclodecane

7. Write the IUPAC name for the following compounds:

   a.

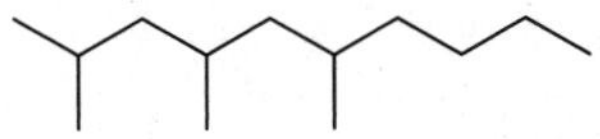

   b.

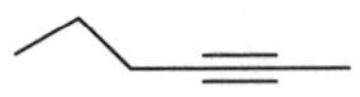

   c.

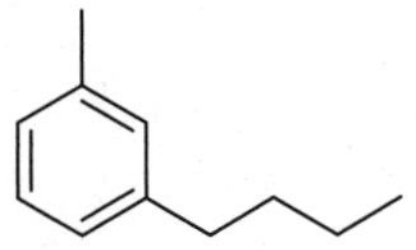

   d.

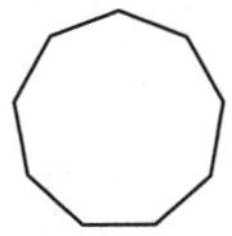

   e.

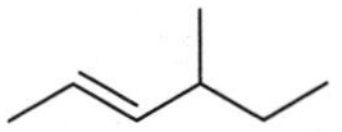

8. Draw the Lewis structure, the condensed structure, and the skeletal line formula for C_3H_8.

9. Draw the Lewis structure, the condensed structure, and the skeletal line structure for 3-ethyl-4-methyloctane.

10. Write the IUPAC name for the following compounds:

    a. $CH_3(CH_2)_7CH_3$

b.

$$\begin{array}{c} CH_3 \\ | \\ CH_3CHCH_2CH_3 \end{array}$$

c.

$$\begin{array}{c} CH_3 \\ | \\ CH_3CHCHCH_3 \\ | \\ CH_3 \end{array}$$

Chapter 6

Answers to Additional Exercises

37 "Good" fats are unsaturated hydrocarbons that contain one or more carbon-carbon double bond and the carbon-carbon double bond creates a kink in the overall shape of the molecule. "Bad" fats have either a long saturated hydrocarbon chain that does not contain a carbon-carbon double bond or are trans fats.

39 Polyunsaturated fats have more double bonds that monounsaturated fats.

41 Catalytic hydrogenation reactions convert carbon-carbon double bonds into carbon-carbon single bonds. The food industry uses this method to produce products that have a longer shelf life and a consistency that consumers prefer.

43 Heteroatoms are nonmetals elements (other than hydrogen and carbon), such as oxygen, nitrogen, phosphorus, and sulfur, that form covalent bonds with carbon.

45 a. False. Organic compounds can be prepared in the laboratory. b. True. Pharmaceuticals can be isolated from plants and animals. c. True. Pharmaceuticals can be prepared in the chemical laboratory. d. False. Organic compounds can be prepared in the laboratory.

47 Hydrocarbons that follow the formula C_nH_{2n+2} are saturated hydrocarbons. C_3H_8 and C_4H_{10} follow this formula. When 3 is substituted for *n* in the formula C_nH_{2n+2}, H = 8; when 4 substituted for *n* in the formula C_nH_{2n+2}, H = 10. C_5H_{10} and C_2H_2 are unsaturated hydrocarbons; they do not follow the formula.

49 a., c., and d. Hydrocarbons are hydrophobic—insoluble in water and soluble in other hydrocarbons, and have low boiling points.

51 Water has a higher boiling point than methane. Water molecules form hydrogen bonds with other water molecules. Hydrogen bonds are stronger than the dispersion forces holding the methane molecules together, so heat is needed to separate water molecules and form a gas.

53 The three parts of an IUPAC name are the prefix, the root, and the suffix.

55 c. hexane. There are six contiguous carbon atoms, so according to Table 6-2, the root is *hexane*. The compound is an alkane, so the ending remains the same.

57 a. There are four contiguous carbon atoms, so according to Table 6-2, the root is *butane*. The compound is an alkane, so the ending remains the same.

b. There are four contiguous carbon atoms, so according to Table 6-2, the root is *butane*. Since there is a carbon-carbon double bond in the molecule, the compound is an alkene and the ending is changed from *-ane* to *-ene*, giving *butene*. A locator number is added to indicate where the double bond first appears when numbering the chain of carbons starting from the right side because that is the side closer to the double bond. Since the carbon-carbon double bond is between C(1) and C(2), the locator number 1- is inserted between the root and the ending: but-1-ene.

c. There are four contiguous carbon atoms, so according to Table 6-2, the root is *butane*. Since there is a carbon-carbon triple bond in the molecule, the compound is an alkyne and the ending is changed from *-ane* to *-yne*, giving *butyne*. A locator number is added to indicate where the triple bond first appears when numbering the chain of carbons starting from the right side because that is the side closer to the triple bond. Since the carbon-carbon triple bond is between C(1) and C(2), the locator number 1- is inserted between the root and the ending: but-1-yne.

59 Alkanes are hydrocarbons that contain only carbon-carbon single bonds and carbon-hydrogen bonds. A cycloalkane is an alkane whose chain of carbon atoms is joined in a way that forms a ring structure. Cycloalkanes are unsaturated hydrocarbons because they contain fewer than the maximum number of hydrogen atoms for a given number of carbon atoms, *n*.

61 Every carbon atom in an alkane must have a tetrahedral geometry; therefore, the overall shape of the molecule takes on a zigzag appearance when the chain has three or more carbon atoms. An alkane can have a non-zigzag appearance, but the non-zigzag conformations are less stable.

63 a. Branched-chain alkane because there is a carbon with bonds to more than two carbons. There is a CH_3 group branching from the third carbon.

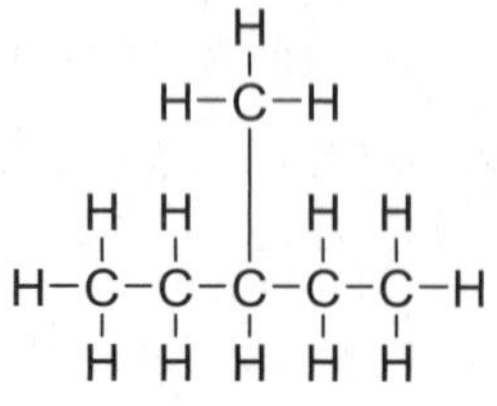

```
        H
        |
      H-C-H
        |
  H H   |   H H
  | |   |   | |
H-C-C---C---C-C-H
  | |   |   | |
  H H   H   H H
```

b. Branched-chain alkane because there is a carbon with bonds to more than two carbons. There is a CH_3CH_2 group branching from the second carbon.

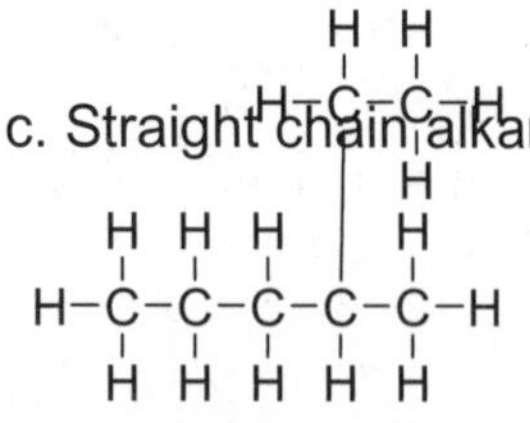

```
          H H
          | |
        H-C-C-H
          |
          | H
  H H H   | H
  | | |   | |
H-C-C-C---C-C-H
  | | |   | |
  H H H   H H
```

c. Straight chain alkan

c. Straight-chain alkane because each carbon in the chain has bonds to no more than two carbons. The subscript 3 in the condensed structure indicates that there are three consecutive CH_2 groups. CH_3 groups are always on each end of the chain:

```
   H H H H H
   | | | | |
 H-C-C-C-C-C-H
   | | | | |
   H H H H H
```

65 a.

b.

c.

d.

e.

67 Oxygen is found in menthol.

```
        3
 2    1    2          1
 2    1  1
            OH
 3    1    3
```

69 a.

```
            H
            |
          H-C-H
            |
          H-C-H
            |
     H H    |  H H                 CH2CH3
     | |    |  | |                    |
   H-C-C----C--C-C-H          CH3CHCHCH2CH3
     | |    H  | |                 |
     H |       H H                 CH3
       |
     H-C-H
       |
       H
```

b.

```
       H
       |
     H-C-H
       |
     H |   H H H                 CH3
     | |   | | |                  |
   H-C-C---C-C-C-H          CH3CCHCH2CH3
     | |   | | |                 ||
     H /    \ H H                |CH3
      /      \                   CH3
   H-C-H    H-C-H
     H        H
```

c.

```
            H
            |
          H-C-H
            |
          H-C-H
            |
  H  H  H   |  H
  |  |  |   |  |
H-C--C--C---C--C-H
  |  |  |   |  |
  H  H  |   H  H
        |
      H-C-H
        |
      H-C-H
        |
        H
```

$$\begin{array}{c} \quad\quad\quad CH_2CH_3 \\ CH_3CH_2CHCHCH_3 \\ \quad\quad CH_2CH_3 \end{array}$$

71 a.

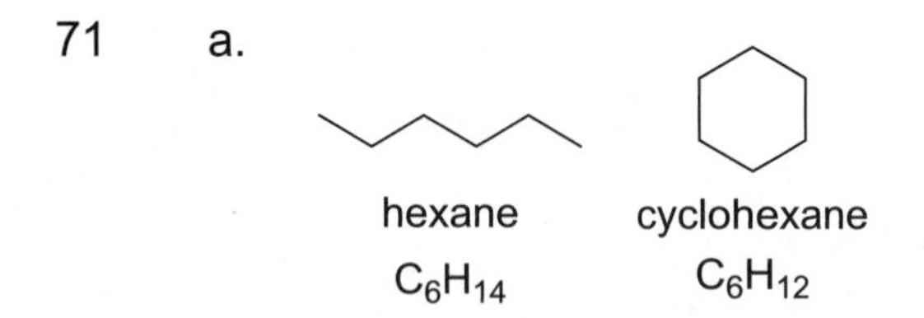

The molecular formulas have the same number of carbon atoms (6), but different numbers of hydrogen atoms (hexane, 14, and cyclohexane, 12). Hexane is a straight-chain alkane; in cyclohexane, the end carbons are bonded together to form a ring.

b.

propane C_3H_8 cyclopropane C_3H_6

The molecular formulas have the same number of carbon atoms (6), but different numbers of hydrogen atoms (propane, 8, and cyclopropane, 6). Propane is a straight-chain alkane; in cyclopropane, the end carbons are bonded together to form a ring.

73 a. Cyclooctane. There are eight carbon atoms in the ring.

b. Cycloheptane. There are seven carbon atoms in the ring.

c. Cyclopentane. There are five carbon atoms in the ring.

75 Three carbons in a ring . Five carbons in a ring . The ring with five carbons in a ring is more commonly found in nature. The three-carbon ring is strained; the bond angles are forced to be 60° when they should be 109.5°.

77 Yes, muscone contains a cycloalkane. There are 15 carbon atoms in the ring. Oxygen is the heteroatom in muscone.

79

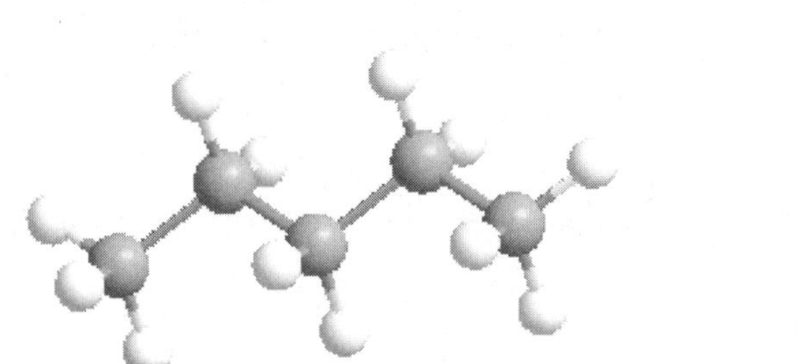

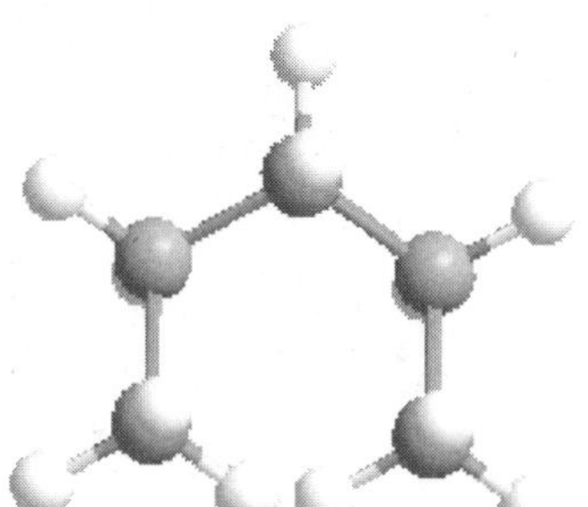

81

```
                                   H
                                   |
                                 H-C-H
   H  H  H  H                    H |   H
   |  |  |  |                    | |   |
 H-C--C--C--C-H               H--C-C---C-H
   |  |  |  |                    | |   |
   H  H  H  H                    H H   H
```

straight chain butane   branched chain

83 The structure on the left has five carbons in the main chain. The structure on the right has three carbons in the main chain and two CH_3 groups branching off the middle carbon. All structural isomers have the same chemical formula.

85 The molecular shape of a carbon atom in a triple bond of an alkyne is linear. The bond angle is 180°.

87

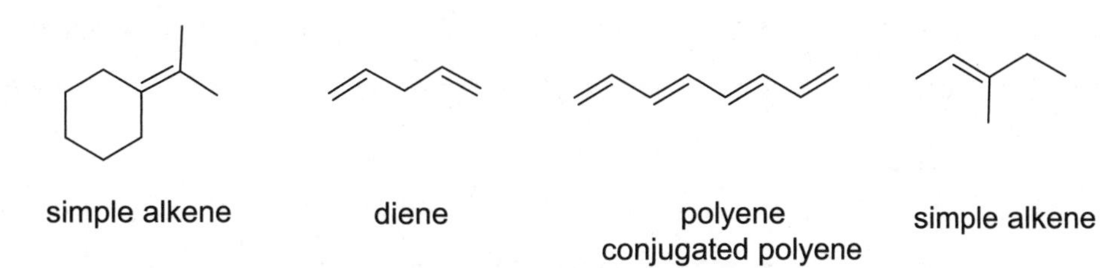

89 a. The two molecules shown are the same molecule.

b.

c.

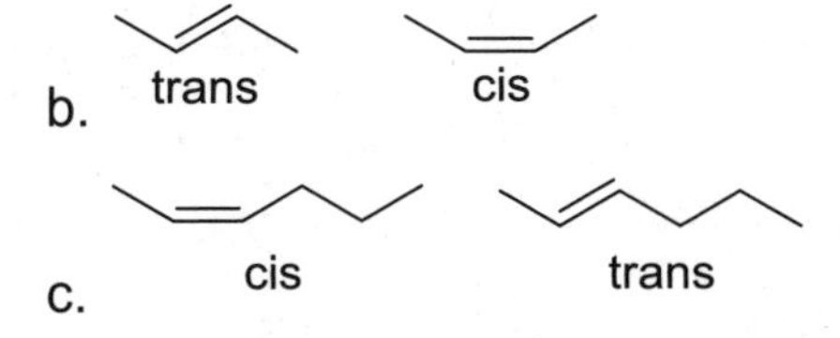

91 The double bonds are cis double bonds. All the groups are on the same side of the double bond. The molecule shown is the natural cis form of the fatty acid.

93 a. The root name hexene indicates a six-carbon chain. The suffix *-ene* indicates a double bond that starts at carbon 3 and is cis.

b. 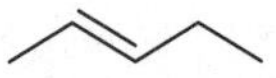 The root name pentene indicates a five-carbon chain. The suffix *-ene* indicates a double bond that starts at carbon 2 and is trans.

c. 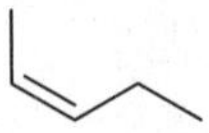 The root name pentene indicates a five-carbon chain. The suffix *-ene* indicates a double bond that starts at carbon 2 and is cis.

d. 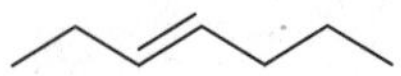 The root name heptene indicates a seven-carbon chain. The suffix *-ene* indicates a double bond that starts at carbon 3 and is trans.

95 a. There are seven contiguous carbon atoms, so according to Table 6-2, the root is *heptane*. Since there is a carbon-carbon triple bond in the molecule, the compound is an alkyne, and the ending is changed from *-ane* to *-yne*, giving *heptyne*. A locator number is added to indicate where the triple bond first appears when numbering the chain of carbons starting from the left side because that is the side closer to the triple bond. Since the carbon-carbon triple bond is between C(1) and C(2), the locator number 1- is inserted between the root and the ending: hept-1-yne.

b. There are six contiguous carbon atoms, so according to Table 6-2, the root is *hexane*. Since there is a carbon-carbon triple bond in the molecule, the compound is an alkyne and the ending is changed from *-ane* to *-yne*, giving *hexyne*. A locator number is added to indicate where the triple bond first appears when numbering the chain of carbons starting from the left side because that is the side closer to the triple bond. Since the carbon-carbon triple bond is between C(3) and C(4), the locator number 3- is inserted between the root and the ending: hex-3-yne.

c. There are six contiguous carbon atoms, so according to Table 6-2, the root is *hexane*. Since there is a carbon-carbon triple bond in the molecule, the compound is an alkyne and the ending is changed from *-ane* to *-yne*, giving *hexyne*. A locator number is added to indicate where the triple bond first appears when numbering the chain of carbons starting from the left side because that is the side closer to the triple bond. Since the carbon-carbon triple bond is between C(2) and C(3), the locator number 3- is inserted between the root and the ending: hex-2-yne.

97 a. The IUPAC name is 2,2,5-trimethyloctane. The longest contiguous chain of carbons is eight carbons, so the root for the main chain is octane. The octane main chain is numbered from the end closer to the substituents, numbering starts from the left as drawn. The multiplier *tri-* is used because there are three methyl groups as substituents. The prefix is 2,2,5 trimethyl because there are two CH_3 groups on carbon 2 and one CH_3 group on carbon 5. Assembling the prefix, main chain, and

ending gives 2,2,5-trimethyloctane. Note: The numbers are separated by commas, and numbers and letters are separated by hyphens.

b. The IUPAC name is 3,4-dimethylheptane. The longest contiguous chain of carbons is seven carbons, so the root for the main chain is *heptane*. The heptane main chain is numbered from the end closer to the substituents, numbering starts from the left as drawn. The multiplier *di-* is used because there are two methyl groups as substituents. The prefix is 3,4 dimethyl because there is one CH_3 group on carbon 3 and one CH_3 group on carbon 4. Assembling the prefix, main chain, and ending gives 3,4-dimethylheptane. Note: The numbers are separated by commas, and numbers and letters are separated by hyphens.

c. The IUPAC name is 3,4-diethylheptane. The longest contiguous chain of carbons is seven carbons, so the root for the main chain is *heptane*. The heptane main chain is numbered from the end closer to the substituents, numbering starts from the left as drawn. The multiplier *di-* is used because there are two ethyl groups as substituents. The prefix is 3,4-diethyl because there is one CH_3CH_2 group on carbon 3 and one CH_3CH_2 group on carbon 4. Assembling the prefix, main chain, and ending gives 3,4-diethylheptane. Note: The numbers are separated by commas, and numbers and letters are separated by hyphens.

d. The IUPAC name is 4-ethyl-5-methyl-6-propylnonane. The longest contiguous chain of carbons is nine carbons, so the root for the main chain is *nonane*. The nonane main chain is numbered from the end closer to the substituents, numbering starts from the right as drawn. The prefix is 4-ethyl-5-methyl-6-propyl because there is one CH_3CH_2 group on carbon 4, one CH_3 group on carbon 5, and one $CH_3CH_2CH_2$ group on carbon 6. Assembling the prefix, main chain, and ending gives 4-ethyl-5-methyl-6-propylnonane. Note: The numbers are separated by commas, and numbers and letters are separated by hyphens. The ethyl group was assigned to carbon 4 because ethyl comes before propyl alphabetically.

e. The IUPAC name is 3-ethyloct-4-yne. The longest contiguous chain of carbons is eight carbons, so the root for the main chain is *octane*. Since there is a carbon-carbon triple bond in the molecule, the compound is an alkyne and the ending is changed from *-ane* to *-yne*, giving *octyne*. A locator number is added to indicate where the triple bond first appears when numbering the chain of carbons starting from the right side because that is the side closer to the triple bond. Since the carbon-carbon triple bond is between C(4) and C(5), the locator number 4- is inserted between the root and the ending: oct-4-yne.The prefix is 3-ethyl because there is one CH_3CH_2 group on carbon 3. Assembling the prefix, main chain, and ending gives 3-ethyl-oct-4-yne. Note: The numbers are separated by commas, and numbers and letters are separated by hyphens.

f. The IUPAC name is 2-methyldec-5-yne. The longest contiguous chain of carbons is ten carbons, so the root for the main chain is *decane*. Since there is a carbon-carbon triple bond in the molecule, the compound is an alkyne and the ending is changed from *-ane* to *-yne*, giving *decyne*. A locator number is added to indicate where the triple bond first appears when numbering the chain of carbons starting from the right side because that is the side closer to the triple bond. Since the carbon-carbon triple bond is between C(5) and C(6), the locator number 5- is inserted between the root and the ending: dec-5-yne.The prefix is 2-methyl because there is one CH_3 group on carbon 2. Assembling the prefix, main chain, and ending gives 2-methyl-dec-5-yne. Note: The numbers are separated by commas, and numbers and letters are separated by hyphens.

99

hexane 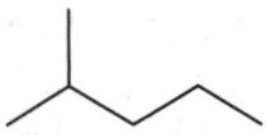 2-methylpentane 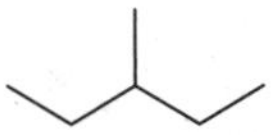 3-methylpentane

2,2-dimethylbutane 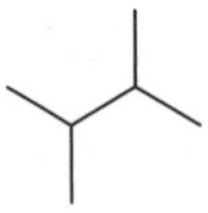 2,3-dimethylbutane

If two compounds end up having the same name, then they were conformational isomers.

101 a. The IUPAC name is 2,3-dimethylpent-2-ene. The longest contiguous chain of carbons is five carbons, so the root for the main chain is *pentane*. Since there is a carbon-carbon double bond in the molecule, the compound is an alkene and the ending is changed from *-ane* to *-ene*, giving *pentene*. A locator number is added to indicate where the double bond first appears when numbering the chain of carbons starting from the left side because that is the side closer to the double bond. Since the carbon-carbon double bond is between C(2) and C(3), the locator number 2- is inserted between the root and the ending: pent-2-ene.The prefix is 2,3-dimethyl because there is one CH_3 group on carbon 2 and one CH_3 group on carbon 3. Assembling the prefix, main chain, and ending gives 2,3-dimethylpent-2-ene. Note:

The numbers are separated by commas, and numbers and letters are separated by hyphens.

b. The IUPAC name is 3-ethylhex-2-ene. The longest contiguous chain of carbons is six carbons, so the root for the main chain is *hexane*. Since there is a carbon-carbon double bond in the molecule, the compound is an alkene and the ending is changed from *-ane* to *-ene*, giving *hexene*. A locator number is added to indicate where the double bond first appears when numbering the chain of carbons starting from the bottom because that is the side closer to the double bond. Since the carbon-carbon double bond is between C(2) and C(3), the locator number 2- is inserted between the root and the ending: hex-2-ene.The prefix is 3-ethyl because there is one CH_3CH_2 group on carbon 3. Assembling the prefix, main chain, and ending gives 3-ethylhex-2-ene. Note: The numbers are separated by commas, and numbers and letters are separated by hyphens.

c. The IUPAC name is cyclopentene. The longest contiguous chain of carbons is five carbons, so the root for the main chain is *pentane*. Since there is a carbon-carbon double bond in the molecule, the compound is an alkene and the ending is changed from *-ane* to *-ene*, giving *pentene*. Since the chain is in a ring, the prefix *cyclo* is added. Assembling the prefix, main chain, and ending gives cyclopentene.

d. The IUPAC name is cycloheptene. The longest contiguous chain of carbons is seven carbons, so the root for the main chain is *heptane*. Since there is a carbon-carbon double bond in the molecule, the compound is an alkene and the ending is changed from *-ane* to *-ene*, giving *heptene*. Since the chain is in a ring, the prefix *cyclo* is added. Assembling the prefix, main chain, and ending gives cycloheptene.

103

The H—C—C bond angles are 120° at every carbon in the molecule

105 Delocalization minimizes the electron-electron repulsions so that benzene is more stable and less likely to undergo chemical reactions.

107 a. True. Aromatic rings are stable and found in many compounds throughout nature. b. True. Benzene is a carcinogen. c. True. Benzene is a flat molecule. d True. The skeletal line structure is shown.

109 All except (c) are acceptable ways to represent benzene. (a) shows the delocalization of the electrons in benzene. (b) shows the delocalization of the electrons in benzene. (d) shows that benzene has alternating carbon-carbon single bonds and carbon-carbon double bonds. (e) is the condensed structural formula for benzene.

111

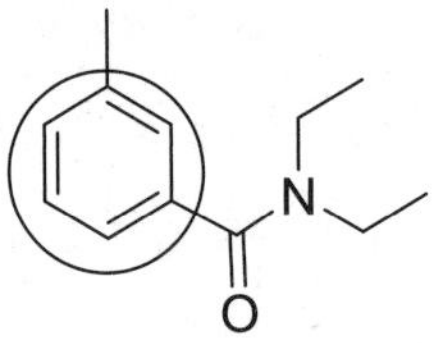

DEET

The substitution is 1,3-.

113

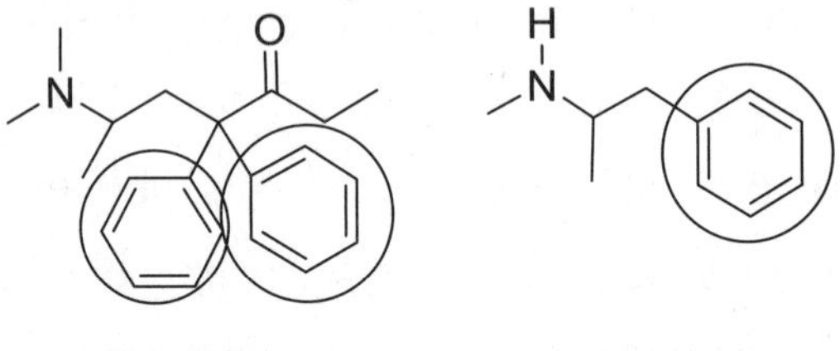

methadone

methamphetamine

They both have a nitrogen atom attached to a carbon atom and an aromatic ring and a methyl group attached to the carbon backbone.

115 a. This double bond is trans. The groups are on opposite sides of the double bond. b. This double bond is cis. The groups are on the same side of the double bond. c. Yes, the molecule would have a different shape if this double bond is isomerized. The change in shape of the protein-retinal complex initiates a nerve impulse that travels along the optic nerve to the brain. This impulse is interpreted as a visual image. d. Double bond (b) undergoes isomerization in the chemical process of vision. The energy comes from light.

117 The protein rhodopsin is found in the surface of the cell membranes of the rods and cones.

119 Rhodopsin is the cis isomer and bathrhodopsin is the trans isomer.

Chapter 7
Organic Chemistry and Biomolecules

Chapter Summary

In this chapter, you continued your focus on the fundamentals of organic chemistry. You have been introduced to all the major functional groups in organic chemistry and biochemistry. You have learned about organic compounds that contain oxygen, nitrogen, and phosphorus atoms and how to name them. You have learned to identify these functional groups within natural products and pharmaceuticals. Knowing the structure of functional groups is key to understanding the way they behave in a chemical reaction.

7.1 Alcohols and Ethers

The presence of heteroatoms oxygen and nitrogen in an organic molecule impacts the physical and chemical properties of the molecule because C—O, C—N, and in particular H—O and H—N, are polar covalent bonds. The permanent dipole created by these heteroatoms causes these compounds to have stronger intermolecular forces of attraction: hydrogen bonding and/or dipole-dipole interactions. To understand the physical and chemical properties of the vast number of organic compounds, they are categorized by the type of functional group they contain. Functional groups are groups of atoms and covalent bonds that "function" in a predictable way in certain chemical reactions.

There are two ways that functional groups can be derived from water: by replacing one of the hydrogen atoms with carbon to make an alcohol, R—O—H, or replacing both hydrogen atoms with carbon atoms to create an ether, R—O—R. The notation —R is used to signify a carbon atom with bonds to other carbon atoms or a chain of carbons of any length. When two R groups are part of the definition, —R and —R' are used to indicate that the carbon chains are not necessarily the same.

Alcohols and ethers have the same molecular shape as water at the oxygen center: bent with a bond angle of approximately 109.5° because the oxygen atom has two bonding and two nonbonding pairs of electrons. An OH group is sometimes called a hydroxyl group. When writing the skeletal line structure of an alcohol or ether, the oxygen atoms must always be written in in order to distinguish the oxygen atom from a carbon atom. In addition, hydrogen atoms bonded to a heteroatom must always be written in; only those hydrogen atoms attached to carbon atoms may be omitted.

In a Nutshell: Alcohols

Alcohols are subdivided into three classes depending on the number of R groups attached to the carbon bearing the hydroxyl group. A primary (1°) alcohol has the hydroxyl group on a carbon atom with one R group and two hydrogen atoms. A secondary (2°) alcohol has the hydroxyl group on a carbon atom with two R groups and one hydrogen atom. A tertiary (3°) alcohol has the hydroxyl group on a carbon atom with three R groups and zero hydrogen atoms.

In a Nutshell: Naming Alcohols

Follow these guidelines to name an alcohol: 1) assign the root, 2) assign the ending, 3) assign a locator number to the root indicating the location of the alcohol, and 4) assign a prefix. To assign the root, select the longest contiguous chain of carbon atoms that also contains the hydroxyl group and assign the rood name based on the number of carbon atoms in this chain To assign the ending, the *-e* at the end of the root is dropped and *-ol* is added to signify that the molecule is an alcohol rather than an alkane. To assign a locator number, start numbering the carbon atoms from the end the chain closer to the hydroxyl group. Then insert a locator number, separated by hyphens, between the root and the ending: *alkan-#-ol*. Assign a prefix according to the rule for naming branched hydrocarbons if there are alkyl substituents on the main chain.

Alcohols containing two hydroxyl groups are known as diols, those containing three hydroxyl groups are known as triols, and those containing many hydroxyl groups are known as polyols.

In a Nutshell: Phenols

A hydroxyl group directly attached to an aromatic ring carbon is not an alcohol, but a related functional group known as phenol. The phenol functional group is defined as the aromatic ring together with the hydroxyl group. The aromatic ring together with the hydroxyl group act as one unit, which causes phenols to have a different chemical reactivity than alcohols. The phenol functional group is responsible for the antioxidant properties in some compounds. A phenol can quench (remove by reacting with) free radicals—unstable atoms or molecules with an odd number of nonbonding electrons that cause damaging oxidation reactions.

Phenols are named according to the rules for naming aromatic compounds, except that the root is phenol rather than benzene, and numbering always begins at the carbon atom bearing the hydroxyl group. Number in the direction, counterclockwise or clockwise, that assigns the other substituents to the lowest set of locator numbers.

In a Nutshell: Ethers

Ethers are functional groups containing an oxygen with two single bonds to carbon atoms: R—O—R'. Ethers may be symmetrical, with identical R groups, or they may be unsymmetrical, with two different R groups. In contrast to alcohols, the —R groups in an ether can be aromatic and still classified as an ether. Ethers are relatively unreactive functional groups; they often serve as solvents in chemical reactions.

In a Nutshell: Naming Ethers

Follow these rules for naming ethers: 1) assign the root, 2) assign the ending, 3) assign a prefix, and 4) assign a locator number. In an ether, the main chain is assigned to the R group with the greater number of contiguous carbons. The root name is assigned based on the number of carbons in the main chain. The ending remains *-ane* since an ether is treated as a substituted alkane. The other R group is named as a substituent and appears in the prefix. An —OR substituent is named *alkoxy*. Name the —OR (R with shorter chain) according to the number of carbon atoms in the —OR group, and change the ending from *-ane* to *-oxy*. When the main chain has three or more carbon atoms, assign a locator number indicating the location of the alkoxy substituent on the main chain. Number the main chain starting from the end closer to the —OR group and place the locator number followed by a hyphen in front of the alkoxy substituent name.

Worked Example #1

Identify the following compounds as alcohols, phenols, or ethers. If the compound is an alcohol, indicate whether it is a primary, secondary, or tertiary alcohol.

a.
```
  H H   H
  | |   |
H-C-C-O-C-H
  | |   |
  H H   H
```

b.

```
   OH
   |
 (benzene ring)
        \   H
         C/
        / \
       H   H
```

c.

```
    H  H  H
    |  |  |
  H-C--C--C-O-H
    |  |  |
    H  H  H
```

d.

```
        H
        |
    H   O   H  H
    |   |   |  |
  H-C---C---C--C-H
    |   |   |  |
    H   |   H  H
        |
      H-C-H
        |
        H
```

a. *Ether. There is a methyl group and an ethyl group attached to the oxygen.*

b. *Phenol. There is an aromatic ring attached to the hydroxyl group.*

c. *Primary alcohol. There is a hydroxyl group present indicating an alcohol group. There is only one carbon atom and two hydrogen atoms attached to the carbon with the hydroxyl group, so it is a primary alcohol.*

d. *Tertiary alcohol. There is a hydroxyl group present indicating the alcohol group. There are three carbon atoms attached to the carbon bearing the hydroxyl group, so it is a tertiary alcohol.*

Try It Yourself #1

Identify the following compounds as alcohols, phenols, or ethers. If the compound is an alcohol, indicate whether it is a primary, secondary, or tertiary alcohol.

a.

```
            H
            |
    H   H   O   H   H
    |   |   |   |   |
  H-C---C---C---C---C-H
    |   |   |   |   |
    H   H   H   H   H
```

b.

```
    H  H  H     H  H
    |  |  |     |  |
H - C -C -C -O -C -C - H
    |  |  |     |  |
    H  H  H     H  H
```

c.

```
                   H
                   |
   H  H  H  H  H   O
   |  |  |  |  |   |
H -C -C -C -C -C  -C - H
   |  |  |  |  |   |
   H  |  H  H  H   H
      |
   H -C - H
      |
   H -C - H
      |
      H
```

d.

```
    H  H
    |  |
H - C -C -        -OH
    |  |
    H  H
```

a. *Atoms bonded to oxygen:* ________________

Alcohol, phenol, or ether: ____________________

If it is an alcohol, number of carbon atoms attached to the carbon atom bearing the hydroxyl group: ______________

Type of alcohol: ________________

b. *Atoms bonded to oxygen:* ________________

Alcohol, phenol, or ether: ____________________

If it is an alcohol, number of carbon atoms attached to the carbon atom bearing the hydroxyl group: ______________

Type of alcohol: ________________

c. *Atoms bonded to oxygen:* ________________

Alcohol, phenol, or ether: ____________________

If it is an alcohol, number of carbon atoms attached to the carbon atom bearing the hydroxyl group: ______________

Type of alcohol: ________________

d. *Atoms bonded to oxygen:* ________________

Alcohol, phenol, or ether: ____________________

If it is an alcohol, number of carbon atoms attached to the carbon atom bearing the hydroxyl group: ______________

Type of alcohol: ________________

Worked Example #2

Assign an IUPAC name to the alcohols, phenols, and ethers in the Worked Example #1.

a. *The longest continuous carbon chain attached to the oxygen has two carbons. The main chain is an ethane. The other group attached to the oxygen has one carbon so it is methoxy. The IUPAC name for this ether is methoxyethane.*

b. *3-methylphenol.*

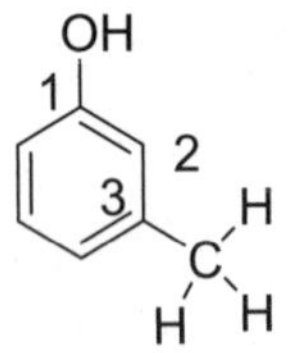

```
    OH
   1|
   ring  2
     3   H
       C
     H   H
```

The root name is phenol. Numbering clockwise from the hydroxyl group puts the methyl group on carbon-3.

c. *propan-1-ol.*

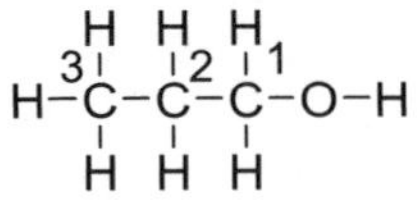

```
   H  H  H
  3|  |2 |1
H-C--C--C-O-H
   |  |  |
   H  H  H
```

The main chain that contains the alcohol group has three carbon atoms. The root name is propane. Since the molecule is an alcohol, the ending is changed from -e to -ol. The hydroxyl group is located on carbon 1.

d. *2-methylbutan-2-ol.*

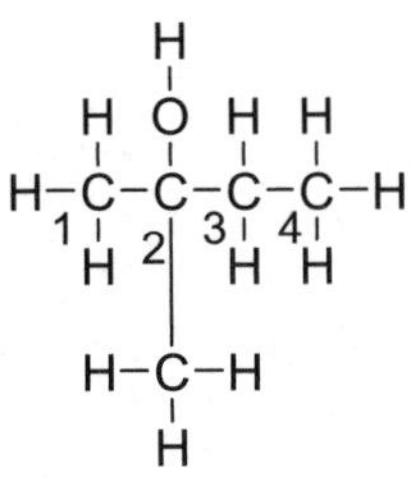

```
      H
      |
   H  O  H  H
   |  |  |  |
H-C--C--C--C-H
  1|  2| 3| 4|
   H  |  H  H
      |
    H-C-H
      |
      H
```

The main chain that contains the alcohol group has four carbon atoms. The root name is butane. The ending is changed from -e to -ol. The hydroxyl group is located on carbon 2. There is a methyl group also located on carbon 2.

Try It Yourself #2

Assign an IUPAC name to the alcohols, phenols, and ethers in the Try It Yourself #1.

a. *Number of carbon atoms in the main chain:* __________

Root name: ______________

Ending for an alcohol: ______________

Locator number for the alcohol: ____________

IUPAC name: _____________________________

e. *Longest contiguous carbon attached to the oxygen atom:* ________________________

Root name for the main chain: ______________

Number of carbon atoms in the other chain: ______________

Name of substituent: ______________

If there are three or more carbon atoms in the main chain, locator number for the substituent: ____________________________

IUPAC name: ____________________________

c. *Number of carbon atoms in the main chain:* _________________

Root name: ________________________

Suffix ending for an alcohol: _____________

Locator number for the alcohol: _______________

Substituent name: ____________________

Locator number for the substituent: _________________________

IUPAC name: _____________________________________

f. *Root name:* __________________

Number of carbon atoms in substituent: _______________________

Name of substituent: _____________________________________

Locator number of substituent: ________________________________

*IUPAC name:*____

In a Nutshell: Physical Properties of Alcohols, Phenols, and Ethers

Alcohols and phenols form hydrogen bonds, the strongest of the intermolecular forces of attraction. Hydrogen boding accounts for the higher boiling points of alcohols and phenols compare to that of other molecules of the same size. Ethers cannot hydrogen bond with other ether molecules because they lack an O—H bond. Ethers interact through dipole-dipole intermolecular forces of attraction as a result of the weakly polar C—O bond. They have slightly higher boiling points than alkanes of comparable size.

Practice Problems for Alcohols and Ethers

1. Indicate whether the following molecules are alcohols, phenols, or ethers. If it is an alcohol, indicate whether it is a primary, secondary, or tertiary alcohol.

a.
```
  H H   H H H H
  | |   | | | |
H-C-C-O-C-C-C-C-H
  | |   | | | |
  H H   H H H H
```

b.
```
          H
          |
  H H H H O H H
  | | | | | | |
H-C-C-C-C-C-C-C-H
  | | | | | | |
  H H H H H H H
```

c.
```
          H
          |
  H H H H O H H
  | | | | | | |
H-C-C-C-C-C-C-C-H
  | | | | | | |
  H H H H | H H
          |
        H-C-H
          |
        H-C-H
          |
          H
```

d.

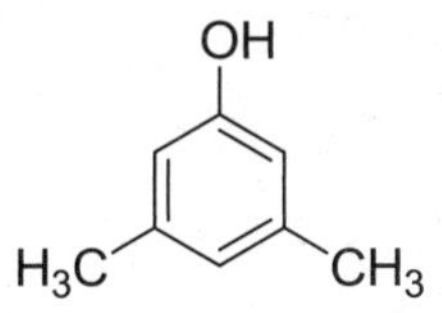

OH

H_3C CH_3

e.

```
   H H H H H H H
   | | | | | | |
 H-C-C-C-C-C-C-C-H
   | | | | | | |
   O H H H H H H
   |
   H
```

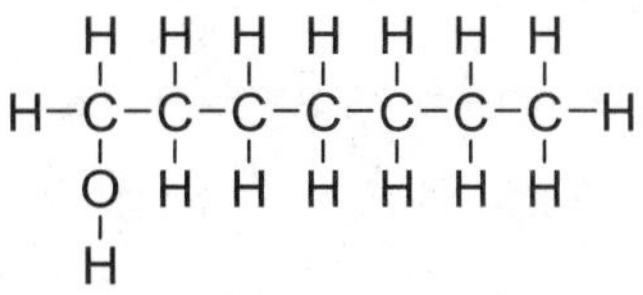

2. Assign an IUPAC name to the molecules shown in Question 1.

3. What intermolecular forces are present for the molecules in Question 1?

7.2 Amines

An amine is a functional group derived from the inorganic compound ammonia (NH_3) by replacing one, two, or three of the hydrogen atoms with an —R group. The —R group can be either alkyl or aryl (aromatic) groups. Amines are further subdivided into three classes depending on the number of —R groups attached to the nitrogen atom. A primary (1°) amine one R group and two hydrogen atoms bonded to the nitrogen atom, RNH_2. A secondary (2°) amine has two R groups and one hydrogen atom bonded to the nitrogen atom, R_2NH. A tertiary (3°) amine has three R groups and zero hydrogen atoms bonded to the nitrogen atom, R_3N.

Since amines are derived from ammonia, they have the same molecular shape as ammonia at the nitrogen center: trigonal pyramidal with bond angles of approximately109.5° because nitrogen has three bonds and one nonbonding pair of electrons.

Tertiary amines do not undergo many of the reactions that primary and secondary amines undergo. Primary and secondary amines, but not tertiary amines, are capable of hydrogen bonding. Consequently, tertiary amines have lower boiling points than primary or secondary amines of similar mass.

Amines are a common functional group in drug molecules such as antidepressants and opioid analgesics. The amine functional group is also present in many natural brain chemicals, such as dopamine and adrenaline. They are also present in compounds that are found in nature are known as alkaloids.

In a Nutshell: Naming Amines

Follow these rules for naming amines: 1) assign the root, 2) assign the ending, 3) assign a locator number, and 4) assign the prefix. Since an amine has up to three R groups, the main chain is assigned to the R group with the greater number of contiguous carbons. The root name is assigned based on the number of carbons in the main chain. Since the main chain is an alkane containing an amine, the *-e* is dropped from the root name and the ending *-amine* is added to signify the molecule is an amine. A locator number is used to indicate which carbon along the main chain contains the bond to the nitrogen atom. If the main chain has more than two carbon atoms, number the chain starting from the end closer to the nitrogen atom. Then insert a locator number separated by hyphens, between the root and the ending: *alkan-#-amine*. For secondary and tertiary amines, the other R group(s) attached to the nitrogen atom are named as though they were substituents; however, instead of a locator number, the prefix *N-* is used to signify that the substituent is bonded to the nitrogen atom and not to a carbon atom in the main chain. For a tertiary amine, the R groups are listed in alphabetical order. If the R groups are identical, the multiplier *di-* is used. If there are also alkyl substituents along the main chain, they are named following the rules for alkanes and place them after the *N*-substituents in the prefix name.

In a Nutshell: Ionic and Neutral Forms of an Amine

One of the most important characteristics of amines is there chemical behavior as organic bases. A base is a compound that accepts a hydrogen ion, H^+, from water. Amines become polyatomic cations when they accept a hydrogen ion, to form an N—H bond. The formation of an N—H bond converts an amine, a neutral compound, into its ionic form. Whether the amine exists in its neutral or ionic form depends on its environment. In the cell, amines are usually in their ionic form.

Worked Example #3

Wellbutrin® is an antidepressant.

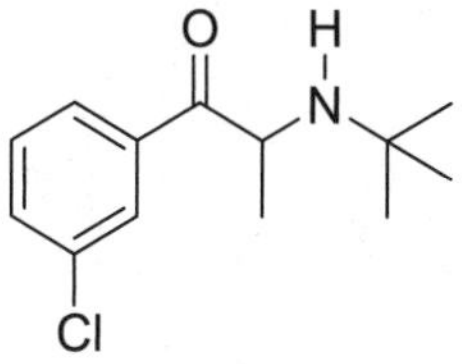

O H N Cl

a. Circle and label the amine functional group in Wellbutrin.
b. Is the amine a primary, secondary, or tertiary amine?
c. What other functional groups are present in this molecule?
d. Draw the ionic form of Wellbutrin.

Solutions

a.

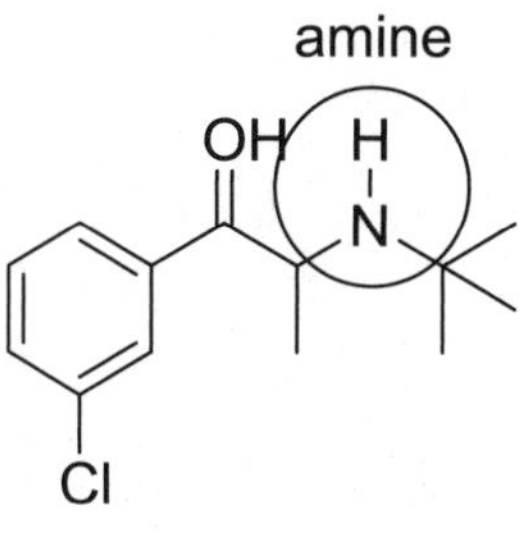

amine

O H N Cl

b. The amine is a secondary amine. There are two carbon groups attached to the nitrogen atom.

c. *There are a ketone and an aromatic ring in the structure of Wellbutrin.*

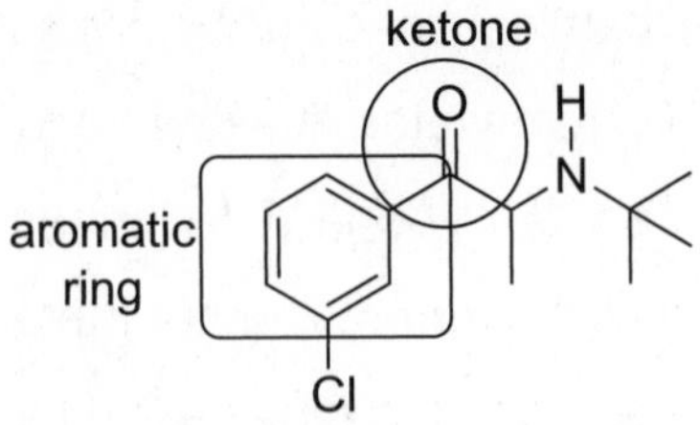

ketone

aromatic ring

Cl

d.

Cl

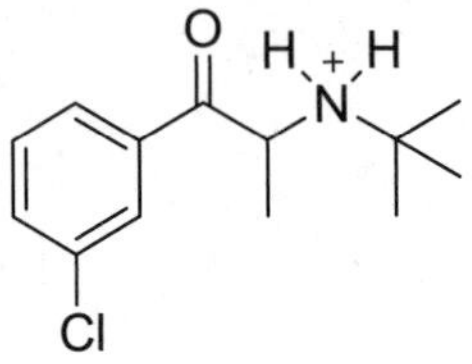

Try It Yourself #3

Paxil ® (paroxetine hydrochloride) is an orally administered psychotropic drug.

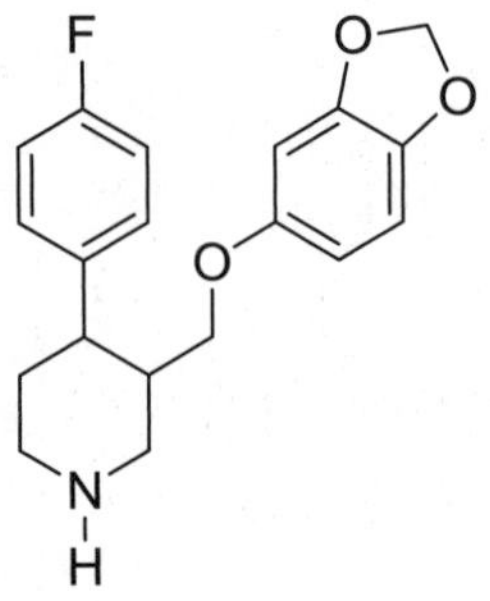

a. Circle and label the amine functional group in Paxil.

b. Is the amine a primary, secondary, or tertiary amine?

c. Draw the ionic form of Paxil.

Solutions

a.

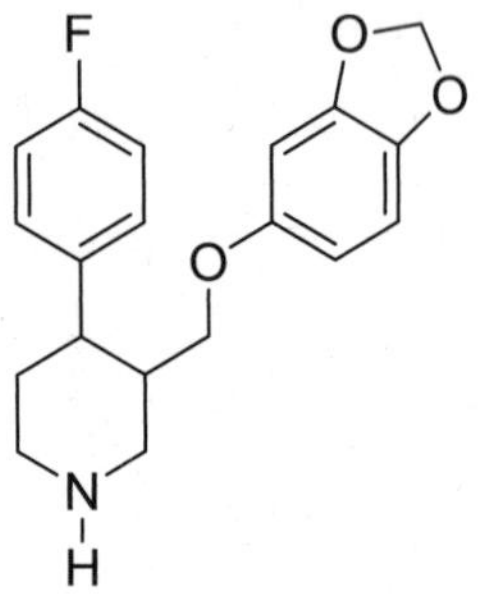

b. *Number of carbon groups attached to nitrogen:* __________

Amine is: ________________________

c. *Ionic form of Paxil:*

Worked Example #4

Write the structure of *N,N*-ethylmethylbutan-2-amine.

Solution

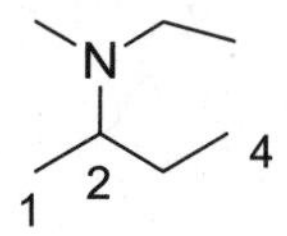

The root name is butane. There are four carbons in the chain containing the amine. The amine is located on carbon 2. There are an ethyl group and a methyl group attached to the nitrogen.

Try It Yourself #4

Write the structure of *N*-ethyl-3-methylpentan-2-amine.

Solution

Number of carbons in the chain containing the amine: _________

Position of the amine: ____________

Substituents on the nitrogen atom: _______________________________

Structure of N-ethyl-3-methylpenta-2-namine:

Worked Example #5

Write the IUPAC name for the following amine.

Solution

N,N-dimethyl-3-methylbutan-2-amine. There are four carbon atoms in the main chain.

The root name is butane. The ending changes from -e to -amine. There is a methyl group attached to carbon 3 of the main chain, so methyl is added as a prefix to the amine name. There are two methyl groups attached to the nitrogen atom; they are indicated by the N,N-dimethyl.

Try It Yourself #5

Write the IUPAC name for the following amine.

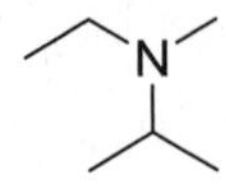

Solution

Number of carbon atoms in the main chain containing the amine: ___________

Root name: ____________________

Ending change: ______________________________

Locator number of the amine: ______________

Substituents attached to the nitrogen atom: ______________________________

IUPAC name: __________________________________

Practice Problems for Amines

1. Are the amines shown the ionic form or the neutral form?

   a.

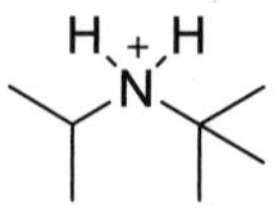

   b.

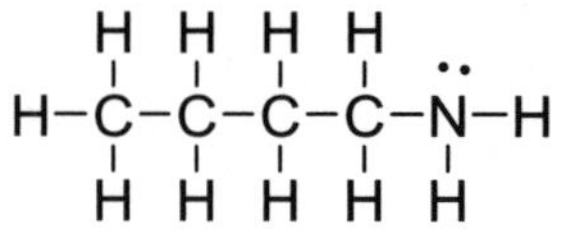

   c.

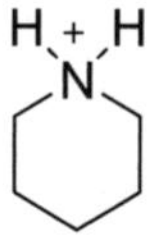

   d.

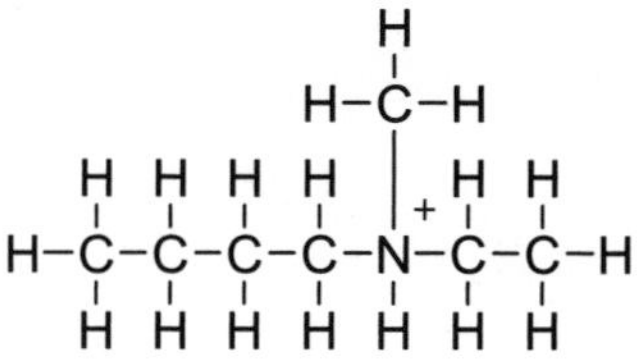

2. Label the amines in Question 1 as primary, secondary, or tertiary amines.

3. Avelox is a broad spectrum antibiotic. Circle and label the amines in the structure of avelox.

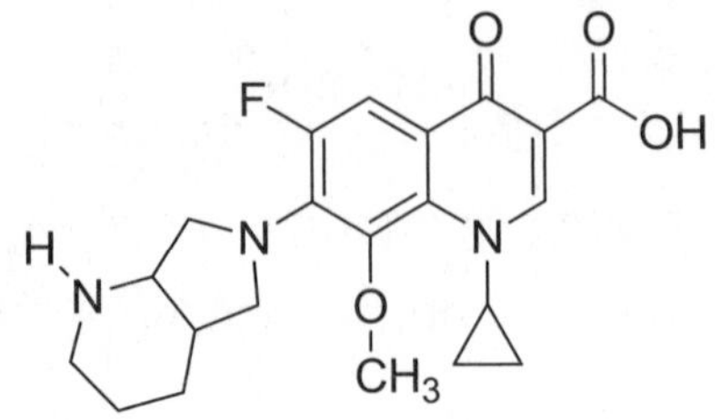

4. Write the IUPAC name for the following amines:

a.

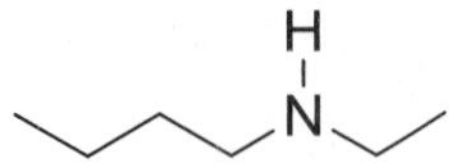

b.

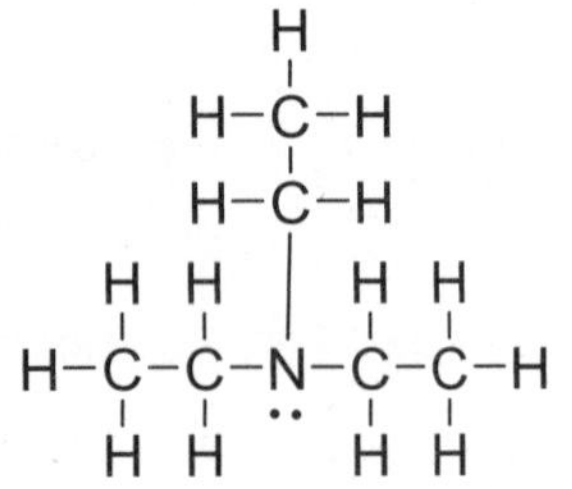

7.3 Carbonyl-Containing Functional Groups

Several functional groups contain a C═O double bond, known as a carbonyl group. The molecular shape around the carbonyl carbon is trigonal planar with bond angles of 120°. Therefore, al the carbonyl-containing functional groups have a trigonal planar molecular shape. The various carbonyl-containing functional groups are distinguished by the two groups or atoms with a bond to the carbonyl carbon. One of these groups is almost always an —R group—a carbon atom, unless the molecule contains only one carbon, in which case it can be a hydrogen atom (H). The other group or atom attached to the carbonyl carbon determines the identity of the functional group. If the other group is —H, it is an aldehyde. If the other group is —R, it is a ketone. If the other group is —OH, it is a carboxylic acid. If the other group is —OR', it is an ester. If it is —NR'R", then it is an amide.

Since oxygen is more electronegative than carbon, the carbonyl group has a polar covalent bond. All functional groups containing carbonyl group have a permanent dipole and dipole-dipole intermolecular forces of attraction. Carboxylic acids and some amides also have the more polar O—H and N—H bond giving them the capacity to hydrogen bond.

In a Nutshell: Aldehydes

An aldehyde is a carbonyl group with a bond to a hydrogen atom and a bond to an R group. Methanal (formaldehyde, H_2CO) is the sole exception because it has two H atoms bonded to the carbonyl group. In condensed structural notation an aldehyde is written as RCHO. When writing the skeletal line structure for an aldehyde, the carbon atom of the carbonyl group is not written in, but the double bond and the oxygen atom must be written in. Sometimes, the hydrogen atom on the carbonyl group is also written in to emphasize the H-atom characteristic of an aldehyde.

In a Nutshell: Naming Aldehydes

Aldehydes can often be recognized by the *-al* ending in their name. Follow these rules for naming aldehydes: 1) assign the root, 2) assign the ending, and 3) assign the prefix. Select the longest contiguous chain of carbon atoms that also contains the carbonyl carbon and assign the root name based on the number of carbon atoms. When counting, begin with the carbonyl carbon. Change the ending on the root by dropping the *-e* and adding *-al* to signify that the molecule is an aldehyde. No locator name is needed for straight-chain aldehydes because the aldehyde is always on carbon number 1. Name any alkyl substituents along the main chain by numbering the main chain beginning with the carbonyl carbon.

In a Nutshell: Ketones

In a ketone, the carbonyl carbon is bonded to two R groups (no hydrogen atoms). When writing condensed structural formulas, a ketone is written as RCOR. In a skeletal line structure for a ketone, the carbonyl carbon does not need to be written in, but the oxygen atom and the double bond do.

In a Nutshell: Naming Ketones

A ketone can be identified by the *-one* ending in its name. Follow these rules for naming ketones: 1) assign the root, 2) assign the ending, 3) assign a locator number, and 4) assign the prefix. Select the longest contiguous chain of carbon atoms that also contains the carbonyl carbon and assign the root name based on the number of carbon atoms it contains. Change the ending on the root by dropping the *-e* and adding *-one* to signify that the molecule is a ketone. Number the main chain from the end closer to the ketone. The insert a locator number, separated by hyphens, between the root and the ending: *alkan-#-one*. Name any alkyl substituents along the main chain.

Worked Example #6

Identify the following compounds as aldehydes or ketones. Write the IUPAC name for them.

a.

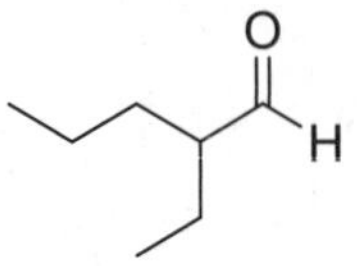

O

H

b.

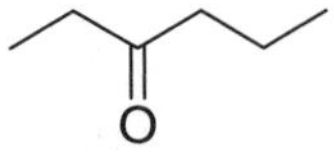

O

a. This molecule is an aldehyde. It has a hydrogen atom bonded to the carbonyl carbon.

O

5 2

1 H

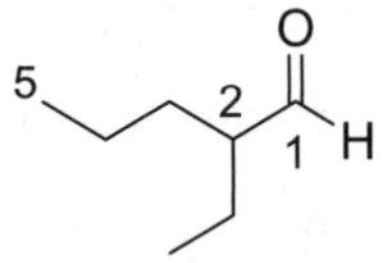

The IUPAC name is 2-ethylpentanal. There are five carbon atoms in the chain with the aldehyde. The root is pentane. The ending changes from -e to -al. The ethyl group is attached to carbon 2 and is added as a prefix.

b. This molecule is a ketone. It has two —R groups attached to the carbonyl carbon.

6

3

1

O

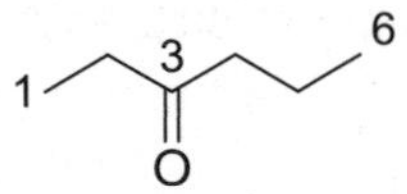

The IUPAC name is hexan-3-one. There are six carbon atoms in the chain. The root is hexane. The ending changes from -e to -one. The locator number is 3 since the carbonyl group is located at carbon 3.

Try It Yourself #6

Identify the following compounds as aldehydes or ketones. Write the IUPAC name for them.

a.

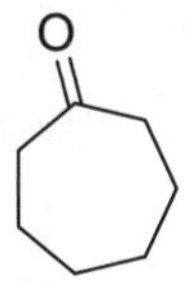

b.

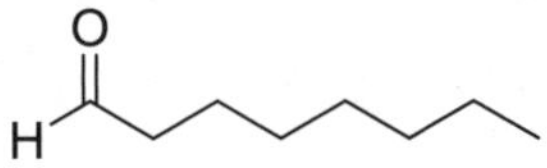

a. *Atoms attached to carbonyl carbon:* ______________

Aldehyde or ketone: ________________________

Number of carbon atoms in ring: ______________________

Root name: ______________________________________

Ending: ________________________________

IUPAC name: __________________________________

b. *Atoms attached to carbonyl carbon:* ______________

Aldehyde or ketone: ________________________

Number of carbon atoms in chain: ______________________

Root name: ______________________________________

Ending: ________________________________

IUPAC name: __________________________________

In a Nutshell: Carboxylic Acids

A carboxylic acid is a functional group containing a carbonyl group attached directly to a hydroxyl group. The condensed structural formula for a carboxylic acid is written as either an RCO_2H or RCOOH. Despite the presence of an O—H group, the chemical reactivity of a carboxylic acid is significantly different from an alcohol. In a carboxylic acid, the carbonyl group and the O—H together act as a single unit.

There are several functional groups with a hydroxyl group. A hydroxyl group attached to a carbonyl group is a carboxylic acid. A hydroxyl group attached to an aromatic ring is a phenol. A hydroxyl group attached to an alkyl with one, two, or three carbon atoms is a primary, secondary, or tertiary alcohol, respectively.

In a Nutshell: Naming Carboxylic Acids

Carboxylic acids can often be recognized by the *-oic acid* ending in their name. Follow these rules for naming carboxylic acids: 1) assign the root, 2) assign the ending, and 3) assign the prefix. Select the longest contiguous chain of carbon atoms that also contains the carbonyl carbon and assign the root name based on the number of carbon atoms. When counting, begin with the carbonyl carbon. Change the ending on the root by dropping the *-e* and adding *-oic acid* to signify that the molecule is a carboxylic acid. No locator name is needed for carboxylic acids because it is always on carbon number 1. Name any alkyl substituents along the main chain by numbering the chain beginning with the carbonyl carbon.

Carboxylic acids are characterized by chemical behavior as organic acids: capable of donating a hydrogen ion, H^+ to water to form hydronium ion, H_3O^+. Upon donating a hydrogen ion, the carboxylic acid becomes a carboxylate ion, $RCOO^-$, a polyatomic anion. In a biological cell, equilibrium favors the ionized form of a carboxylic acid, the carboxylate.

Carboxylic acids are capable of hydrogen-bonding intermolecular forces of attraction because they contain the strong O—H bond dipole and a partial negative charge on both oxygen atoms of the carboxylic acid. The ionized form, the carboxylate has very different physical properties than its neutral form because of the negative charge. Carboxylic acid is a molecule, while a carboxylate ion is an ion. The ion is soluble in water, whereas carboxylic acids with five or more carbon atoms do not dissolve in water.

Worked Example #7

Indicate whether the following compounds are the neutral form or the ionic form of the carboxylic acid. Write the IUPAC name for the carboxylic acid.

a.
```
   H H H H H H O
   | | | | | | ||
 H-C-C-C-C-C-C-C-OH
   | | | | | |
   H H H H H H
```

b.
```
     O H H H
  -  || | | |
   O-C-C-C-C-H
       | | |
       H H H
```

a. *This molecule is a neutral carboxylic acid. The IUPAC name is heptanoic acid. There are seven carbon atoms in the chain, so the root name is heptane. The ending is changed by dropping* -e *and adding* -oic acid.

b. *This molecule is the ionic form of the carboxylic acid. The IUPAC name for the carboxylic acid is butanoic acid. There are four carbon atoms in the chain, so the root name is butane. The ending is changed by dropping* -e *and adding* -oic acid.

Try It Yourself #7

Indicate whether the following compounds are the neutral form or the ionic form of the carboxylic acid. Write the IUPAC name for the carboxylic acid.

a.

```
    H H H H H O
    | | | | | ||  -
  H-C-C-C-C-C-C-O
    | | | | |
    H H H H H
```

b.

```
      O H H H H H H H
      || | | | | | | |
   HO-C-C-C-C-C-C-C-C-H
        | | | | | | |
        H H H H H H H
```

a. *Ionic form or neutral form:* ____________________

Number of carbon atoms in chain: ____________________

Root name: ________________________________

Suffix ending: ____________________________

IUPAC name: ________________________________

b. *Ionic form or neutral form:* ____________________

Number of carbon atoms in chain: ____________________

Root name: ________________________________

Suffix ending: ____________________________

IUPAC name: ________________________________

In a Nutshell: Esters

An ester is similar to a carboxylic acid, except that in an ester the carbonyl carbon has a bond to an alkoxyl group (—OR) rather than a hydroxyl group (—OH). Esters can be prepared from carboxylic acids in a chemical reaction. An ester is also different from an ether because it contains a carbonyl group directly attached to the —OR group. The carbonyl group and the alkoxyl group act together as one unit. An ester is written as RCO_2R' or RCOOR' in condensed notation, where R' means that the two R groups are not necessarily the same. All molecules, including esters, can be written from left to right or right to left. Esters contain a permanent dipole but they do not have the more polar N—H or O—H bond. They have dipole-dipole intermolecular forces of attraction, which are weaker than hydrogen bonding.

In a Nutshell: Naming Esters

Follow these rules for naming esters: 1) assign the root, 2) assign the ending, and 3) assign the prefix. There are two R groups in an ester: The one containing the carbonyl carbon is the main chain, and the other is in the —OR' group, which is treated as a substituent. Assign the root name based on the number of contiguous carbon atoms in the main chain. Change the ending on the root by dropping the *-e* and adding *-oate* to signify that the molecule is an ester. No locator number is needed for an ester because the carbonyl carbon is C(1). Name the R' group attached to the oxygen as you would a substituent by counting the number of carbon atoms it contains and changing the ending to *-yl*. Insert the name of the R' group as a prefix before the root name followed by a space. If there are alkyl substituents on the main chain, name them.

Worked Example #8

Americaine® is an anesthetic lubricant. Circle and label the ester functional group. Circle and label the aromatic ring functional group.

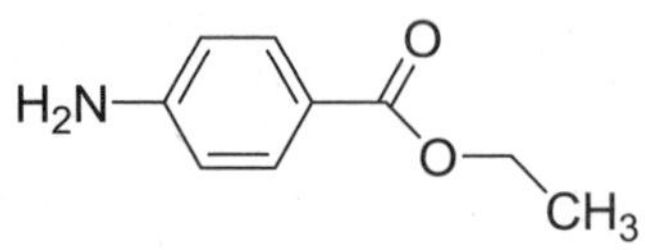

O
H_2N—
O
CH_3

Solution

ester

H_2N — aromatic ring — ester — CH_3

Try It Yourself #8

The structure of aspirin (acetylsalicylic acid) is shown below. Circle and label the ester functional group. What other functional groups are present in the molecule?

O, OH, O, O, CH_3

Worked Example #9

Write the IUPAC name for the following ester.

```
  H H H H O     H H H
  | | | | ||    | | |
H-C-C-C-C-C-O-C-C-C-H
  | | | |       | | |
  H H H H       H H H
```

Propyl pentanoate. There are five carbon atoms in the chain containing the carbonyl carbon. The root name is pentane. The suffix changes from -ane to -anoate. There are three carbon atoms in the chain attached to the oxygen atom. The substituent name is propyl.

Try It Yourself #9

Write the IUPAC name for the following ester.

O, O

Number of carbon atoms in the main chain containing the carbonyl carbon: ______

Root name: __________________

Suffix change: ____________________

Number of carbon atoms in the chain attached to the oxygen atom: ____________

Substituent name for the chain: ____________________________

IUPAC name for the ester: _________________________________

Worked Example #10

Write the skeletal line structure for the 1-methylethyl butanoate.

Solution

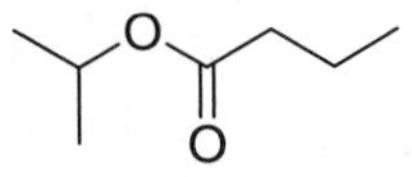

The root name is butane. There are four carbon atoms in the chain containing the carbonyl carbon. The substituent attached to the oxygen atom is an ethyl group with a methyl group on carbon 1.

Try It Yourself #10

Write the skeletal line structure for cyclopentyl propanoate.

Solution

Number of carbon atoms in the main chain: _______________

Structure of the substituent attached to the oxygen atom: ________________

Structure of the ester:

In a Nutshell: Amides

An amide functional group contains a nitrogen atom attached to carbonyl carbon. The nitrogen atom will have one bond to the carbonyl carbon and two bonds to either H atoms or R groups.

Amides are carboxylic acid derivatives because they can be prepared from carboxylic acids in a chemical reaction.

An amide differs from an amine in having a carbonyl carbon directly bonded to the nitrogen atom. The carbonyl group together with the nitrogen atom act as one unit. Amides and amines have distinctively different chemical properties. Amides are generally less reactive than amines. The condensed structural formula for an amide appears as $RCONH_2$, RCONHR', $RCONR'_2$, depending on whether the nitrogen atom has zero, one, or two R groups attached. When writing a skeletal line structure for an amide, the nitrogen atom and hydrogen atoms bonded to nitrogen must be written in.

In a Nutshell: Naming Amides

Follow these rules for naming amides: 1) assign the root, 2) assign the ending, and 3) assign the prefix. Select the longest contiguous chain of carbon atoms that also contains the carbonyl carbon and assign the root name based on the number of carbon atoms. Change the ending of the root by dropping the *-e* and adding *-amide* to signify that the molecule is an amide. No locator number is needed for an amide because the carbonyl carbon is C(1). If the amide nitrogen is bonded to two H atoms, there is no *N*- prefix. If the amide nitrogen has one or two R' groups, treat the R groups as substituents, but instead of a locator number, use the prefix *N*- to indicate that the substituent is bonded to the nitrogen atom and not a carbon atom in the main chain. If there are two R groups, list them in alphabetical order. If the R groups are identical, use the multiplier *di-*. If there are also alkyl substituents along the main chain, name those and list them after the *N*-substituents.

Worked Example #11

The structure for Tylenol®, acetaminophen, follows. Circle and label the amide functional group. How many R groups are attached to the nitrogen atom? What other functional groups are present in acetaminophen?

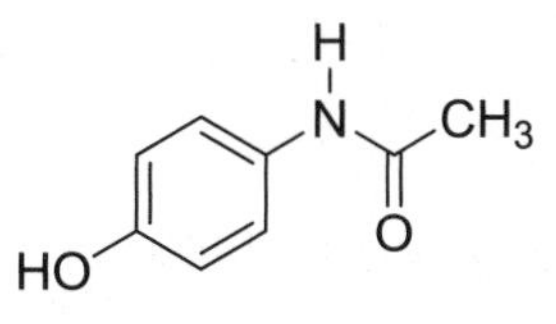

Solution

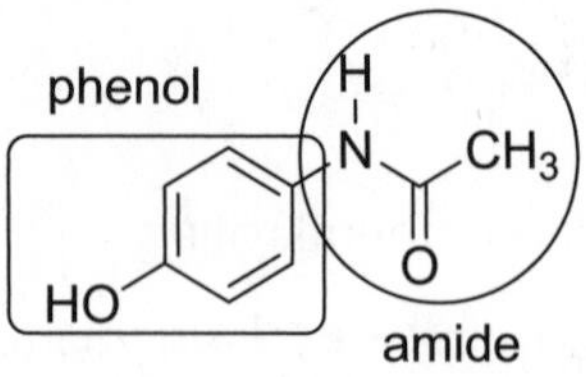

There is one R group attached to the nitrogen atom; it is the aromatic ring.

Try It Yourself #11

Keflex® is an antibiotic. Circle and label the amide functional groups. What other functional groups are present in Kelfex?

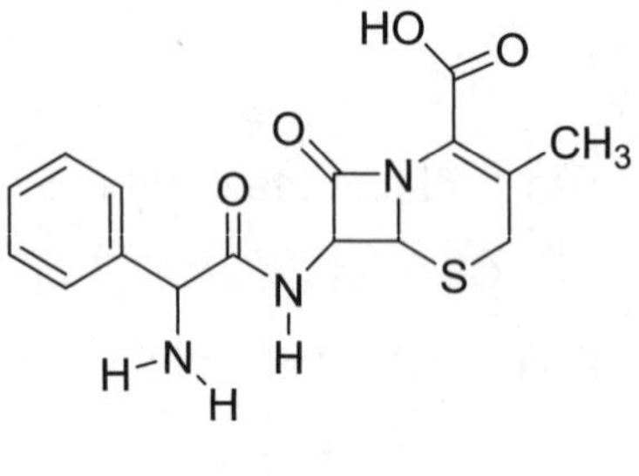

Worked Example #12

Write the IUPAC name for the following amide.

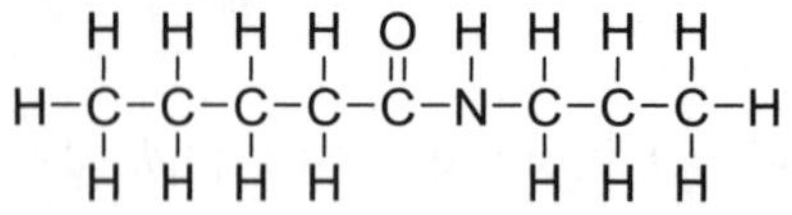

N-propylpentanamide. There are five carbon atoms in the chain containing the carbonyl carbon. The root name is pentane. The suffix changes from -e to -amide. There are three carbon atoms in the chain attached to the nitrogen atom. The substituent name is propyl.

Try It Yourself #12

Write the IUPAC name for the following ester.

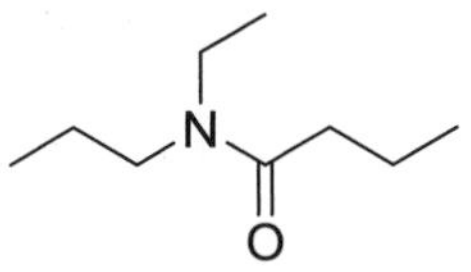

Number of carbon atoms in the main chain containing the carbonyl carbon: ______

Root name: ____________________

Ending change: ______________________

Number of carbon atoms in the chain attached to the nitrogen atom: ______________

Substituent name for the chain: ________________________________

Number of carbon atoms in the chain attached to the nitrogen atom: ______________

Substituent name for the chain: ________________________________

IUPAC name for the amide: ______________________________________

Worked Example #13

Write the skeletal line structure for the *N,N*-ethylmethylbutanamide.

Solution

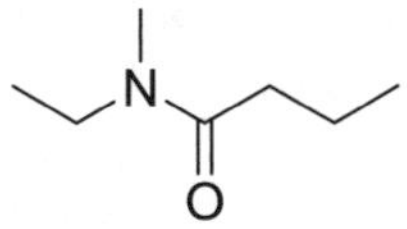

The root name is butane. There are four carbon atoms in the chain containing the carbonyl carbon. One substituent attached to the nitrogen atom is an ethyl group, and the other substituent is a methyl group.

Try It Yourself #13

Write the skeletal line structure for *N*-cyclopentylhexanamide.

Solution

Number of carbon atoms in the main chain: _________________

Structure of the substituent attached to the nitrogen atom: __________________

Structure of the amide:

Practice Problems for Carbonyl-Containing Functional Groups

1. Identify the following carbonyl-containing functional groups:

   a. 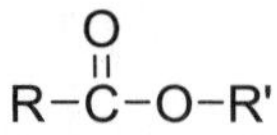 $R{-}\overset{\overset{\large O}{||}}{C}{-}O{-}R'$

   b. $R{-}\overset{\overset{\large O}{||}}{C}{-}R'$

   c. $R{-}\overset{\overset{\large O}{||}}{C}{-}\underset{\underset{\large H}{|}}{N}{-}R'$

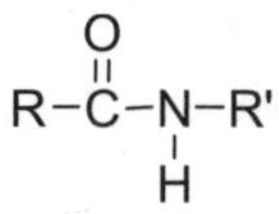

   d. 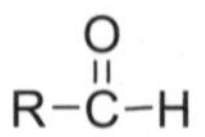 $R{-}\overset{\overset{\large O}{||}}{C}{-}H$

   e. 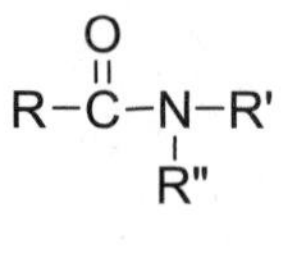 $R{-}\overset{\overset{\large O}{||}}{C}{-}\underset{\underset{\large R''}{|}}{N}{-}R'$

   f. 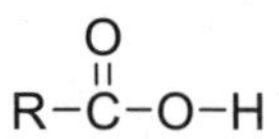 $R{-}\overset{\overset{\large O}{||}}{C}{-}O{-}H$

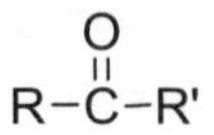

2. Depo Provera (medroxyprogesterone acetate) is used as a contraceptive. Its structure is shown below. Circle and label the carbonyl-containing functional groups.

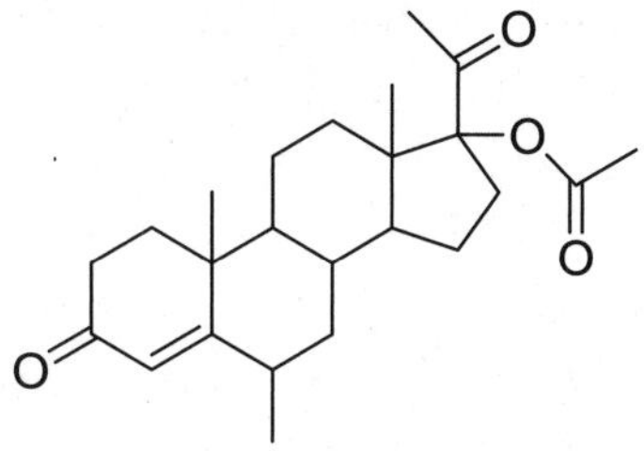

3. Write the IUPAC name for the following molecules:

a.

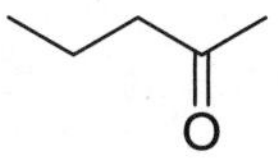

b.

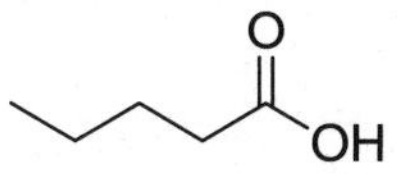

c.

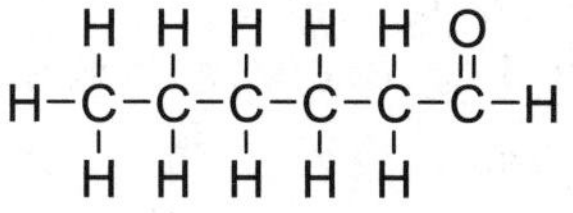

```
   H H H H H O
   | | | | | ||
 H-C-C-C-C-C-C-H
   | | | | |
   H H H H H
```

d.

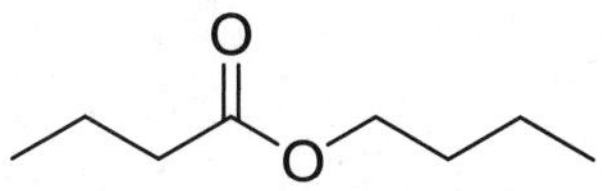

e.

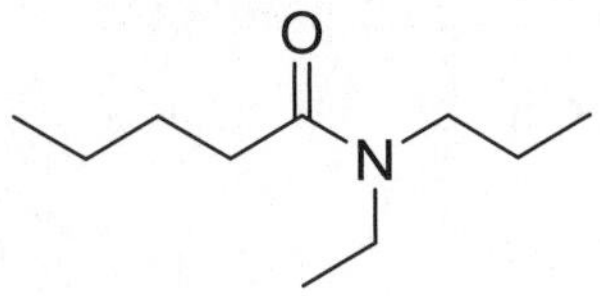

7.4 Stereoisomers

Stereoisomers are molecules with the same chemical formula and the same bonding arrangement of atoms but a different three-dimensional spatial arrangement of the atoms as a result of chirality. Chirality is a symmetry property present of some objects and molecules.

In a Nutshell: Chirality

An object or molecule is chiral if it is non-superimposable on its mirror reflection. Two objects are non-superimposable when you can never get all the components to perfectly overlay. An object or molecule is achiral if it is superimposable on its mirror image because it is identical to its mirror image.

In a Nutshell: Enantiomers

A pair of non-superimposable mirror image stereoisomers is known as a pair of enantiomers. In the IUPAC naming system, the prefixes D- and L- and *R*- and *S*- are used to distinguish two enantiomers. A pair of enantiomers have identical physical and chemical properties, such as boiling point, melting point, and solubility. Consequently, enantiomers are very difficult to separate. When enantiomers are placed in a chiral environment, such as the body, they can exhibit profoundly different chemical properties. Often, particularly in man-made drugs, a chiral substance will be produced in a 50:50 mixture of enantiomers, known as a racemic mixture.

Chirality in a molecule exists when there is one or more centers of chirality in a molecule. A center of chirality is a tetrahedral carbon atom with four single bonds to four different atoms or groups of atoms. A molecule containing one center of chirality is by definition a chiral molecule. A molecule with two or more centers of chirality is chiral unless is it superimposable on its mirror image and therefore achiral.

Worked Example #14

Indicate whether the following objects are chiral or achiral.

a. a square
b. a pair of scissors
c. a mitten
d. a golf ball

a. A square is achiral. Its mirror image is superimposable.
b. A pair of scissors is chiral.
c. A mitten is chiral.
d. A golf ball is achiral. Its mirror image is superimposable.

Try It Yourself #14

Indicate whether the following objects are chiral or achiral.

a. your left foot
b. a circle
c. a triangle
d. a hammer

a. Is the mirror image superimposable? ____________
Your left foot is: ______________.
b. Is the mirror image superimposable? ____________
A circle is: ______________.
c. Is the mirror image superimposable? ____________
A triangle is: _____________.
d. Is the mirror image superimposable? ____________
A hammer is: ______________.

Worked Example #15

L-methionine, a natural amino acid, is shown below.

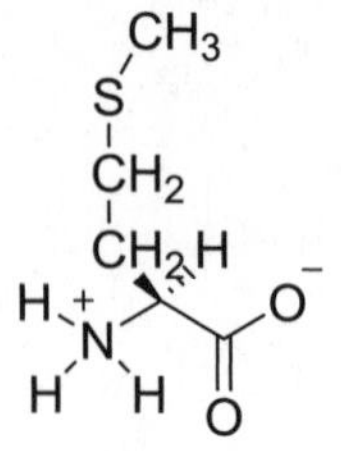

a. Is this molecule chiral?

b. Identify the chiral center in the molecule.

c. Write the structure of the enantiomer of this amino acid.

a. The molecule is chiral because it is non-superimposable on its mirror image.

b.

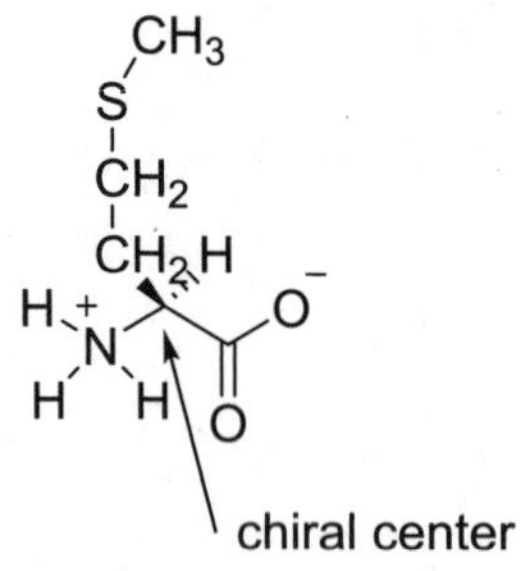

chiral center

c.

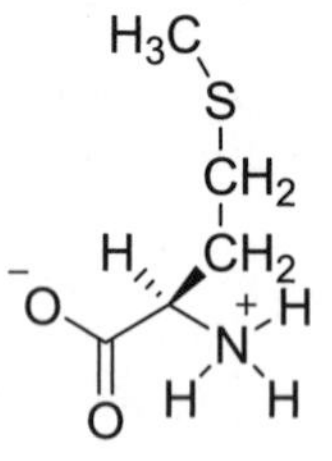

Try It Yourself #15

The structure of L-isoleucine, one of the natural amino acids, is shown below.

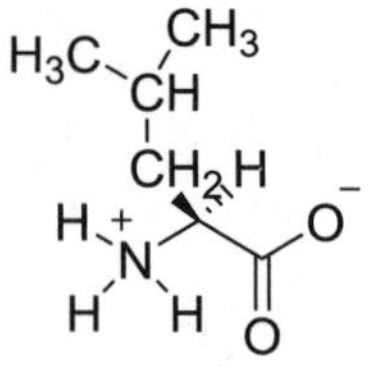

a. Is isoleucine chiral?

b. Identify the chiral center.

c. Draw the enantiomer of this isoleucine.

a. *The mirror image of this amino acid is (superimposable or non-superimposable). The amino acid is: _________________.*

b.

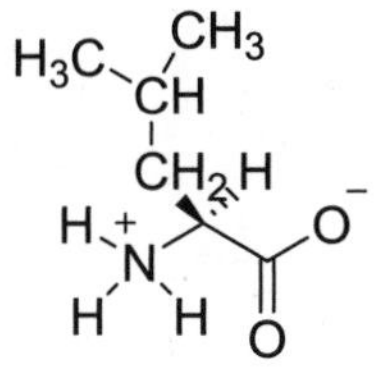

H₃C CH₃ CH CH₂ H H N⁺ H H O⁻ O

c.

Practice Problems for Stereoisomers

1. The structure of phenylalanine, an amino acid is shown below.

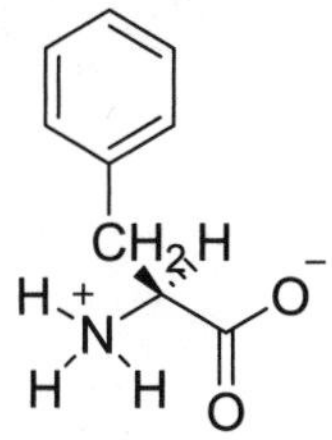

CH₂ H H N⁺ H H O⁻ O

a. Is phenylalanine chiral?

b. Draw the enantiomer of this amino acid.

2. The structure of glutamic acid, an amino acid, is shown below.

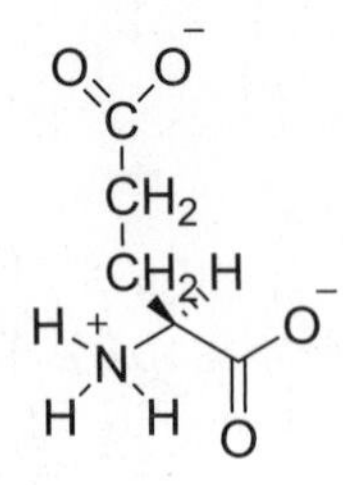

a. Is glutamic acid chiral?

b. Place an arrow pointing to the chiral center.

b. Draw the structure of the enantiomer of this amino acid.

3. Are the following objects achiral or chiral?

a. your right ear

b. a ping pong ball

c. a shoe

c. a corkscrew

7.4 Phosphate Ester Functional Groups

The final functional groups to consider are those derived from phosphoric acid (H_3PO_4). These functional groups have one phosphorous-oxygen double bond, P=O, and three phosphorus-oxygen single bonds, P—O. The functional groups derived from phosphoric acid are found in important biomolecules, such as DNA, RNA, ATP, and coenzyme A.

Phosphoric acid has an expanded octet; it contains a central phosphorus atom surrounded by 10 bonding electrons. Phosphoric acid can donate one, two, or all three of its hydrogen atoms. The number of hydrogen atoms lost depends on the environment. The most abundant form in the cell is the ion that has lost two hydrogen atoms. This form, HPO_4^{2-}, is called monohydrogen phosphate or inorganic phosphate and abbreviated P_i.

If the hydrogen atom is monohydrogen phosphate is substituted with an R, the result is an organic molecule with a functional group known as a phosphate ester. When two hydrogen atoms in phosphoric acid are substituted by R groups, the result is a phosphate diester—a molecule with two phosphate ester functional groups.

One of the unique characteristics of phosphate esters is that the phosphate group ($—PO_3^{2-}$) can form a bond to a second and even a third phosphate group, P_i, to form a diphosphate ester and a triphosphate ester, respectively. Phosphoanhydride bond join phosphate groups; these bonds are very important in biochemistry because they are high energy bonds as a result of the unstable arrangement created when two negatively charged oxygen atoms are in close proximity. The cell stores energy by forming phosphoanhydride bonds.

Worked Example #16

Identify the structures below as mono-, di-, or triphosphate esters. Place a box around each phosphate group.

a.

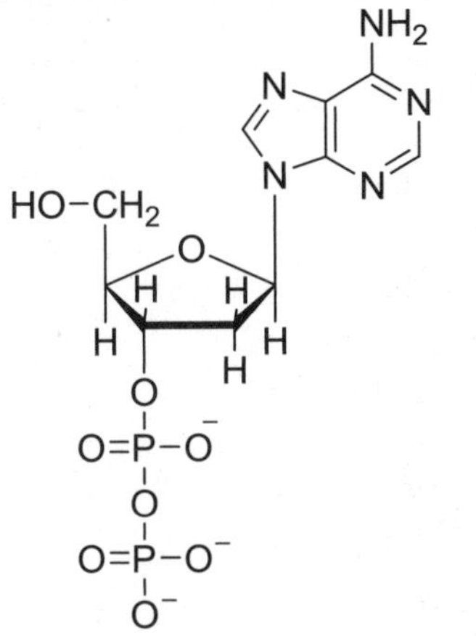

b.

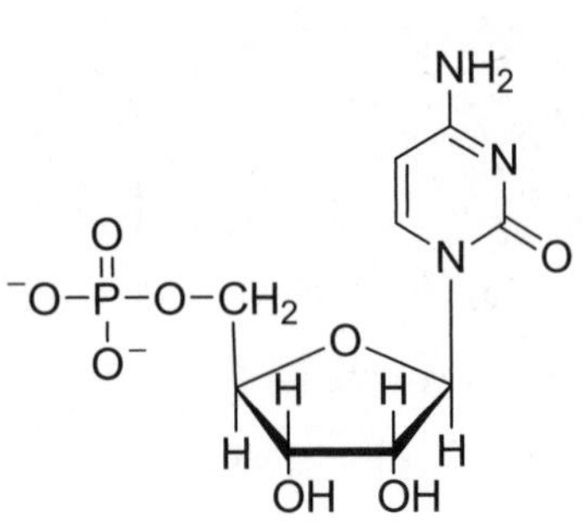

Solution

a. *This molecule is a diphosphate ester. There are two phosphate groups.*

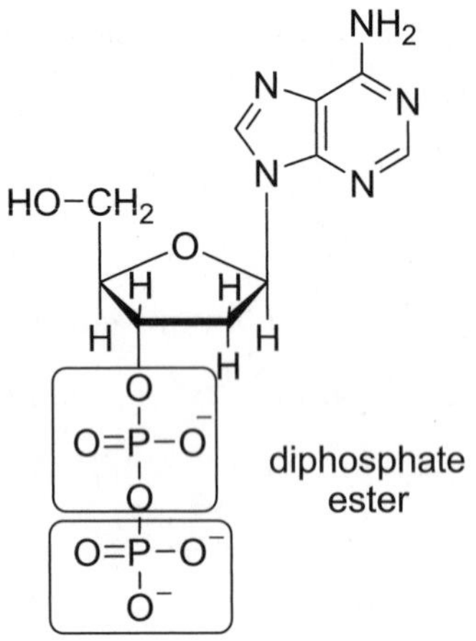

diphosphate ester

b. *This molecule is a monophosphate ester. There is one phosphate group.*

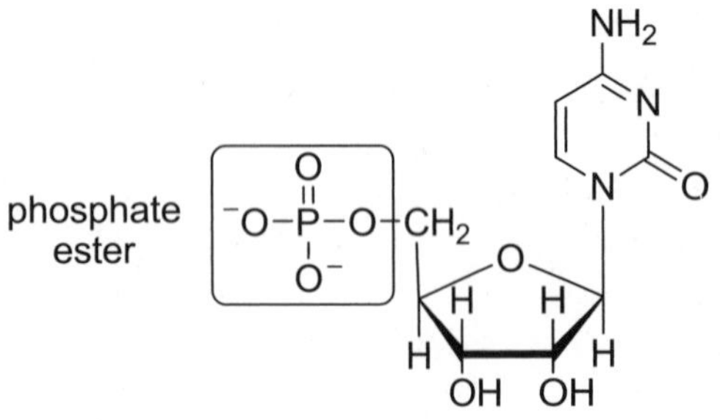

phosphate ester

Try It Yourself #16

Identify the structures that follow as mono-, di-, or triphosphate esters. Place a box around each phosphate group.

a.

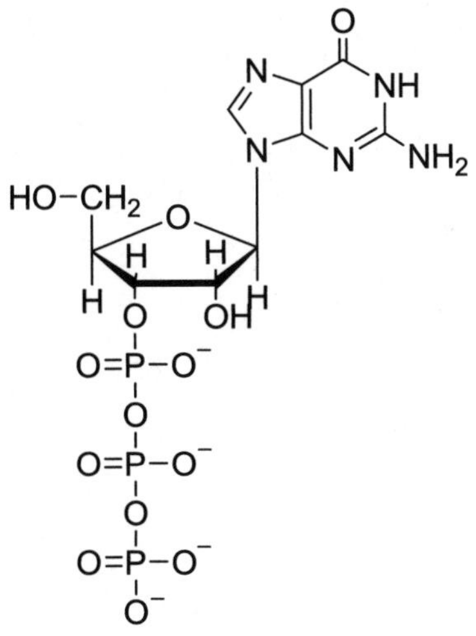

b.

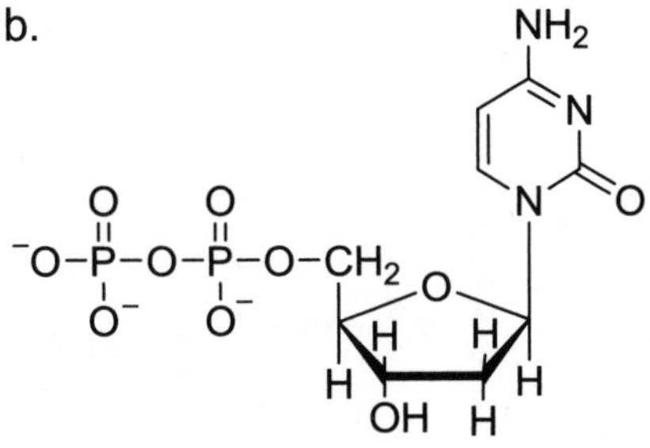

Solutions

a. Number of phosphate groups: ____________________

Type of phosphate: ____________________

Draw a box around each phosphate group.

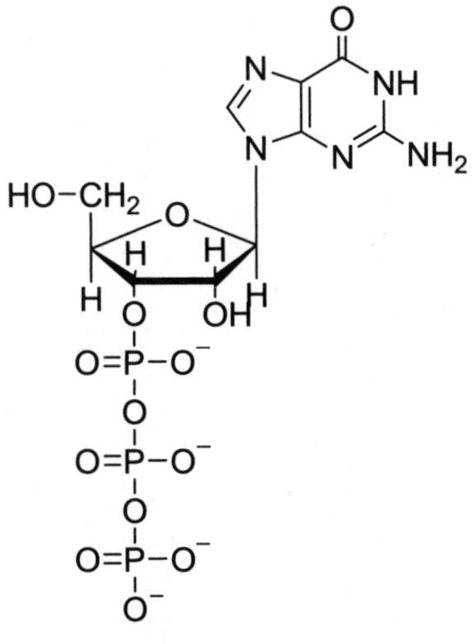

b. Number of phosphate groups: ____________________________

Type of phosphate: ____________________________________

Draw a box around each phosphate group.

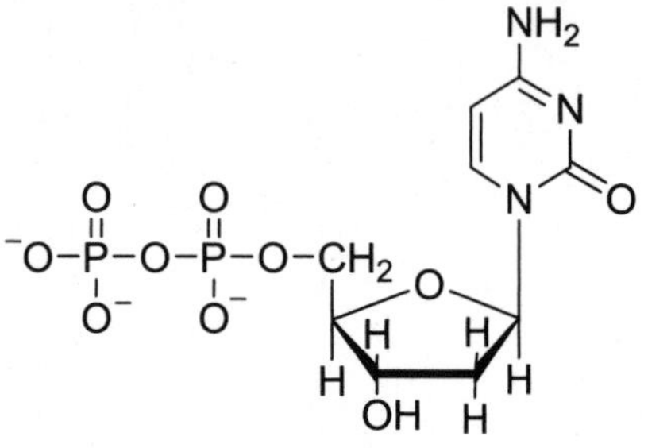

Practice Problems for Phosphate Ester Functional Groups

1. Identify the structures below as mono-, di-, or triphosphate esters. Place a box around each phosphate group.

   a.

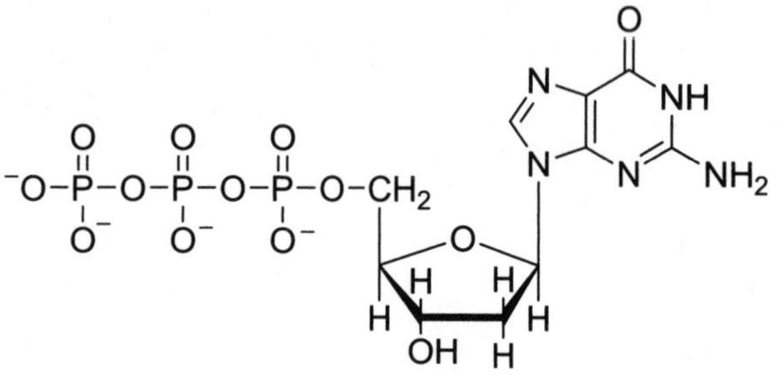

   b.

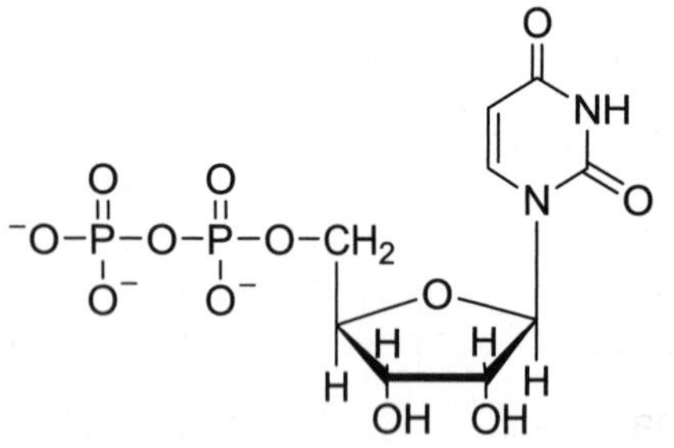

2. What is the total charge on the molecules shown in Question 1?

Chapter 7 Quiz

1. Doxycycline is a broad spectrum antibiotic. Circle and label the functional groups in the structure shown below.

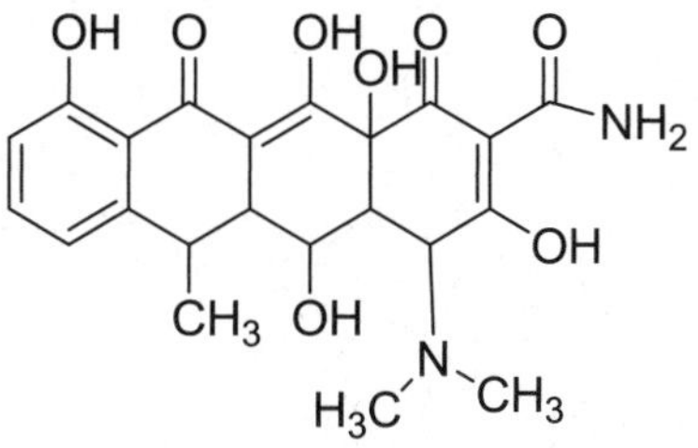

2. Inderal is a beta-adrenergic receptor blocking agent. It is used in the management of hypertension. Circle and label the functional groups in the structure shown below.

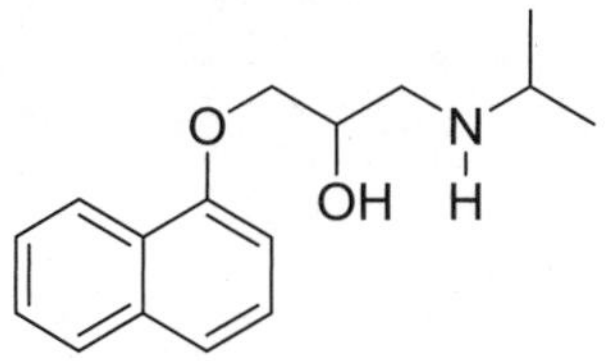

Inderal

3. Ritalin is a mild central nervous system stimulant. It is used in the treatment of attention deficit disorders. Circle and label the functional groups in the structure shown below. Is the amine in its neutral or ionic form?

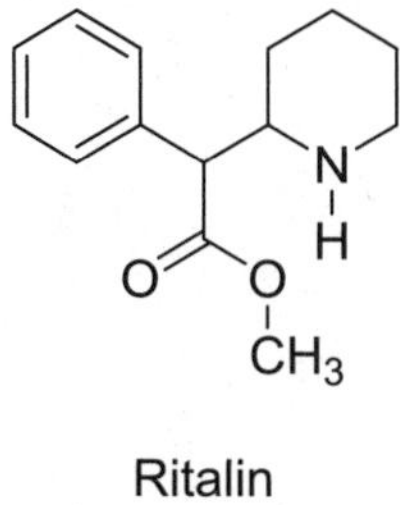

Ritalin

4. Glyceraldehyde 3-phosphate is an intermediate of glycolysis, a metabolic pathway. Circle and label the functional groups in the structure shown below.

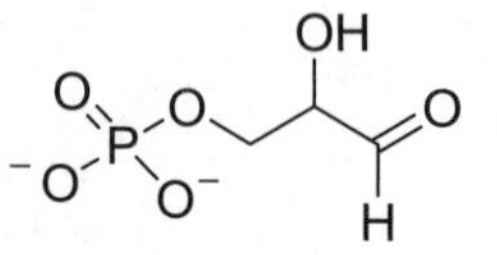

Glyceraldehyde 3-phosphate

5. Lotensin® (Benzapril hydrochloride) is used in the treatment of hypertension. It is an ACE (angiotension-converting enzyme) inhibitor.

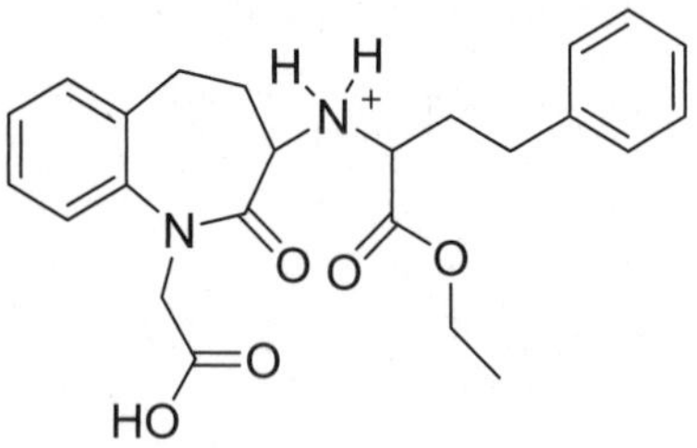

Lotensin

   a. Circle and label the following functional groups in Lotensin: two aromatic rings, an ester, an amide, an amine, and a carboxylic acid.
   b. Is the amine in its ionic or neutral form?

   c. Is the carboxylic acid in its neutral or ionic form?

6. Aldactone® (Spironolactone) is a diuretic used to treat edema in patients with congestive heart failure, hypertension, and hypokalemia (low blood potassium). Circle and label the functional groups in Aldactone.

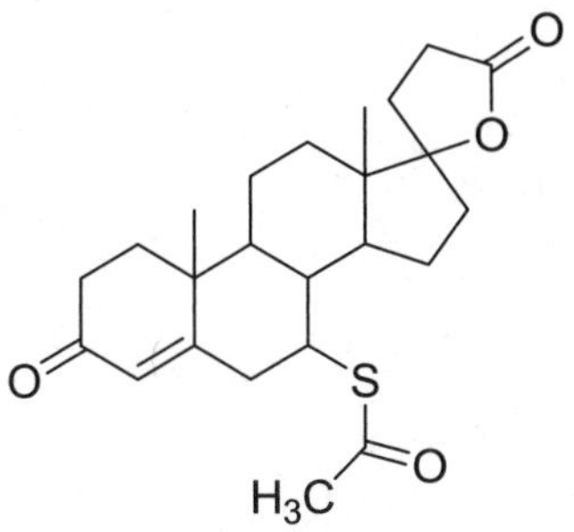

H_3C

Aldactone

7. Assign an IUPAC name to the molecules shown below:

   a.

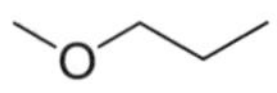

   b. $CH_3CH_2CH_2CH_2CH_2CH_2CH_2CHO$

   c.

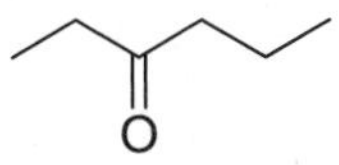

   d. $CH_3CH_2CH_2CH_2CH_2CH_2CH_2CH_2OH$

e.

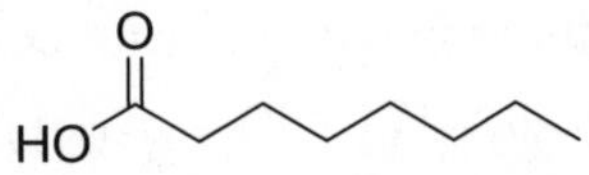

HO

O

f.

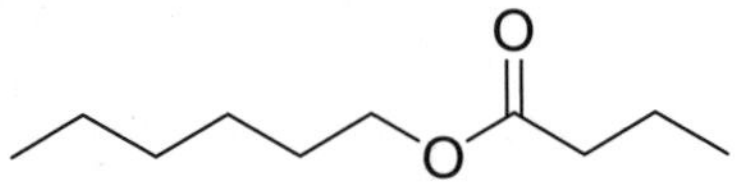

O

O

g.

O

N

H

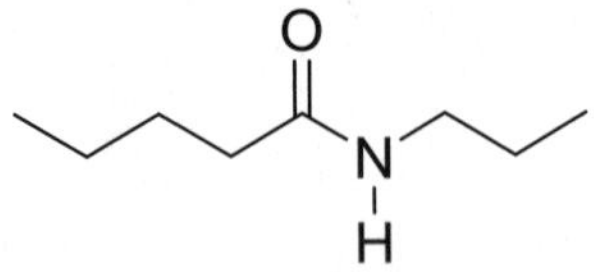

8. Write a skeletal line structure for the following molecules:

a. heptanal

b. 3-heptanone

c. pentanoic acid

d. methoxymethane

e. 2-methyl-3-pentanol

f. *N*-propylbutanamide

g. ethylhexanoate

9. Classify the following amines and alcohols as primary, secondary, or tertiary amines or alcohols.

   a. 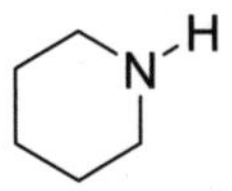 N–H (six-membered ring)

   b. 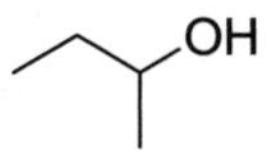 OH

c. HO

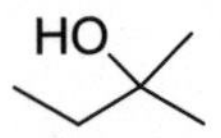

d. H
H N

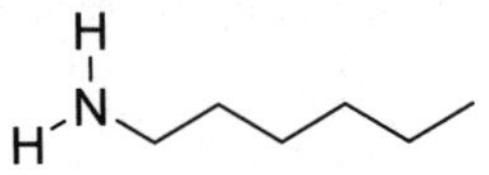

10. Are the neutral forms or the ionic forms of the following amines and carboxylic acids shown?

a. O^- O

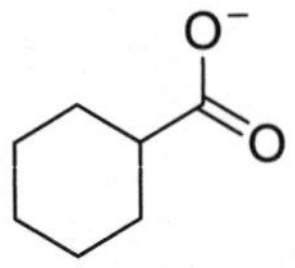

b. N

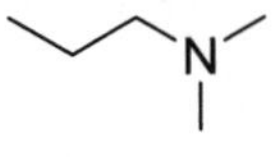

c. H + N H
H

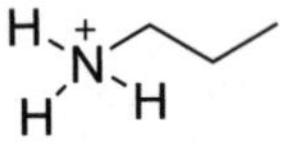

d. O
HO

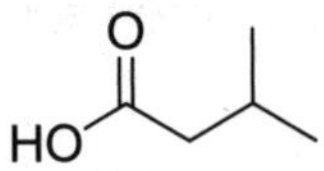

Chapter 7

Answers to Additional Exercises

45 Functional groups in a molecule determine its chemical and physical properties.

47 The distribution of morphine, heroin, and hydrocodone can lead to dependence and, with repeated use, addiction.

49 When opioids bind to opioid receptors, it initiates a sequence of biological events that leads to reduced pain sensations.

51 Endorphins and opioids bind to opioid receptors.

53 An ether has an oxygen atom bonded to two carbon atoms. An alcohol has oxygen atom bonded to a hydrogen atom and a carbon atom.

55 a. 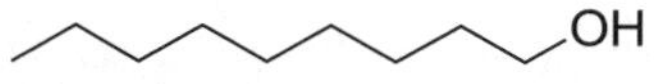 OH There are nine carbon atoms in the main chain. The OH group is located on carbon 1. The alcohol is a primary alcohol.

b. 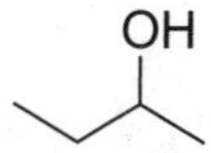 OH There are four carbon atoms in the main chain. The OH group is located on carbon 2. The alcohol is a secondary alcohol.

c. 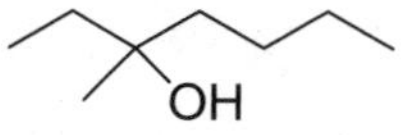 OH There are seven carbon atoms in the main chain. The OH group is located on carbon 3. There is also a methyl group located on carbon 3. The alcohol is a tertiary alcohol.

57 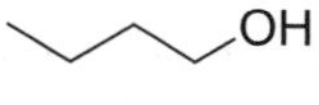 OH The molecular shape around the oxygen is bent tetrahedral. The C—O—H bond angle is 109°.

59

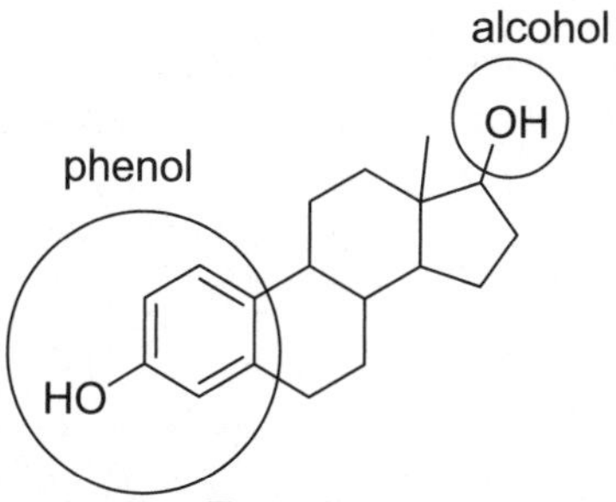

Estradiol

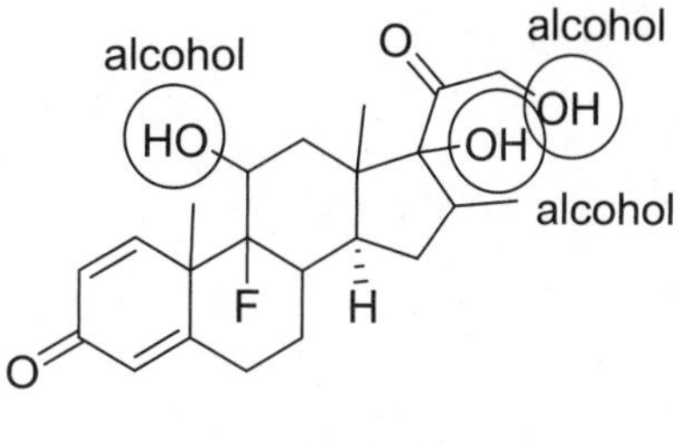

Betamethasone

The phenol is an alcohol attached to an aromatic ring. They both have three six-membered rings and a five-membered ring fused together.

61

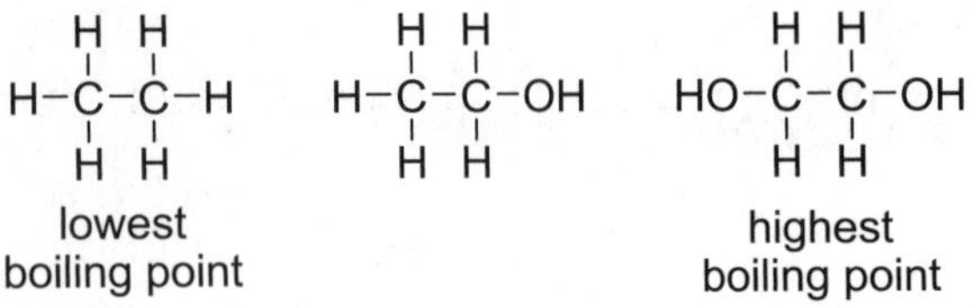

```
   H H          H H           H H
   | |          | |           | |
 H-C-C-H      H-C-C-OH     HO-C-C-OH
   | |          | |           | |
   H H          H H           H H
  lowest                    highest
boiling point            boiling point
```

63

alcohol

HO

ether

O

O

ONO_2

ether

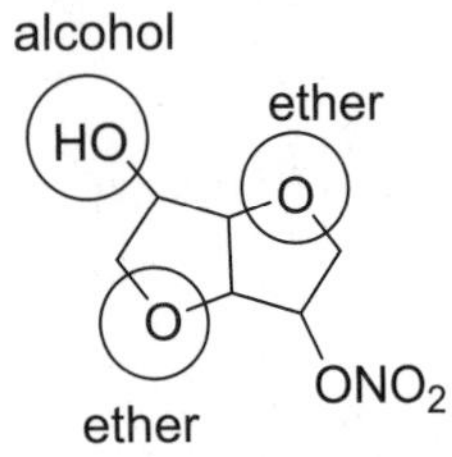

65. a. 1-ethoxyethane. Both carbon chains have the same number of carbon atoms—two . The root name is ethane. The alkoxy substituent has two carbon atoms, so it is ethoxy.

b. 2-methoxypentane. There are five carbons in the main chain. The alkoxy substituent has one carbon, so it is methoxy. The methoxy substituent is on carbon number 2.

c. 3-ethoxypentane. There are five carbons in the main chain. The alkoxy substituent has two carbon atoms, so it is ethoxy. The ethoxy substituent is on carbon number 3.

67 a. ether b. primary alcohol c. ether d. tertiary alcohol e. ether f. secondary alcohol g. primary alcohol

69 a. Butan-2-amine. The main chain has four carbon atoms. The nitrogen atom is on carbon 2. Primary amine—there is one carbon atom attached to the nitrogen atom. b. Butan-1-amine. The main chain has four carbon atoms. The nitrogen atom is on carbon 1. Primary amine—there is one carbon atom attached to the nitrogen atom c. *N*-methylbutan-1-amine. The main chain has four carbon atoms. The nitrogen atom is on carbon 1. There is one methyl group attached to the nitrogen atom. Secondary amine—there are two carbon atoms attached to the nitrogen atom.

71 (c) is a primary amine. There is one carbon atom attached to the nitrogen atom. Only structure (c) is capable of hydrogen bonding.

73 a. and b.

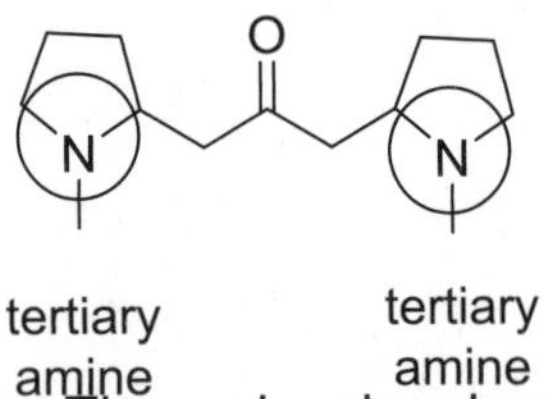

O

N N

tertiary amine

tertiary amine

c. The molecular shape around the nitrogen atom is trigonal pyramidal.

d. The C—N—C bond angle is 109.5°.

e. It is an alkaloid because it contains a nitrogen atom and it is produced by a plant.

75 a.

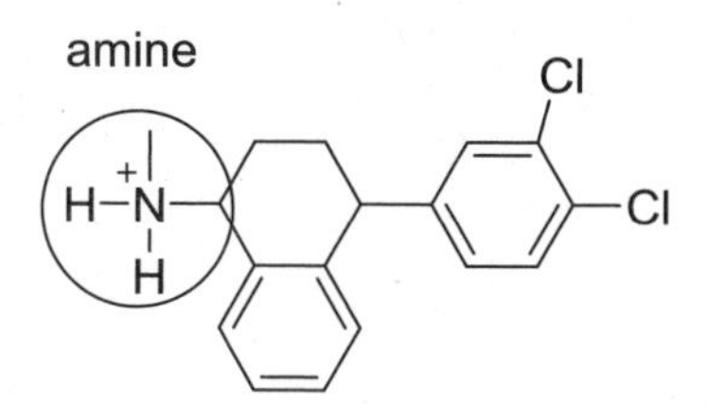

b. The amine is in the ionized form.

c. The neutral form of the amine is a secondary amine.

77 a. Aldehyde. The IUPAC name is propanal. There are three carbon atoms in the main chain. The ending is changed to *-al* to reflect that it is an aldehyde.

b. Ketone. The IUPAC name is pentan-3-one. There are five carbon atoms in the main chain. The ending is changed to *-one* to reflect that it is a ketone. The carbonyl carbon is located on carbon 3.

c. Aldehyde. The IUPAC name is pentanal. There are five carbon atoms in the main chain. The ending is changed to *-al* to reflect that it is an aldehyde.

d. Ketone. The IUPAC name is hexan-2-one. There are six carbon atoms in the main chain. The ending is changed to *-one* to reflect that it is a ketone. The carbonyl carbon is located on carbon 2.

e. Aldehyde. The IUPAC name is 3-methyl pentanal. There are five carbon atoms in the main chain. The ending is changed to *-al* to reflect that it is an aldehyde. There is a methyl group on carbon 3.

79 a.

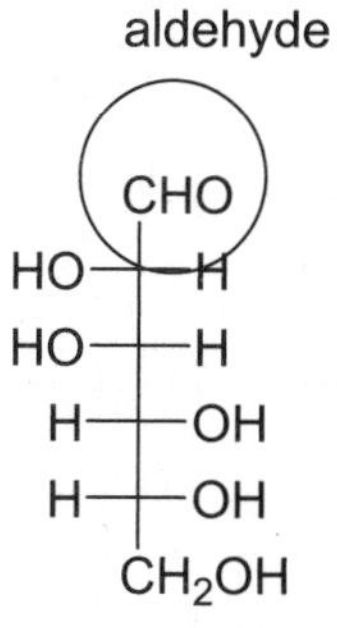

b. The molecular shape around the carbonyl carbon is trigonal planar.

c.

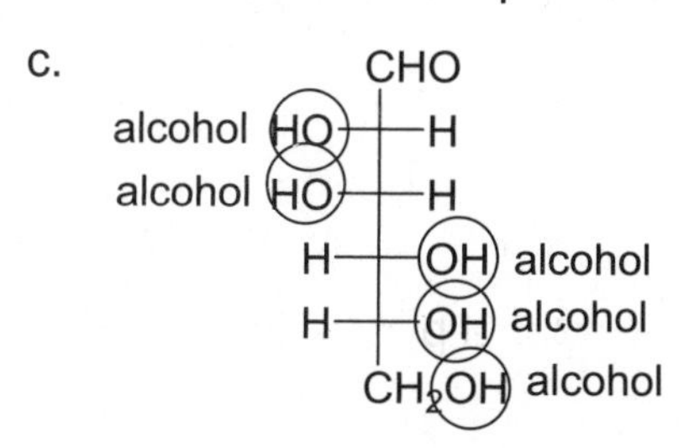

d. Mannose is a carbohydrate.

81

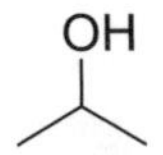

OH O

propan-2-ol 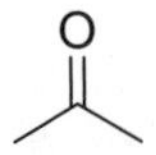 propan-2-one

Propan-2-ol has the higher boiling point. Propan-2-ol has hydrogen bonding and dipole-dipole interactions, while propan-2-one only has dipole-dipole interactions.

83 a.

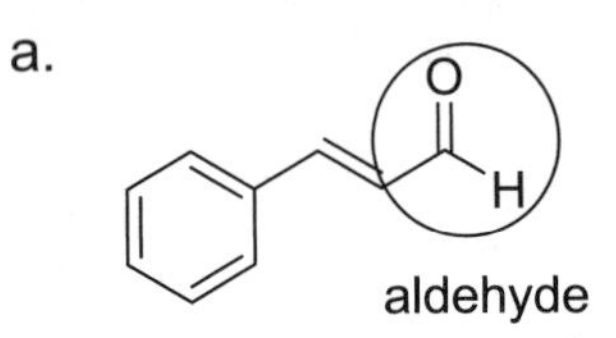

b. The molecular shape around the carbonyl carbon is trigonal planar.

c.

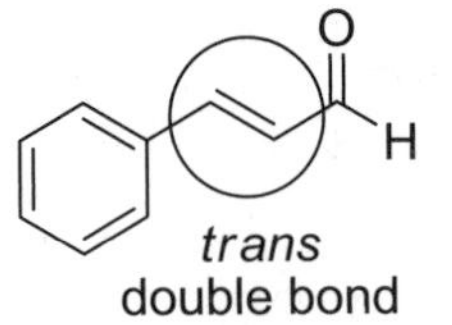

O
H
trans
double bond

85 a. The IUPAC name is pentanoic acid. The main chain has five carbon atoms. The ending is changed to *-oic acid* to signify that there is a carboxylic acid present.

b. The IUPAC name is hexanoic acid. The main chain has six carbon atoms. The ending is changed to *-oic acid* to signify that there is a carboxylic acid present.

c. The IUPAC name is propanoic acid. The main chain has three carbon atoms. The ending is changed to *-oic acid* to signify that there is a carboxylic acid present.

d. The IUPAC name is 3-methylpentanoic acid. The main chain has five carbon atoms. The ending is changed to *-oic acid* to signify that there is a carboxylic acid present. There is a methyl group on carbon 3.

87 a.

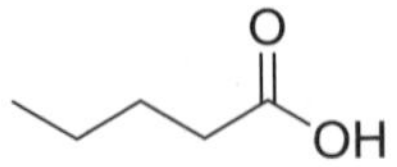

O
OH

b.

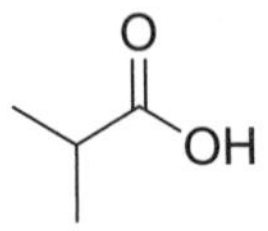

O
OH

c.

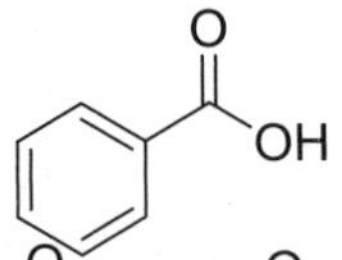

O
OH

89

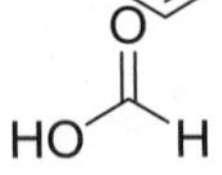

O O
HO H OH
formic acid 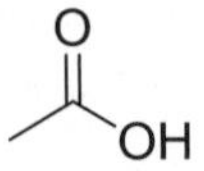 acetic acid

91 a.

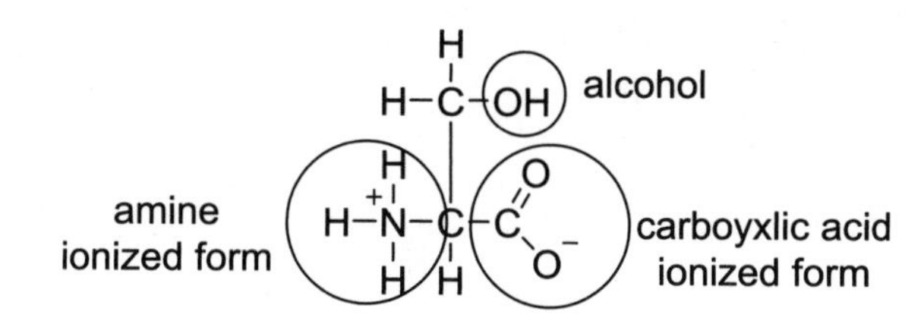

b. The amine is in its ionic form.

c. The carboxylic acid is in its ionic form.

d. The overall charge on serine is zero.

93 a.

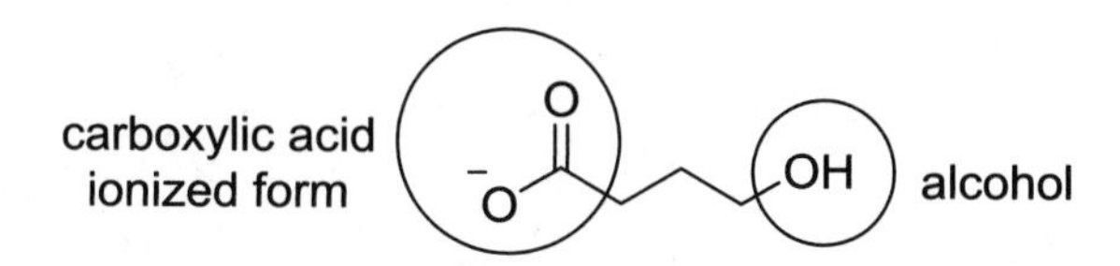

b. The carboxylic acid is in its ionic form.

95 a. and b.

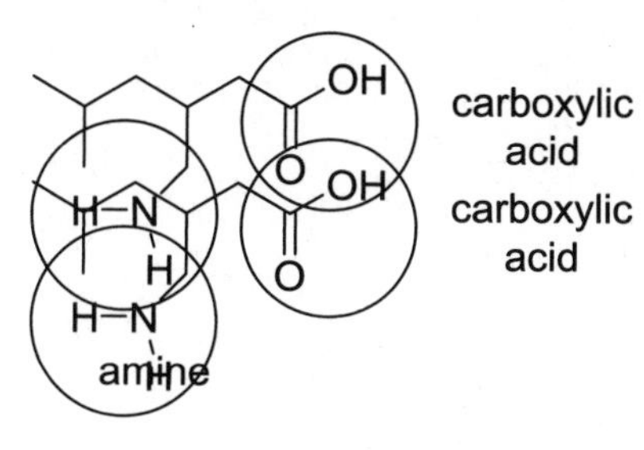

amine

c. The carboxylic acid is in its neutral form.

d. The amine is in its neutral form.

97 a.

b.

c.

99 The IUPAC name is butyl ethanoate. The main

chain with the carbonyl carbon has two carbon atoms. The ending is changed to *-oate* to signify that the molecule is an ester. The R' group has four carbons in it, so it is a butyl group.

101

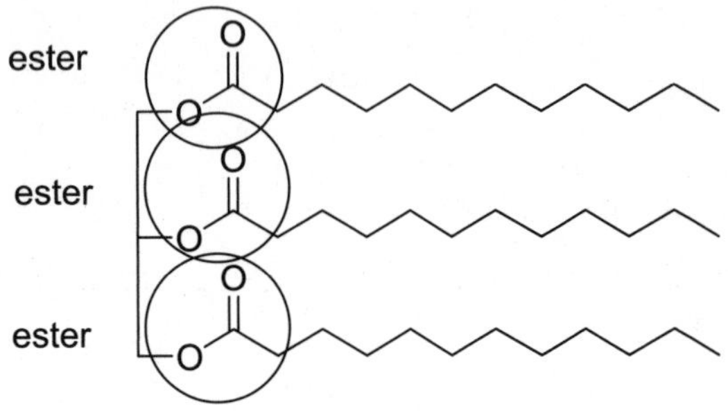

ester

ester

ester

Triglycerides are more commonly known as fats and oils.

103

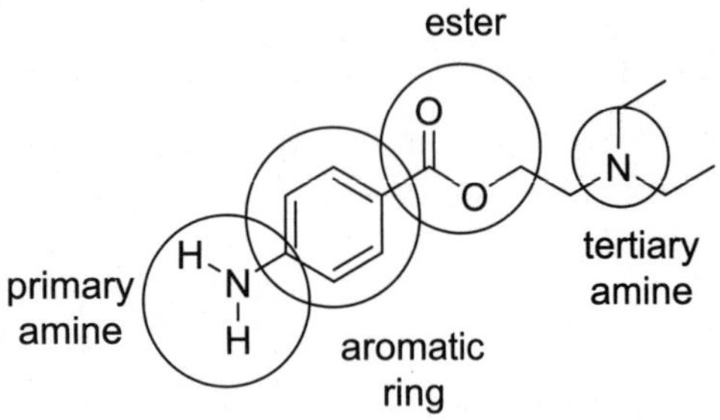

ester

primary amine

aromatic ring

tertiary amine

105

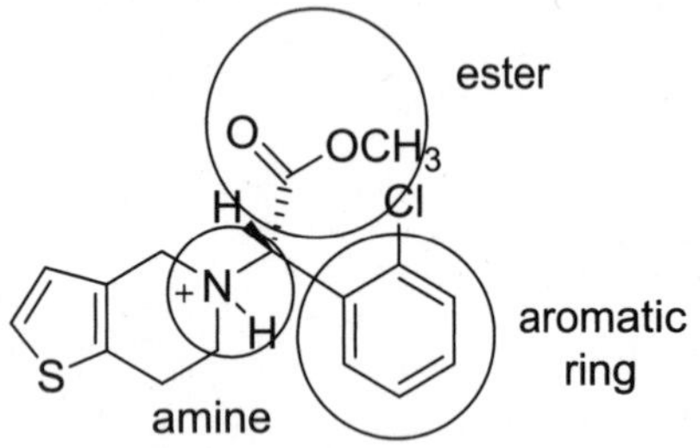

ester

aromatic ring

amine

The amine is in its ionic form.

107 (b) and (c) are amides

109

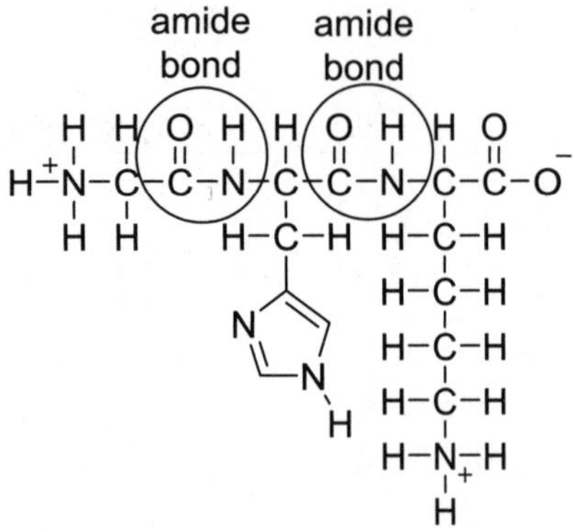

amide bond

amide bond

111

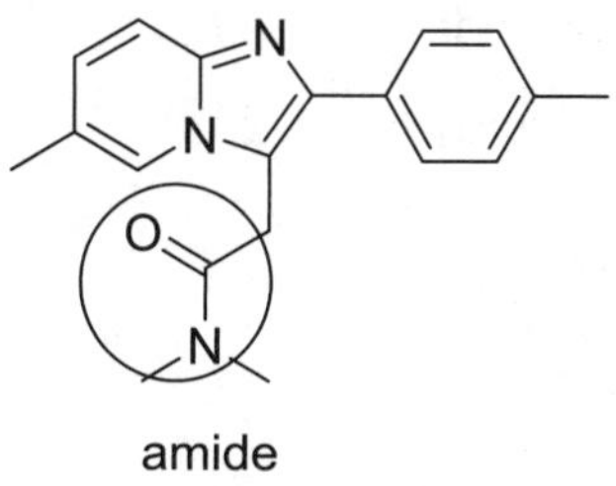

amide

113 Esters and amides can be prepared from carboxylic acids.

115 An achiral molecule is superimposable on its mirror image—it is identical to its mirror image.

117 a. achiral b. chiral c. chiral d. chiral

119 a. The two molecules are identical. b. Enantiomers. The two molecules are non-superimposable on each other.

121 No, ethambutol should not be sold as a racemic mixture. One of the enantiomers causes harm (blindness).

123

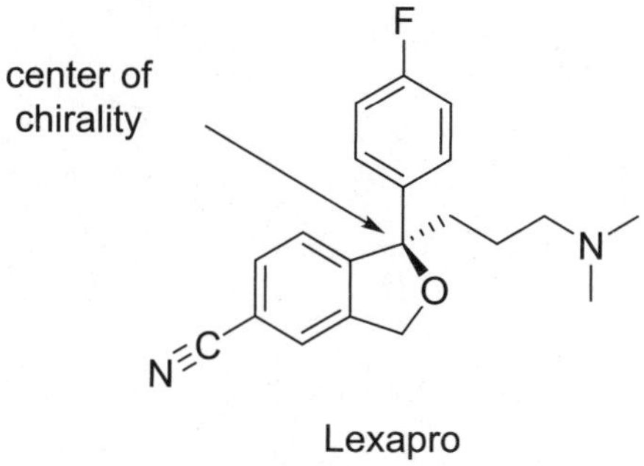

center of chirality

Lexapro

The center of chirality is tetrahedral, and the bond angles are 109.5°.

125

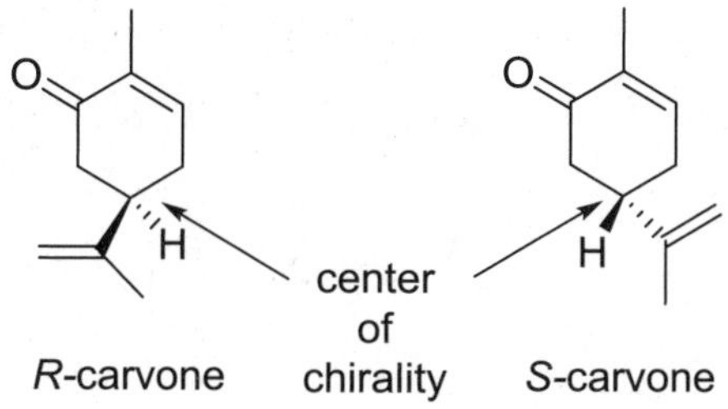

R-carvone center of chirality *S*-carvone

The hydrogen atom attached to the center of chirality projects toward you in S-carvone and away from you in R-carvone.

127

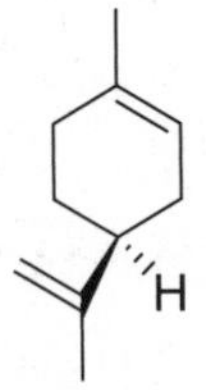

S-limonene

The receptors for smell are a chiral environment. When the two enantiomers are placed in a chiral environment, they exhibit different physical properties, such as smell.

129

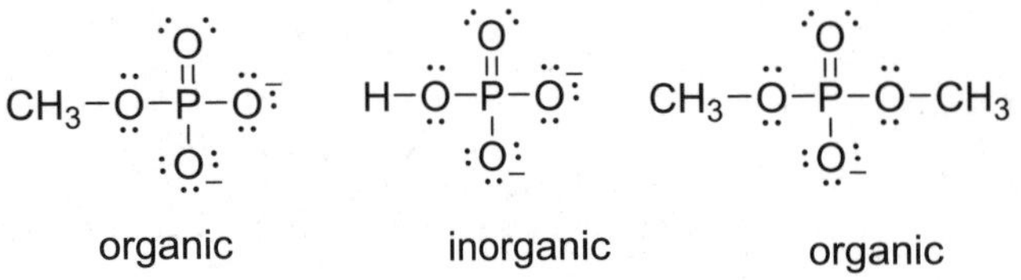

$CH_3-O-P(=O)(-O^-)-O^-$ — organic

$H-O-P(=O)(-O^-)-O^-$ — inorganic

$CH_3-O-P(=O)(-O^-)-O-CH_3$ — organic

The organic compounds contain carbon atoms, while the inorganic ones do not.

131 a.

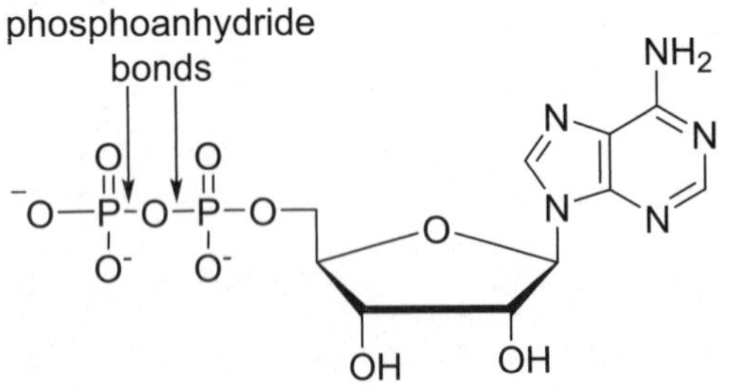

phosphoanhydride bonds

NH_2

OH OH

b. The cell stores potential energy by forming phosphoanhydride bonds.

c. The compound is a phosphate diester. There are two phosphate groups joined together with phosphoanhydride bonds.

133 A person who has schizophrenia has an excess amount of dopamine in the brain; a person who has Parkinson's disease has a decreased amount of dopamine in the brain.

135 neurotransmitters; amine functional group

137 Dopamine cannot pass through the blood-brain barrier to reach the brain where it is needed.

139 Parkinson's disease is the loss of dopamine producing neurons along the neuronal pathway controlling movement.

141 Atypical antipsychotics are the medications that do not affect movement.

Chapter 8
Mixtures, Solution Concentrations, and Diffusion

Chapter Summary

In this chapter, you have learned about how scientists and medical professionals classify and quantify the compounds and elements that make up a mixture. You have also learned how to distinguish mixtures and different types of solutions. Knowing how to interpret concentrations of solutions and understanding how to use concentrations play an important role in the health field, from preparing IV solutions to delivering correct drug dosages to patients. Understanding concentrations will provide insight into dialysis, osmosis, and other biochemical pathways that use membranes.

8.1 Mixtures

A combination of two or more elements and/or compounds, in any proportion, is known as a mixture. Mixtures can be separated into their pure components through physical separation techniques. There are two basic types of mixtures: heterogeneous and homogeneous. In a homogeneous mixture, the components are evenly distributed throughout the mixture. There are two types of homogenous mixtures: a solution and a colloidal dispersion. In a heterogeneous mixture, the components are unevenly distributed throughout the mixture.

Solutions, colloidal dispersions, and suspensions differ at the atomic level in the size of their particles. Solutions contain the smallest particles, known as solute, that measure less than 1 nm in diameter. Colloidal dispersions contain particles intermediate in size, known as colloids, which range from 1 nm to 1 μm in size. Suspensions contain the largest particles, greater than 1 μm.

In a Nutshell: Colloidal Dispersions

The component of the colloid dispersion that is present in the greatest amount is known as the medium. The minor component(s) are called colloids. Colloids range in size from 1 nm to μm. Colloidal particles do not dissolve in their medium because they are significantly larger

than the molecules of the medium. Colloids can be large molecules or many small molecules clumped together, known as aggregates.
Colloidal dispersions can be found in all three states of matter: aerosols (solid or liquid colloids dispersed in a gas medium), sols (solid colloids dispersed in a liquid medium), emulsions (liquid colloids dispersed in a liquid medium), and gels (liquid colloids in a solid medium).

In a Nutshell: Suspensions

The particles in a suspension are larger than 1 μm and can often be seen with the naked eye or under a light microscope. In the absence of constant stirring, particles in a suspension eventually settle due to gravity and the large size and mass of their particles. Particles in a suspension can be separated from the medium by a filter or a centrifuge.

A centrifuge contains receptacles for several test tubes, which are spun at high speed, causing the suspended particles to collect at the bottom of the test tube in what is known as a pellet, with the remainder of the mixture, known as the supernatant, on top. The supernatant is then easily separated from the pellet.

Worked Example #1

Classify each mixture below as homogenous or heterogeneous:

a. a glass of orange juice with no pulp
b. a glass of orange juice with pulp
c. balsamic vinaigrette salad dressing
d. soapy water

Solutions

a. *Homogeneous mixture because the components are even distributed throughout*
b. *Heterogeneous mixture because the pulp can settle or be filtered out*
c. *Heterogeneous mixture because the oil and water will separate over time*
d. *Homogeneous mixture because the soap and water are evenly distributed*

Try It Yourself #1

Classify each mixture below as homogenous or heterogeneous:

a. hot chocolate
b. a glass of beer
c. a bowl of cereal in milk
d. an oatmeal-raisin cookie

Solutions

a. Hot chocolate is: ______________________________.
b. A glass of beer is: ______________________________.
c. A bowl of cereal in milk is: __________________________________.
d. An oatmeal raisin cookie is: __________________________________.

Worked Example #2

Do the following samples represent solutions, colloids, or suspensions?

a. a glass of soda
b. cement
c. mud

Solutions

e. A glass of soda is a solution. There is a gas (carbon dioxide) dissolved in a liquid in a homogenous mixture.
f. Cement is colloid. The particles (i.e., rock, sand) are large and evenly distributed throughout the medium.
g. Mud is a suspension. The dirt particles will settle after some time.

Try It Yourself #2

Do the following samples represent solutions, colloids, or suspensions?

a. whipped cream
b. milk
c. sugar and cream in coffee
d. paint

Solutions

a. Whipped cream is: ___________________________.

b. *Milk is:* ______________________________.

c. *Sugar and cream in coffee is:* ______________________________.

d. *Paint is:* ______________________________.

Practice Problems for Mixtures

1. Are the following heterogeneous or homogenous mixtures?

   a. pizza

   b. a milkshake

   c. an oral suspension of Tylenol

2. Do the following samples represent solutions, colloids, or suspensions?

   a. flour in water

   b. half-normal saline

   c. mayonnaise

   d. dust particles in the air

   e. barium enema

   f. cheese

g. champagne

8.2 Solutions

A solution is a homogenous mixture containing small molecules, atoms, or ions with a diameter less than 1 nm. The component that is present in the greatest amount is called the solvent and the component(s) present in lesser amounts are known as solute(s). A solution can contain more than one solute in a solvent. One of the most common solvents is water, and a solution containing water as the solvent is known as an aqueous solution and abbreviated *aq*.

A solution can have a solid solute in a liquid solute, or a gas solute in a gas solvent (air). There are also solutions with liquid solutes/liquid solvents (alcoholic beverages), gas solutes/liquid solvents (carbonated beverages), solid solutes/solid solvents (alloys), or liquid solute/solid solvent (dental amalgam). The most common types of solutions in medicine are aqueous solutions.

The term dissolved means that each of the solute atoms, ions, or molecules is completely surrounded by solvent molecules and evenly distributed throughout the solvent. Dissolution is a physical change.

In a Nutshell: Polarity of Solute and Solvent: "Like Dissolves Like"

The solubility of a solute in a solvent depends mainly on the polarity of the solute and the solvent. Polar solvents will dissolve polar solutes, and nonpolar solvents will dissolve nonpolar solutes, as characterized by the phrase "like dissolve like."

In contrast, a polar and a nonpolar liquid do not form a homogeneous mixture. When nonpolar molecules encounter polar molecules such as water, they cannot participate in the hydrogen bonding network between the water molecules. Instead, they form a separate layer that "avoids" the water molecules. Nonpolar molecules are hydrophobic. The tendency to form a separate layer is known as the hydrophobic effect and is driven by the natural tendency of all systems to attain the lowest energy.

Common polar solvents include water, small alcohols, and small carboxylic acids. These are all liquids that have strong hydrogen bonding forces of attraction. Common nonpolar solvents include: liquid hydrocarbons, carbon dioxide, elements and carbon tetrachloride. Nonpolar compounds interact through dispersion forces.

In a Nutshell: Dissolution of Ionic Compounds in Aqueous Solvent

When an ionic compound dissolves in water, the electrostatic attractions between ions—the ionic bonds—are broken and substituted with electrostatic interactions between each ion and several water molecules. Every cation is completely surrounded by the partial negatively charged pole of several water molecules and every anion is similarly surrounded by the partial positively charged pole of several water molecules. The electrostatic attraction between an ion and several polar solvent molecules is known as an ion-dipole interaction. Ionic compounds that dissolve in water do so because the solution is lower in energy than the undissolved lattice in water.

Although dissolution is a physical process, we can use reaction equations to represent the dissolution process. The formula unit for the ionic compound and its physical state—solid—are on the left side of the arrow and the separate ions, including their charges, followed by the abbreviation *aq,* enclosed in parentheses to signify the ions are surrounded by water molecules, are on the right side of the arrow. The subscripts in the formula unit of an ionic compound appear as coefficients on the right side of the equation. When writing an equation to show the dissolution of an ionic compound containing a polyatomic ion, remember to keep the formula unit for the polyatomic ion intact.

Some ionic compounds do not dissolve in water. Insoluble ionic compounds have stronger electrostatic interactions between the cation and anion in the solid ionic lattice than as dissolved ions. Insoluble ionic compounds form suspensions, a heterogeneous mixture consisting of large aggregates of ionic lattice suspended in water.

Even for ionic compounds that are soluble in water, there is a limit to the amount of solute that will dissolve in a given volume of solvent. A solution in which the solute no longer dissolves in the solvent is known as a saturated solution, a heterogeneous mixture. A solution becomes saturated when there are no more solvent molecules left to surround—dissolve—added solute.

Dissolved ions are referred to as electrolytes, especially in biological applications. Solutions containing electrolytes conduct electricity. The most important electrolytes that are critical to cellular function are Na^+, K^+, Ca^{2+}, Mg^{2+}, and Cl^-. Proper electrolyte balance is regulated by hormones in the body. Ionic compounds that do not dissolve in water are nonelectrolytes.

In a Nutshell: Dissolution of Molecular Compounds in Solution

When a covalent compound or diatomic element dissolves in a solvent, the covalent bonds remain intact, and only the intermolecular forces of attraction between solute molecules are disrupted as new solute-solvent intermolecular forces of attraction are formed.

Polar molecules dissolve in polar solvents ("like dissolves like"). When a polar solute molecule dissolves in a polar solvent, the intermolecular forces of attraction between the solute molecules are disrupted and new intermolecular forces of attraction, usually hydrogen-bonding forces are formed between the solute molecules and the solvent molecules. Polar solute molecules dissolve in water because the hydrogen-bonding interactions between solute and solvent are stronger than those in the lattice surrounded by water. The dissolved solution is lower in energy than the undissolved heterogeneous mixture. Solutions that contain dissolved molecules are classified as nonelectrolytes because the aqueous solution does not conduct electricity.

Nonpolar solutes dissolve in nonpolar solvents ("like dissolves like"). Nonpolar solutes interact with nonpolar solvents through weak dipole-dipole interactions or dispersion forces. When a nonpolar solute is dissolved in a nonpolar solvent, the solute is distributed uniformly throughout the mixture and dispersion forces of attraction form between the solute and solvent.

Worked Example #1

Identify the solute and solvent in each solution.

a. 500 g of ethanol containing 2 g of iodine
b. 50 mL of CO_2 and 245 mL of H_2O
c. 3 g KCl and 235 g of water

In each case, identify which substance is present in the smallest amount and which substance is present in the greatest amount. The substance present in the greatest amount is the solvent, while the substance present in the smallest amount is the solute.

a. *Substance present in the smallest amount: iodine*

Substance present in the greatest amount: ethanol

Solute: iodine

Solvent: ethanol

b. *Substance present in the smallest amount:* CO_2

Substance present in the greatest amount: H_2O

Solute: CO_2

Solvent: H_2O

c. *Substance present in the smallest amount: KCl*

Substance present in the greatest amount: water

Solute: KCl

Solvent: water

Try It Yourself #1

Identify the solute and solvent in each solution.

a. 125 g of Cu and 7 g of Sn
b. 150 mL of ethanol and 25 mL of CO_2
c. 27 g O_2 and 235 g of water

a. *Substance present in the smallest amount:* ____________

Substance present in the greatest amount: ____________

Solute: ____________

Solvent: ____________

b. *Substance present in the smallest amount:* ____________

Substance present in the greatest amount: ____________

Solute: ____________

Solvent: ____________

c. *Substance present in the smallest amount:* ____________

Substance present in the greatest amount: ____________

Solute: ____________

Solvent: ____________

Worked Example #2

For the following solutions, indicate whether the solute is a molecule or an electrolyte.

a. $CaCl_2$

b. sucrose

Identify the type of bond in type of bond in the solute; then determine the type of solute.

a. *$CaCl_2$ has an ionic bond. It produces ions when dissolved in water. Therefore, it is an electrolyte.*

$$CaCl_2\ (s) \longrightarrow Ca^{2+}\ (aq) + 2\ Cl^-\ (aq)$$

*(**1** calcium ion)* *(**2** chloride ions)*

b. *Sucrose has covalent bonds. Therefore, it is a molecule.*

Try It Yourself #2

For the following solutions, indicate whether the solute is a molecule or an electrolyte.

a. O_2

b. dextrose ($C_6H_{12}O_6$)

c. KCl

a. *Type of bond:* ________________

Molecule or electrolyte: ________________

b. *Type of bond:* ________________

Molecule or electrolyte: ________________

c. *Type of bond:* ________________

Molecule or electrolyte: ____________________

Worked Example #3

How many ions of each type will be produced when $Ca(HCO_3)_2$ is dissolved in aqueous solution? Would you describe the solutes in $Ca(HCO_3)_2$ (aq) as molecules or electrolytes? Explain.

The formula unit for calcium hydrogen carbonate ($Ca(HCO_3)_2$) indicates that there is one calcium ion for every two polyatomic hydrogen carbonate anions (HCO_3^-) released:

$$Ca(HCO_3)_2\ (s) \longrightarrow Ca^{2+}\ (aq) + 2\ HCO_3^-\ (aq)$$

*(**1** calcium ion)* *(**2** hydrogen carbonate ions)*

Formula unit for calcium hydrogen carbonate: $Ca(HCO_3)_2$
Cation: Ca^{2+}
Number of cations produced: one
Anion: HCO_3^-
Number of anions produced: two

$Ca(HCO_3)_2$ produces ions when dissolved in water; therefore, it is an electrolyte.

Try It Yourself #3

How many ions of each type will be produced when $Ca_3(PO_4)_2$ is dissolved in aqueous solution?

Formula unit for calcium phosphate: ____________________
Cation: ______________
Number of cations produced: ________________
Anion: ______________________
Number of anions produced: __________________

Practice Problems for Solutions

Identify the solute and solvent for the following solutions. Is the solute a molecule or an electrolyte, or a gas? If the solute is an electrolyte, what are the ions produced and how many of each ion are produced?

a. a teaspoon of sugar and a teaspoon of cream dissolved in a cup of coffee

b. 25 mL of water and 50 L of air

c. 1.23 g Na_3PO_4 and 235 g of water

d. 500 mL ethanol and 34 mL of CO_2

e. 175 mL water and 7 g $Mg(HCO_3)_2$

8.3 Concentrations

The amount of solute in a given volume of solutions is known as its concentration. Several methods are used to measure the concentration of a solution. One method, known as the colorimetric technique measures the intensity of the color of the solution to determine its concentration. Quantitative measurements of concentration are routine in science and medicine. The concentration of a solution is a quantitative measure the amount of a particular solute is dissolved in a given volume of solution. Thus, concentration is always expressed as a ratio, where the numerator represents the amount of solute and the denominator represents the volume of the total solution (solutes + solvent). Units of concentration are also a ratio: a unit representing the amount of solute in the numerator and a unit of volume for the solution in the denominator. Concentrations are usually based on the mass of solute or the moles of solute. The units of concentration most commonly encountered in the medical field are mass/volume (m/v), % mass/volume (%m/v), moles/volume (mol/L), and equivalents/volume (eq/L).

In a Nutshell: Preparing a Solution with a Specified Concentration

All concentration units are defined in terms of volume of total solution, not the volume of the solvent. A solution with a specified concentration can be prepared using a volumetric flask. First, calculate and then weigh out the amount of solute needed. Next, transfer the solute to an appropriately sized volumetric flask. Volumetric flasks come with a mark on the next of the flask representing one precise volume. Finally, add solvent to the volumetric flask up to the mark on the neck of the flask and mix it.

In a Nutshell: Mass/Volume Concentration

The concentrations of many important solutes in a blood test are reported as mass per volume (m/v), a common unit of concentration in medicine. The mass per volume of a solution gives the ratio of the mass of the solute as a metric unit of mass, per volume of solution, in a metric unit of volume. The general form of the equation for determining a mass/volume concentration follows:

$$\frac{\text{mass}}{\text{volume}} = \frac{\text{mass of solute}}{\text{volume of solution}}$$

Worked Example #4

Auralgan is used as a topical decongestant and analgesic. It comes in solution form and contains 54 mg of antipyrine per mL. What is the concentration of this solution in g/dL?

You are asked to convert the concentration into another metric unit, so two metric conversions are required—one for the numerator (mg to g) and one for the denominator (mL to dL). Thus, you need to multiply the supplied concentration by two metric conversions, in any order.

$$\frac{54\ \cancel{\text{mg}}}{1\ \cancel{\text{mL}}} \times \frac{1\ \text{g}}{10^3\ \cancel{\text{mg}}} \times \frac{1000\ \cancel{\text{mL}}}{10\ \text{dL}} = 5.4\ \text{g/dL}$$

The first ratio represents the supplied concentration, the second is the metric mg-to-g conversion factor, and the third is the metric mL-to-dL conversion factor. Always use the form of the conversion factor that allows the supplied unit to cancel, thereby leaving the requested unit.

Try It Yourself #4

Amprenavir, used to treat HIV-1, comes as oral solution with a concentration of 15 mg of amprenavir per mL. What is this concentration in μg/L?

Conversion factors needed:

Set up the calculation so that the supplied units cancel.

In a Nutshell: % mass/volume concentration

Solutions used in intravenous (IV) therapy often have concentrations given in units of percent mass per volume, %m/v. To calculate the concentration of a solution in %m/v, the units of the solute must be in grams (g) and the units of solution must be in milliliters (mL). The resulting ratio is then solved and multiplied by 100 to obtain a percentage:

$$\%\frac{\text{mass}}{\text{volume}} = \frac{\text{g of solute}}{\text{mL of solution}} \times 100$$

A % symbol always stands for "per 100," which means "divided by 100." The terms ppm (parts per million) and ppb (parts per billion) are analogous to % mass/volume, except instead of "per 100," they stand for "per million" and "per billon," respectively. The mass must be in grams and the volume must be in milliliters.

$$\text{ppm} = \frac{\text{g of solute}}{\text{mL of solution}} \times 1{,}000{,}000$$

$$\text{ppb} = \frac{\text{g of solute}}{\text{mL of solution}} \times 1{,}000{,}000{,}000$$

Worked Example #5

What is the % m/v of sucrose in a carbonated beverage that contains 39 g of sucrose in 355 mL of beverage?

The supplied units are g/mL. We will use the equation for % mass/volume.

$$\%\frac{\text{mass}}{\text{volume}} = \frac{\text{g of solute}}{\text{mL of solution}} \times 100$$

$$\%\frac{\text{mass}}{\text{volume}} = \frac{\text{39 g sucrose}}{\text{355 mL of beverage}} \times 100 = \text{11 \% (m/v) sucrose}$$

Try It Yourself #5

What is the % m/v of I_2 in a solution prepared by dissolving 2.35 g I_2 in enough ethanol to make 155 mL of solution?

Tools: equation for % mass/volume

Worked Example #6

You have been asked to prepare 1.75 L of 0.90% NaCl (% m/v) solution for IV therapy. How many g of NaCl should you weigh out?

In this problem, you are supplied with the concentration and the volume of the solution and asked to calculate the mass of solute. Using dimensional analysis, first convert the supplied volume unit, 1.75 L, into mL. Then multiply by the correct form of the concentration conversion factor that allows mL to cancel, leaving only grams:

$$1.75\,\cancel{\text{L}} \times \frac{10^3\ \text{mL}}{1\,\cancel{\text{L}}} \times \frac{0.90\ \text{g}}{100\,\cancel{\text{mL}}} = 16\ \text{g}$$

Note that the concentration 0.90% has been written as the ratio 0.90g/100 mL because % means "per 100."

Try It Yourself #6

You have been asked to prepare 2.5 L of a 5% dextrose solution. How many g of dextrose should you weigh out?

Supplied units: ______________________________

Requested units: ______________________________

Set up the calculation, so that the supplied units cancel.

In a Nutshell: Molar Concentration

The most common units of concentration based on moles of solute are molarity and equivalents/liter. The unit of concentration most often used in chemistry is molar concentration, also known as molarity and abbreviated *M*. The molar concentration of a solution is defined as the number of moles of solute. The unit of volume in the denominator must be in liters.

$$\text{Molarity (M)} = \frac{\text{moles of solute}}{\text{L of solution}}$$

Blood tests typically report concentrations of calcium, sodium, potassium, and chloride ions and also carbon dioxide in units of millimoles per liter (mmol/L = mM). In calculations involving molarity as a conversion factor, you should always replace the abbreviation M with moles/L so that units can cancel.

Worked Example #7

A solution having a volume of 5.8 L contains 34 mmol of CO_2 (carbon dioxide). What is the concentration of carbon dioxide in the solution in mol/L?

You are asked to calculate a molar concentration—moles per liter given the mmols of solute and the liters of solution. First, mmol needs to be converted to mol. Then use the molarity equation.

Set up the calculation, so that the supplied units cancel:

$$\text{M} = \frac{34\ \cancel{\text{mmol}}}{5.8\ \text{L}} \times \frac{1\ \text{mol}}{10^3\ \cancel{\text{mmol}}} = 0.0059\ \text{mol/L}$$

Try It Yourself #7

A solution having a volume of 2.4 L contains 253 mmol of O_2. What is the concentration of oxygen in the solution in mol/L?

Supplied units: ________________________

Requested unit: ________________________

Set up the calculation, so that the supplied units cancel.

Worked Example #8

How many moles of potassium ions (K^+) are there in 3.2 L of 45 mM KCl?

Using dimensional analysis, multiply the supplied unit by the correct form of the supplied conversion factor that allows volume to cancel. Then convert mmol to moles.

$$3.2\ \cancel{\text{L}} \times \frac{45\ \cancel{\text{mmol}}}{\cancel{\text{L}}} \times \frac{1\ \text{mol}}{10^3\ \cancel{\text{mmol}}} = 0.14\ \text{moles}$$

Try It Yourself #8

How many moles of sodium ions (Na^+) are there in 7.2 L of 5.1 mM Na_3PO_4?

Supplied units: _________________

Requested units: _________________

Set up the calculation, so that the supplied units cancel.

In a Nutshell: Equivalents per Liter

For an electrolyte, the moles of charge per liter are reported as equivalents per liter, eq/L. An equivalent, abbreviated eq, is the moles of charge, which depends on the magnitude of charge on the ion and the number of moles of ion. Equivalents per liter (eq/L) are simply calculated by multiplying the molar concentration of the solution, M, by the numerical value of the charge on the ion:

$$\frac{\text{eq}}{\text{L}} = \frac{\text{mol}}{\text{L}} \times |\text{charge}| = \text{M} \times |\text{charge}|$$

where, the symbol | | stands for "absolute value," which means we remove the sign. Thus, for ions with a +1 or −1 charge, the equivalent per liter is equal to the moles per liter, M.

Worked Example #9

A patient's blood test comes back with an iron level of 15 mmol/L. Convert this value to units of meq/L Fe^{3+}. Note that meq/L is equal to mmol/L × charge.

Use Table 8-5 to determine the number of equivalents per mole. Substitute the molar concentration into the equation for meq/L. The charge on iron is +3, so we use the value without the sign, 3, in the equation:

$$\frac{\text{meq}}{\text{L}} = \text{mM} \times \text{charge}$$

$$\frac{\text{meq}}{\text{L}} = 15\ \text{mM} \times 3 = 45\ \text{meq/L}$$

Try It Yourself #9

A patient's blood test comes back with a calcium level of 1.2 mmol/L. Convert this value to units of meq/L.

Tools: Table 8-5

Supplied units: ___________________

Requested units: _________________

Set up the calculation, so that the supplied units cancel.

Worked Example #10

a. How many equivalents/L of citrate ion ($C_6H_4O_7^{3-}$) are there in a 0.54-M solution of $Na_3C_6H_4O_7$?
b. How many eq/L of Na^+ are there in this solution?

Tools: Table 8-5

a. *In this problem, we must calculate the number of moles of citrate ($C_6H_4O_7^{3-}$) in solution for every 1 mole $Na_3C_6H_4O_7$.*

$Na_3C_6H_4O_7$ (s) → 3 Na^+ (aq) + $C_6H_4O_7^{3-}$ (aq)

The dissolution equation shows that one mole of $C_6H_4O_7^{3-}$ enter solution for every one formula unit of $Na_3C_6H_4O_7$

$$0.54\ \text{M}\ \cancel{Na_3C_6H_4O_7} \times \frac{1\ \text{M}\ C_6H_4O_7^{3-}}{1\text{M}\ \cancel{Na_3C_6H_4O_7}} = 0.54\ \text{M}\ \ C_6H_4O_7^{3-}$$

Next, multiply the molar concentration of $C_6H_4O_7^{3-}$ by the numerical value of the charge on a citrate ion, 3:

$$\frac{\text{eq}}{\text{L}} = \text{M} \times \text{charge} = 0.54\ \text{M} \times 3 = 1.6\ \text{eq/L}$$

b. *Since the solution of $Na_3C_6H_4O_7$ is neutral, the eq/L of Na^+ ions in this solution must be equal to the eq/L of $C_6H_4O_7^{3-}$ ions. Therefore, there are 1.6 eq/L Na^+ ions.*

Try It Yourself #10

a. How many eq/L of calcium ions (Ca^{2+}) are there in a 1.8-M solution of $CaCl_2$?
b. How many eq/L of chloride ions (Cl^-) are there in this solution?

Tools: Table 8-5

Dissolution equation for $CaCl_2$:

Number of moles of Ca^{2+} for every mole of $CaCl_2$ dissolved:

Set up the calculation so that the supplied units cancel (molar concentration × charge).

In a Nutshell: Solution Dilution

A solution with a given concentration can also be prepared by diluting a more concentrated solution. Dilution is a method for preparing a less concentrated solution form a more concentrated solution. Knowing how to prepare a solution by dilution is important because many solutes are not readily available in their pure form and are instead only available as stock solutions-pre-prepared solutions of known concentration.

To prepare a solution with a volume V_2 and concentration C_2 from a more concentrated solution with a concentration C_1 and a volume V_1, use the dilution equation:

$$C_1 \times V_1 = C_2 \times V_2$$

where C_1 = concentration of more concentrated solution (the stock solution)
V_1 = volume of the more concentrated solution (stock solution)
C_2 = concentration of the dilute solution
V_2 = volume of the dilute solution

This equation can be applied to any concentration units: m/v, %m/v. molarity, M, provided C_1 and C_2 have the same units and V_1 and V_2 have the same units.

Follow these guidelines for solving a dilution calculation: 1) Begin by determining the three supplied variables and the variable that needs to be calculated. 2) Rearrange the dilution equation given above so that the variable you are solving for—the unknown—is isolated on one side of the equality. You can do so by dividing both sides of the equation by the appropriate variable (s). 3) Substitute the supplied variables listed in (1) into the algebraically rearranged equation from (2) and solve for the unknown variable. 4) Prepare the dilute solution by transferring the calculated volume of stock solution from (3) to a volumetric flask with volume V_2. Then add water to the mark, with mixing.

Worked Example #11

What volume of a 2.5 M dextrose solution would you need in order to prepare 100. mL of a 0.75 M dextrose solution?

1) Write the supplied variables and determine the variable that needs to be calculated:

$V_1 = ?$

$C_1 = 2.5\ M$

$V_2 = 100.\ mL$

$C_2 = 0.75\ M$

2) Rearrange the dilution equation so that the variable that you are solving for is isolated on one side of the equality, by dividing both sides of the equation by the appropriate variable(s). Begin with the dilution equation:

$$C_1 \times V_1 = C_2 \times V_2$$

Rearrange the equation so that V_1, the unknown variable, is isolated on one side of the equality. This is done by dividing both sides by C_1:

$$V_1 = \frac{C_2 \times V_2}{C_1}$$

3) Substitute the variables from (1) into the rearranged equation from (2) and solve for the unknown variable from (1).

$$V_1 = \frac{0.75\ \cancel{M} \times 100.\ mL}{2.5\ \cancel{M}} = 30.\ mL$$

Try It Yourself #11

What volume of a 0.90% m/v saline solution do you need to prepare 150 mL of 0.60% m/v saline solution?

1) Write the supplied variables and determine the variable that needs to be calculated:

$V_1 =$ _____ $C_1 =$ _____ $V_2 =$ _____ $C_2 =$ _____

2) Rearrange the dilution equation so that the variable that you are solving for is isolated on one side of the equality, by dividing both sides of the equation by the appropriate variable(s). Begin with the dilution equation:

$$C_1 \times V_1 = C_2 \times V_2$$

Rearranged equation:

3) Substitute the variables from (1) into the rearranged equation from (2) and solve for the unknown variable from (1).

Worked Example #12

What is the %m/v of a saline solution prepared by diluting 15 mL of a 0.90 %m/v saline solution to 50. mL in a 50-mL volumetric flask?

1) Write the supplied variables and determine the variable that needs to be calculated:

V_1 = 15 mL

C_1 = 0.90 %m/v

V_2 = 50. mL

C_2 = ?

2) Rearrange the dilution equation so that the variable that you are solving for is isolated on one side of the equality, by dividing both sides of the equation by the appropriate variable(s). Begin with the dilution equation:

$$C_1 \times V_1 = C_2 \times V_2$$

Rearrange the equation so that C_2, the unknown variable, is isolated on one side of the equality. This is done by dividing both sides by V_2:

$$C_2 = \frac{C_1 \times V_1}{V_2}$$

3) Substitute the variables from (1) into the rearranged equation from (2) and solve for the unknown variable from (1).

$$C_2 = \frac{0.90\ \%\text{m/v} \times 15\ \cancel{\text{mL}}}{50.\ \cancel{\text{mL}}} = 0.27\ \%\text{m/v}$$

Try It Yourself #12

What is the molarity, M, of a glucose solution prepared by diluting 35 mL of a 3.1-M solution to 150. mL in a 150-mL volumetric flask?

1) Write the supplied variables and determine the variable that needs to be calculated:

V_1 = _____ C_1 = _____ V_2 = _____ C_2 = ______

2) Rearrange the dilution equation so that the variable that you are solving for is isolated on one side of the equality, by dividing both sides of the equation by the appropriate variable(s). Begin with the dilution equation:

$$C_1 \times V_1 = C_2 \times V_2$$

Rearranged equation:

3) Substitute the variables from (1) into the rearranged equation from (2) and solve for the unknown variable from (1).

In a Nutshell: Oral Medications

Calculating dosages of drugs delivered as solutions is an important part of the health care worker's responsibility. Oral medications are substances given by mouth and administered, especially to children, in solution form.

Worked Example #13

Gantrisin®, an antibiotic used to treat urinary tract infections, is prescribed for a child. The dosage is 2045 mg to be administered in a 24-hour period. The suspension contains 500 mg in every 5 mL. How many mL of the suspension should be administered to the patient in the 24-hour period?

The supplied unit is 2045 mg of the drug. The concentration of the drug is also supplied (500 mg/ 5 mL). Calculate the volume of suspension required using dimensional analysis by multiplying the supplied mass of solute by the correct form of concentration as a conversion factor:

$$2045\ \text{mg} \times \frac{5\ \text{mL}}{500\ \cancel{\text{mg}}}\ 20\ \text{mL}$$

Thus, the patient should receive 20 mL over a 24-hour period.

Try It Yourself #13

Aristospan®, used to treat bursitis, is a suspension that contains 20 mg in every milliliter. A patient requires an injection of 15 mg of aristospan. How many mL should you administer to the patient?

Supplied units: ____________________

Concentration of suspension: _____________

Set up the calculations so that the supplied units cancel.

In a Nutshell: IV Solutions

Intravenous (IV) administration directly into a vein is often used to deliver a pharmaceutical gradually over a period of time because the body metabolizes the drug rapidly, the drug is not stable when given orally, or the patient is unconscious or unable to take medications orally. For IV drug administration, a solution with a known concentration of medicine is infused at specified volume (mL) of solution per unit of time (minutes or hours), known as a flow rate.

$$\text{flow rate} = \frac{\text{solution volume}}{\text{time}}$$

The flow rate selected for the IV depends on the rate that the medicine is prescribed by the physician:

$$\text{rate medicine prescribed} = \frac{\text{solute mass (medicine)}}{\text{time}}$$

The rate at which the drug should be administered is usually given in milligram (mg) per minute, microgram (μg) per minute, or grams (g) per hour. Both the flow rate and the rate at which the medicine is delivered are ratios in which time is in the denominator. Both ratios can be used as conversion factors

Worked Example #14

Levaquin® is a powerful antibiotic used to treat pneumonia. An order is given to administer 250 mg of levaquin by IV over a period of 60 minutes. The IV bag contains 5 mg of levaquin in every 1 mL. What should the flow rate be in mL/min?

You are asked to calculate the flow rate (mL/min) and supplied with the concentration (5 mg/1mL) and the rate the medicine should be administered (250 mg/60 min). The supplied terms are conversion factors that can be used in either of two forms:

$$\text{Rate medicine delivered: } \frac{250\text{ mg}}{60.\text{ min}} \text{ or } \frac{60.\text{ min}}{250\text{ mg}}$$

$$\text{Concentration: } \frac{5\text{ mg}}{1\text{ mL}} \text{ or } \frac{1\text{ mL}}{5\text{ mg}}$$

The flow rate will also be a ratio:

$$\text{flow rate} = \frac{\text{solution volume}}{\text{time}} = \frac{\text{mL}}{\text{minute}}$$

In this type of problem, look at the units required in the answer, flow rate, (mL/min), to determine how to multiply the conversion factors so that units cancel, leaving only mL in the numerator and minutes in the denominator. To do this, select the correct form of the concentration conversion factor that has mL in the numerator and multiply by the correct form of the prescribed rate that has minute in the denominator:

$$\frac{1\text{ mL}}{5\,\cancel{\text{mg}}} \times \frac{250\,\cancel{\text{mg}}}{60\text{ min}} = \frac{0.8\text{ mL}}{\text{min}}$$

inverted concentration × prescribed rate = flow rate

Try It Yourself #14

Retrovir IV is used to treat HIV patients. An order is given to infuse 64 mg of retrovir IV per hour. The concentration of the IV solution is 4 mg per mL. What should the flow rate be in mL/hr?

Supplied concentration: ________________________________

Supplied rate: ________________________________

Write the two forms of the conversion factors.

Rate medicine delivered: $\frac{}{}$ *or* $\frac{}{}$

Concentration: $\frac{}{}$ *or* $\frac{}{}$

Set up the calculation so that the required units appear in the numerator and the denominator.

Worked Example #15

Intravenous gamma globulin is used to treat immunodeficiency orders. The concentration of the intravenous gamma globulin is 100. mg per milliliter. The patient receives the drug at a flow rate of 50.9 mg per minute. How many milliliters of gamma globulin per minute is the patient receiving?

In this exercise, you are asked to calculate the rate the medicine is administered, in milliliters per minute (mL/min) given the flow rate (50.9 mg/min) and the concentration (100. mg/mL). The two conversion factors supplied can be used in either of two forms:

Flow rate: $\frac{50.9\text{ mg}}{1\text{ min}}$ or $\frac{1\text{ min}}{50.9\text{ mg}}$

Concentration: $\frac{100.\text{mg}}{1\text{ mL}}$ or $\frac{1\text{ mL}}{100.\text{mg}}$

To set up the calculation, multiply the conversion factor that has milliliters in the numerator by the conversion factor that has minutes in the denominator so that mg cancel:

$$\frac{1\text{ mL}}{100.\,\cancel{\text{mg}}} \times \frac{50.9\,\cancel{\text{mg}}}{1\text{ min}} = \frac{0.509\text{ mL}}{\text{min}}$$

The drug is administered at a rate of 0.509 mL per minute.

Try It Yourself #15

Brevibloc is used for rapid control of the ventricular rate during surgery. An order is given to administer brevibloc at a rate of 12.8 mg per 4 minute interval. The flow rate of the IV solution is 0.30 mL per minute. What is the concentration in mg per mL of the brevibloc solution?

Supplied flow rate: ______________________________

Supplied rate medicine prescribed: ______________________________

Write the two forms of the conversion factors.

Rate medicine prescribed: —— *or* ——

Flow rate: —— *or* ——

Set up the calculation so that the requested units appear in the numerator and the denominator.

Practice Problems for Concentration

1. Every 5 mL of an oral suspension of amoxicillin contains 200 mg of amoxicillin. What is the concentration of the oral suspension in g/dL?

2. You are asked to prepare 2.75 L of 3.3% m/v dextrose solution for IV therapy from a stock solution of 5.0% m/v dextrose. How many liters of the stock dextrose solution do you need?

3. A patient's blood test shows that there is 0.9 mmol/L of magnesium in her blood. What is this concentration in mol/L? How many meq/L of Mg^{2+} are there in her blood?

4. Indocin IV is used to close the ductus arteriosus, an arterial shunt, in premature infants. An order is given to administer 0.238 mg of indocin by IV over a period of 30 minutes. The concentration of the indocin solution is 0.05 mg/mL. What should the flow rate be in mL/min?

8.4 Osmosis and Dialysis

In our bodies, membranes separate the aqueous mixture on the inside of the membrane from the mixture on the outside of the membrane. Cell membranes separate and control what enters and leaves the cell, a property of a selectively permeable membrane. A selectively permeable membrane allows only certain substances to pass through the membrane at certain times. Colloids and suspended particles are too large to pass through a selectively permeable membrane; thus proteins and cellular organelles stay on the inside of a cell. Whether or not a particular solute can pass through a membrane depends on the nature of the membrane and the charge, polarity, and size of solute.

In a Nutshell: Permeability of the Cell Membrane

The most important factors in determining whether a solute can pass through a cell membrane are the size and polarity of the solute. The interior of the cell membrane is nonpolar, so the most permeable substances are small nonpolar molecules such as oxygen and carbon dioxide, followed by small polar neutral molecules such as water, followed by larger polar molecules such as glucose, and the least permeable are the highly polar charged ions such as chloride, sodium, and potassium. Colloids and suspensions are unable to cross the cell membrane because of their large size. Ions require assistance to cross the membrane, a process that often requires energy and specialized proteins.

It is the selective permeability of the cell membrane that allows the concentration of ions to be significantly different on the inside of the cell compared to the outside of the cell. The relative concentration of two solutions separated by a selectively permeable membrane can be described by the following terms: hypertonic, hypotonic, and isotonic. Hypertonic describes the solution with the higher solute concentration (lower water concentration). Hypotonic describes the solution with the lower solute concentration (higher water concentration). When both solutions have equal solute concentrations, they are isotonic.

In a Nutshell: Dialysis

Osmosis and dialysis describe the movement of solvent and solute across a selectively permeable membrane. Osmosis refers to the flow of solvent (water) though a selectively permeable membrane. Dialysis refers to the flow of small solutes through a selectively permeable membrane.

The solutes in a liquid solvent are in constant random motion as a result of kinetic energy. The movement of solute particles through the solvent is known as diffusion. Two solutions with different concentrations separated by a selectively permeable membrane create a concentration gradient. Diffusion of the solute always occurs from the hypertonic solution to the hypotonic solution (in the direction of the concentration gradient) until both solutions have an equal concentration. The movement of solute particles across a selectively permeable membrane is known as dialysis. Dialysis involves the diffusion of only certain small solute particles. Colloids (proteins) and suspended particles do not cross the cell membrane.

Worked Example #16

Biochemists often use dialysis to purify a protein by separating it from a salt. One compartment contains a solution of salt (small molecule) and protein (large molecule). The other compartment contains pure water. The two compartments are separated by a selectively permeable membrane. Describe how you would separate the salt from the protein, based on the principles of dialysis.

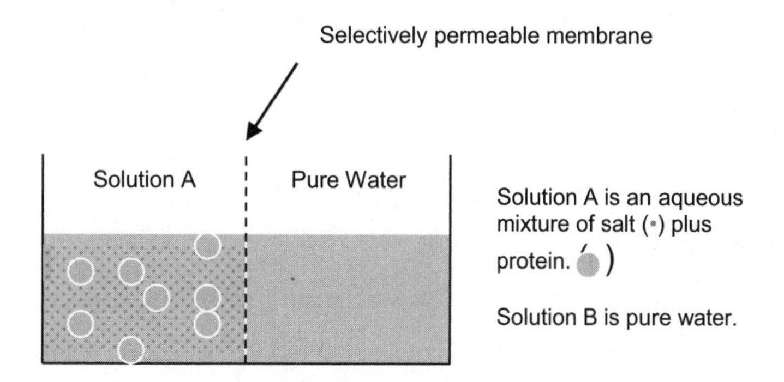

Solution

Since salt is a small molecule, it will diffuse through the membrane from solution A to the pure water side, B, from higher concentration to lower concentration, until the concentrations as equal. The protein will be unable to diffuse through the membrane because of its large size. By periodically replacing solution B with water, the salt can be separated from the protein. In the end, solution A will contain only protein. The solutions of water will contain salt.

Try It Yourself #16

When a patient undergoes kidney dialysis, small molecules such as urea and creatine are separated from the blood. One compartment contains the patient's blood and the other compartment contains the dialysate, a solution similar in composition to blood. The two compartments are separated by a selectively permeable membrane. Describe how urea and creatine are separated from blood by dialysis.

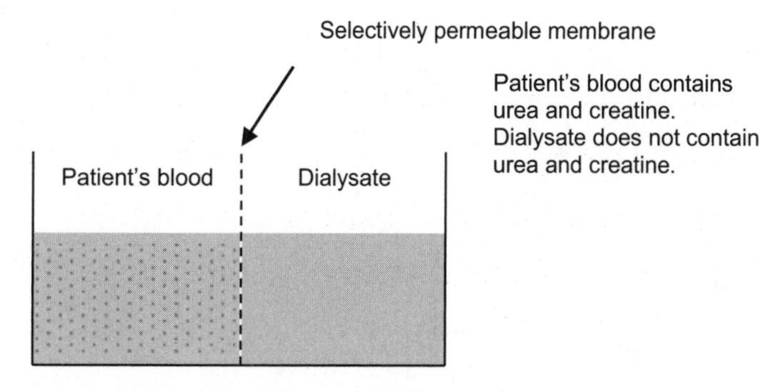

Side with lower concentration of urea and creatine: ______________________________

Side with higher concentration of urea and creatine: ______________________________

Which way will urea and creatine flow? ______________________________

In a Nutshell: Osmosis

Water, the solvent or medium in most biological mixtures can cross a selectively permeable membrane while solutes cannot. The diffusion of water across a selectively permeable membrane is referred to as osmosis. Osmosis occurs when there is a concentration gradient between two solutions separated by a selectively permeable membrane. Osmosis, the diffusion of water across a membrane, always occurs from the solution with the lower solute concentration (higher water concentration), the hypotonic solution to the solution with the higher solute concentration (lower water concentration), the hypertonic solution, until both solutions have the same concentration (isotonic). So instead of the solute diffusions (dialysis) from the more concentrated to the less concentrated solution, water diffuses (osmosis) from the less concentrated to the more concentrated solution, achieving the same end result: solutions with equal concentrations. The hypertonic solution also increases in volume as a result of osmosis.

The flow of water through a membrane is governed by the total number of solute particles, not their mass, size, or identity. Osmosis is governed by the total number of solute particles, not their mass, size, or identity.

Osmosis can be stopped by placing external pressure on the hypertonic solution. The minimum amount of pressure that must be applied to the hypertonic solution to stop water from flowing across the membrane from the hypotonic solution is known as the osmotic pressure. If a pressure greater than the osmotic pressure is applied to the hypertonic solution, reverse osmosis occurs, and in this way water can be forced to diffuse against a concentration gradient.

Worked Example #17

Consider the following solutions separated by a selectively permeable membrane. Indicate the direction of water flow between solution A and solution B, or state if no flow occurs. Which solution is hypertonic?

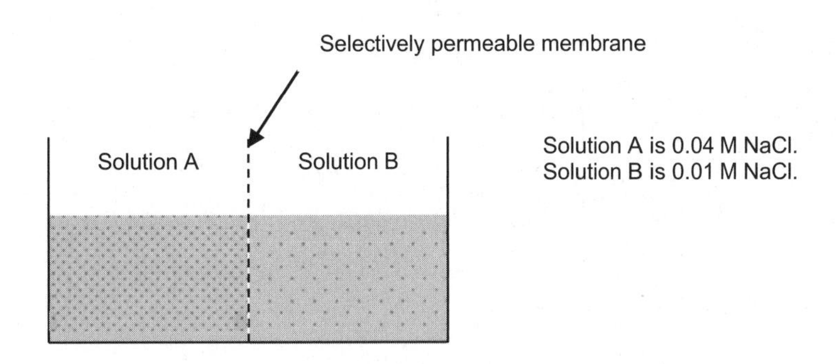

Since the question asks about water flowing across the membrane, the process involved is osmosis. Solution A has a higher solute concentration and lower water concentration. The water will flow from solution A to solution B. Solution A is hypertonic; solution B is hypotonic.

Try It Yourself #17

Consider the following solutions separated by a selectively permeable membrane. Indicate the direction of water flow between solution A and solution B, or state if no flow occurs. Which solution is hypotonic?

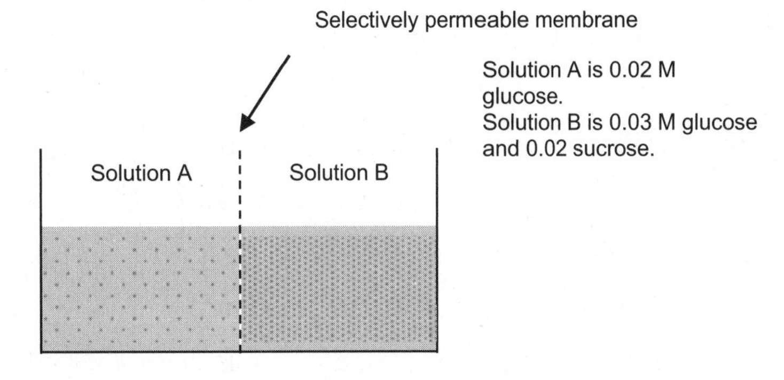

What is flowing across the membrane? ______________________________

What process is involved? ______________________________

Which solution has the higher concentration? ______________________________

Which solution has the lower concentration? ______________________________

Which solution is hypertonic? ________________________________

Which solution is hypotonic? ________________________________

Which way will the water flow? ________________________________

Practice Problems for Osmosis and Dialysis

1. Two solutions are separated by a selectively permeable membrane. Solution A contains 3.6 mmol of glucose and solution B contains 7.2 mmol of glucose. Which solution is hypertonic? In which direction will the water flow between solution A and solution B?

2. Explain how a solution of glucose (a small molecule) and glycogen (a large molecule) can be separated using dialysis.

Chapter 8 Quiz

1. Identify the following as a mixture, solution, colloid, or suspension:
   a. milk
   b. concrete
   c. brass
   d. whipped cream
   e. mud
   f. mist
   g. fog
   h. a glass of champagne
   i. tea
2. An IV solution contains 0.45% m/v NaCl.
   a. What is the solute in this solution?

b. What is the solvent?

c. Is the solute a molecule, an electrolyte, or a gas?

d. What is the concentration of the solution in g/mL?

3. Singulair® is used to treat allergies in infants. The drug is prepared as follows; 4 mg of Singulair® is dissolved in 5 mL of baby formula. What is the m/v concentration of this solution in g/dL?

4. Minocin® is used to treat an infection caused by a microorganism. An order is given to administer 200 mg. The suspension contains 50 mg in every 5 mL. How mL of the suspension should be administered to the patient?

5. Remifentanil hydrochloride is used as an analgesic with anesthesia during surgery. An order is given to administer 15.9 μg/min. The IV solution contains 20 μg/mL. What is the flow rate in mL/min?

6. A patient's blood test shows a Mg^{2+} ion concentration of 0.85 mmol/L. How many meq/L of Mg^{2+} are there?

7. A patient's blood test shows that 3 mL contains 0.0126 mmol of cholesterol. What is the molar concentration of cholesterol in mmol/mL?

8. What is the molarity, M, of a solution prepared by diluting a 135 mL of a 1.2-M solution of citric acid to 250. mL in a 250-mL volumetric flask?

9. Two solutions are separated by a semipermeable membrane. Solution A contains 0.06 M of glucose and solution B contains 0.02 M of glucose and 0.04 M of sucrose. In what direction will water flow between solution A and solution B?

10. The urea from a solution containing urea and platelet cells can be separated by dialysis. Explain.

Chapter 8

Answers to Additional Exercises

45 The kidneys filter blood to remove waste products (ions and small molecules) and water.

47 It filters out ions, small molecules, and water.

49 The millions of tubules present in the kidneys are selectively permeable membranes that filter the blood. The waste and surplus water exits the kidneys through the ureters to the bladder, where it is stored until it is eliminated.

51 Renal failure is when the kidneys stop working altogether. It may be acute or chronic.

53 Chronic renal failure can result from diabetes, high blood pressure, or certain hereditary factors.

55 A kidney transplant is the other alternative for the patient.

57 a. False. A mixture can be separated into its components through physical means. b. False. A mixture has two or more components. c. True. Mixtures can differ in the relative amounts of each component. d. True. A solution is a type of mixture.

59 a. homogeneous b. homogeneous c. heterogeneous d. homogenous e. homogenous f. homogeneous

61 In colloidal dispersions, the particles are known as colloids. The major component is called the medium.

63 Colloidal dispersions have particles ranging from 1 nm to 1 μm in diameter.

65 Aggregates are many small molecules clumped together.

67 No, the kidneys do not remove small proteins. Proteins are too big and cannot pass through the selectively permeable membrane of the kidney.

69 Yes, the kidneys remove urea. Urea is a small molecule that can pass through the selectively permeable membrane of the kidney.

71 a. solution b. colloid c. solution and colloid d. solution e. solution f. colloid

73 A centrifuge can be used to separate particles out of a suspension.

75 The medication is insoluble in all acceptable media, and some people have trouble swallowing capsules or tablets.

77 The two components of a solution are the solute and the solvent.

79 a. Glucose is the solute. It is a solid. The solvent is water. b. Carbon dioxide is the solute and is a gas. The solvent is water. c. Iodine is the solute. It is a solid. The

solvent is ethanol. d. Magnesium sulfate is the solute. It is a solid. The solvent is water.

81 The solubility of a solute in a solvent depends on the polarity of the solute and the solvent. Polar compounds will dissolve in polar solvents and nonpolar compounds dissolve in nonpolar solvents.

83 Hydrocarbons are not soluble in water. The hydrocarbons form a separate layer that avoids the water molecules—the hydrophobic effect.

85 b, c, and d would dissolve in water. They are polar molecules and water is a polar molecule. a and e would not; they are nonpolar molecules.

87 Propan-2-ol and acetic acid are polar solvents. CCl_4, carbon dioxide, and octane are nonpolar.

89 a. $CaCl_2(s) \rightarrow Ca^{2+}(aq) + 2\ Cl^-(aq)$

b. $NH_4Cl(s) \rightarrow NH_4^+(aq) + Cl^-(aq)$

c. $K_3PO_4(s) \rightarrow 3\ K^+(aq) + PO_4^{3-}(aq)$

d. $NaCl(s) \rightarrow Na^+(aq) + Cl^-(aq)$

91 The particles of acetaminophen will settle to the bottom. The bottle needs to be shaken before use so that the particles are distributed throughout the medium.

93 Gout is caused by saturated solutions of calcium salts such as calcium phosphate and calcium oxalate. When the calcium salts precipitate, they form crystals in joints.

95 b and c. NaCl and K_3PO_4 are ionic compounds and form electrolytes when dissolved in water.

97 a. False. When a molecule dissolves in water, the covalent bonds do not break. b. True. Hydrogen bonds aid dissolving polar molecules in water. c. True. Solutions containing electrolytes conduct electricity. d. False. Dissolving an ionic compound in water is a physical process.

99 In the colorimetric technique, a colored dye is added and the intensity of the color of the solution determines the concentration. The darker solutions contain more solute and are more concentrated.

101 Using dimensional analysis, multiply the total volume of the solution given by the correct form of the concentration conversion factor that allows volume to cancel leaving only units of mass. You will also need to convert mL to dL and mg to g.

$$5.0\ \cancel{mL} \times \frac{1\ \cancel{dL}}{100\ \cancel{mL}} \times \frac{78\ \cancel{mg}}{1\ \cancel{dL}} \times \frac{1g}{1000\ \cancel{mg}} = 3.9 \times 10^{-3}g$$

103 Using dimensional analysis, multiply the mass of solute (the medicine) prescribed by the correct form of the concentration conversion factor that allows mass to cancel leaving only units of volume. You will also need to convert mg to g.

$$2.0\ \text{g} \times \frac{1000\ \text{mg}}{1\ \text{g}} \times \frac{1\ \text{mL}}{250\ \text{mg}} = 8.0\ \text{mL}$$

105 Use the equation that defines ppb. The mass needs to be converted from μg to g and the volume needs to be converted from L to mL.

$$\text{ppb}\ {}^{\text{m}}{}_{\text{v}} = \frac{\text{g solute}}{\text{mL of solution}} \times 1{,}000{,}000{,}000$$

$$\text{ppb}\ {}^{\text{m}}{}_{\text{v}} = \frac{400\ \mu\text{g}}{1\ \text{L}} \times \frac{1\ \text{g}}{10^6\ \mu\text{g}} \times \frac{1\ \text{L}}{1000\ \text{mL}} \times 1{,}000{,}000{,}000 = 400\ \text{ppb}$$

107 You are supplied with the concentration and the volume of solution and asked to solve for the mass of solute. Using dimensional analysis, apply the correct form of the concentration conversion factor that allows mL to cancel leaving only grams.

$$500\ \text{mL} \times \frac{0.90\ \text{g}}{100\ \text{mL}} = 4.50\ \text{g}$$

109 You are supplied with the concentration and the mass of solute and asked to solve for the volume of solution. Using dimensional analysis, apply the correct form of the concentration conversion factor that allows grams to cancel leaving only mL.

$$36.0\ \text{g} \times \frac{100\ \text{mL}}{3.3\ \text{g}} = 1.09 \times 10^3\ \text{mL}$$

111 You are supplied with the concentration and the mass of solute and asked to solve for the volume of solution. Using dimensional analysis, apply the correct form of the concentration conversion factor that allows grams to cancel leaving only mL.

$$27.5\ \text{g} \times \frac{100\ \text{mL}}{0.90\ \text{g}} = 3.06 \times 10^3\ \text{mL}$$

113 Using dimensional analysis, perform a metric conversion on the supplied volume of solution, mL to L and then multiply the correct form of the supplied conversion factor that allows volume to cancel. You will also need to convert mmol to mol.

$$8.00\ \text{mL} \times \frac{1\ \text{L}}{1000\ \text{mL}} \times \frac{4.14\ \text{mmol}}{1\ \text{L}} \times \frac{1\ \text{mol}}{1000\ \text{mmol}} = 3.31 \times 10^{-5}\ \text{mol}$$

115 The moles of solute and the volume are supplied. Use the molarity equation to calculate the number of moles. You will need to convert mL to L.

$$\text{M} = \frac{\text{moles of solute}}{\text{liters of solution}}$$

$$\text{M} = \frac{0.24\ \text{moles}}{185\ \text{mL}} \times \frac{1000\ \text{mL}}{\text{L}} = 1.3\ \text{M}$$

117 The number of moles is supplied. Using dimensional analysis, multiply the moles of solute by the correct form of the conversion factor that allows moles to cancel.

$$2.5 \text{ mol} \times \frac{1 \text{ L}}{0.56 \text{ mol}} = 4.5 \text{ L}$$

119 Substitute the molar concentration into the equation for eq/L given. The charge on a magnesium ion is +2.

$$\frac{\text{eq}}{\text{L}} = \frac{\text{mol}}{\text{L}} \times \text{charge}$$

$$\frac{\text{eq}}{\text{L}} = \frac{0.9 \times 10^{-3}\text{mol}}{\text{L}} \times 2 = 1.8 \times 10^{-3}\,\frac{\text{eq}}{\text{L}}$$

121 First, we must calculate the number of moles of Mg^{2+} and SO_4^{2-} in solution for every 1 mole of $MgSO_4$:

$MgSO_4(s) \rightarrow Mg^{2+}(aq) + SO_4^{2-}(aq)$

The dissolution equation shows that one mole of Mg^{2+} and one mole of SO_4^{2-} are in solution for every one formula unit of $MgSO_4$:

$$0.15 \text{ M } \mathrm{MgSO_4} \times \frac{1\text{M } \mathrm{SO_4^{2-}}}{1\text{M } \mathrm{MgSO_4}} = 0.15 \text{ M } \mathrm{SO_4^{2-}}$$

Multiply the molar concentration of SO_4^{2-} by the numerical value of the charge, 2:

$$\frac{\text{eq}}{\text{L}} = \text{M} \times \text{charge} = 0.15 \text{ M} \times 2 = 0.30\frac{\text{eq}}{\text{L}}\ \mathrm{SO_4^{2-}}$$

Multiply the molar concentration of Mg^{2+} by the numerical value of the charge, 2:

$$\frac{\text{eq}}{\text{L}} = \text{M} \times \text{charge}\frac{eq}{L} = M \times charge = 0.15 \text{ M} \times 2 = 0.30\frac{\text{eq}}{\text{L}}\mathrm{Mg^{2+}}$$

123 Use the equation that relates molar concentration to equivalents for eq/L. The charge on a calcium ion is +2. The supplied unit is 0.010 eq/L. The variable that needs to be calculated is mol/L.

$$\frac{\text{eq}}{\text{L}} = \frac{\text{mol}}{\text{L}} \times \text{charge}$$

Algebraically rearrange the equation so that the variable that you are solving for is isolated on one side of the equality, by dividing both sides by charge.

$$\frac{\frac{\text{eq}}{\text{L}}}{\text{charge}} = \frac{\text{mol}}{\text{L}}$$

Substitute the supplied variables into the rearranged equation.

$$\frac{0.010\frac{\text{eq}}{\text{L}}}{2} = 0.005\,\frac{\text{mol}}{\text{L}}$$

The phosphate ion has a charge of +3. Substitute this value into the rearranged equation:

$$\frac{0.010 \frac{\text{eq}}{\text{L}}}{3} = 0.003\ \frac{\text{mol}}{\text{L}}$$

125 Begin by writing the three supplied variables and determine the variable that needs to be calculated:

V_1 = 30. mL

C_1 = 4.5 M

V_2 = 250. mL

C_2 = ?

Algebraically rearrange the dilution equation so that the variable that you are solving for is isolated on one side of the equality.

$$C_2 = \frac{C_1 \times V_1}{V_2}$$

Substitute the supplied variables into the rearranged equation.

$$C_2 = \frac{C_1 \times V_1}{V_2} = \frac{4.5\ \text{M} \times 30.\ \cancel{\text{mL}}}{250.\ \cancel{\text{mL}}} = 0.54\ \text{M}$$

127 Begin by writing the three supplied variables and determine the variable that needs to be calculated:

V_1 = 15 mL

C_1 = 2.0 M

V_2 = 25 mL

C_2 = ?

Algebraically rearrange the dilution equation so that the variable that you are solving for is isolated on one side of the equality.

$$C_2 = \frac{C_1 \times V_1}{V_2}$$

Substitute the supplied variables into the rearranged equation.

$$C_2 = \frac{C_1 \times V_1}{V_2} = \frac{2.0\ \text{M} \times 15.\ \cancel{\text{mL}}}{25\ \cancel{\text{mL}}} = 1.2\ \text{M}$$

129 The supplied unit is the order of 187.5 mg of Betapen-VK (the solute). The concentration of the suspension is also supplied (125 mg/5.0 mL). Calculate the volume of the suspension required by using dimensional analysis by multiplying the supplied mass of solute by the correct form of the concentration as a conversion factor:

$$187.5\ \cancel{\text{mg}} \times \frac{5.0\ \text{mL}}{125\ \cancel{\text{mg}}} = 7.5\ \text{mL}$$

The patient should receive 7.5 mL of Betapen-VK.

131 The supplied units are the rate the medicine is administered, milligrams per hour, (250. mg/6 hr) and concentration (125 mg/0.02 mL). You are asked to find the flow rate in milliliters per hour. To set up the calculation, multiply the conversion factor that has milliliters in the numerator by the conversion factor that has hours in the denominator so that mg cancel:

$$\frac{0.02\ \text{mL}}{125\ \cancel{\text{mg}}} \times \frac{250.\cancel{\text{mg}}}{6\ \text{hr}} = 0.007\ \frac{\text{mL}}{\text{hr}}$$

133 The supplied units are the flow rate, milliliters per hour, (60. mL/1 hr) and concentration (50. mg/250 mL). You are asked to find the rate that the drug is administered in milligrams per hour. To set up the calculation, multiply the conversion factor that has milligrams in the numerator by the conversion factor that has hours in the denominator so that mL cancel:

$$\frac{50.\text{mg}}{250\ \cancel{\text{mL}}} \times \frac{60.\cancel{\text{mL}}}{1\ \text{hr}} = 12\ \frac{\text{mg}}{\text{hr}}$$

135 The supplied units are the flow rate, micrograms per minute, (225 μg/1 min) and concentration (50. mg/250 mL). You are asked to find the rate that the drug is administered in milliliters per hour. To set up the calculation, multiply the conversion factor that has milliliters in the numerator by the conversion factor that has hours in the denominator so that mg cancel: You will need to convert μg to mg and min to hr.

$$\frac{250\ \text{mL}}{50.\cancel{\text{mg}}} \times \frac{225.\cancel{\mu\text{g}}}{1\ \cancel{\text{min}}} \times \frac{1000\ \cancel{\text{mg}}}{10^6\cancel{\mu\text{g}}} \times \frac{60\ \cancel{\text{min}}}{1\ \text{hr}} = 68\ \frac{\text{mL}}{\text{hr}}$$

137 The nature of the membrane and the charge, polarity, and size of the solute determine whether a solute can pass through the membrane.

139 Simple diffusion is the spontaneous movement of a molecule or ion from a region of higher concentration to a region of lower concentration.

141 Dialysis is the movement of solute particles across a selectively permeable membrane.

143 A hypertonic solution is the solution with the higher solute concentration. A hypotonic solution is the solution with the lower solute concentration. Isotonic solutions have equal solute concentrations.

145 The pressure is greater than the osmotic pressure and water is forced to diffuse against the concentration gradient.

147 a. The solution with the higher concentration is solution A. The water will flow from solution B to solution A. The volume of solution A will increase.

b. The solution with higher concentration is solution B. The question asks about water flow (i.e., osmosis), so the identity of the solutes does not matter, just the total concentration of the solutes. The water will flow from solution A to solution B. The volume of solution B will increase.

c. The solutions are isotonic. The water will not flow between the solutions. The volumes will remain the same.

149 In osmosis, water flows from a hypotonic solution to a hypertonic solution.

151 No, you should not expect dialysis to occur. With isotonic solutions, there is no difference in concentration between the two solutions and no driving force for the solutions to move.

153 If you apply pressure to seawater separated from freshwater, the water will flow from the seawater (the hypertonic solution) to the freshwater (the hypotonic solution). This process is called reverse osmosis. If pressure is not applied, the water will flow from the freshwater (higher water concentration) to the seawater (lower water concentration).

155 The major components of blood are the plasma and blood cells.

157 Plasma contains serum and fibrinogen and other clotting factors. Serum is 90% water and does not contain fibrinogen and other clotting factors.

159 Some small molecules found in blood are glucose, amino acids, creatinine, and urea. Some electrolytes found in blood are Na^+, K^+, Ca^{2+}, Cl^-, and HCO_3^-.

161 In order to compare the fasting glucose level from the blood test to the normal range, we need to convert g to mg.

$$\frac{0.085\ \text{g}}{1\ \text{dL}} \times \frac{10^3\ \text{mg}}{1\ \text{g}} = 85\ \text{mg/dL}$$

The glucose level is within the normal range. The patient does not have diabetes.

Chapter 9

Acids and Bases

Chapter Summary

In this chapter, you learned about the unique properties of acids and bases and their role in health and medicine. You learned about the formation of acids and bases, how to determine the difference between strong acids, weak acids, strong bases, and weak bases, the factors that affect acid-base equilibria, and neutralization reactions. You discovered how to determine the concentrations of hydronium ion and hydroxide ion and how to determine the pH. You gained an understanding of how buffers work and how they control the pH within the body.

9.1 Acids and Bases

The reactions of acids and bases in aqueous solution are the simplest, fastest, and most common of all reactions. Acids are known for their sour taste, their ability to neutralize bases, and their ability to turn blue litmus paper red. Bases are known for their bitter taste, slippery feel, ability to neutralize acids, and their ability to turn red litmus paper blue. Acids and bases are defined by their chemical behavior in water. Two definitions are used to define acids and bases: the Arrhenius definition and the Brønsted-Lowry definition.

In a Nutshell: The Arrhenius Definition of Acids and Bases: Hydronium Ion and Hydroxide Ion

According to the Arrhenius definition, an acid is a substance that produces hydrogen ions (H^+) in aqueous solution. A hydrogen ion is a hydrogen atom without an electron; thus, it is simply a proton (H^+). A hydrogen ion always bonds to a water molecule in aqueous solution to form a hydronium ion, H_3O^+. Acids produce hydronium ions (H_3O^+) in solution.

According to the Arrhenius definition, a base is a substance that produces hydroxide ions (OH^-) when dissolved in water. Hydroxide ion–containing salts are bases because in aqueous solution they release hydroxide ions.

When a proton from an acid, H^+, is combined with a hydroxide ion, OH^-, they react to form a molecule of water, H_2O. Because water is a neutral substance, the reaction is called a neutralization reaction:

$$H^+ (aq) + OH^- (aq) \rightarrow H_2O (l)$$

In a Nutshell: Brønsted-Lowry Definition of Acids and Bases: Proton Transfer

Another more widely applicable definition of an acid and a base is taken from the perspective of the proton (H^+) being transferred. According to the Brønsted-Lowry definition, an acid is a proton donor, and a base is a proton acceptor. One advantage of the Brønsted-Lowry definition of a base (a proton acceptor) is that it allows us to see how molecular compounds can act as bases.

According to the Brønsted-Lowry definition, an acid transfers a proton to a base, and the products formed are related to the acid and the base by a proton; hence, the products themselves are an acid and a base. After transferring a proton, the acid becomes what is known as the conjugate base of the acid. Similarly after receiving a proton, the base becomes what is known as the conjugate acid of the base. An acid and its conjugate base and a base and its conjugate acid are known as a conjugate acid-base pair.

Generally, bases are neutral molecules or negatively charged polyatomic anions. When the base is neutral, the conjugate acid is a polyatomic cation. When the base is negatively charged, the conjugate acid is a neutral molecule. Generally, acids are neutral molecules or positively charged polyatomic cations. When the acid is neutral, the conjugate base is an anion. When the acid is positively charged, the conjugate base is neutral.

Worked Example #1

Write the reaction that represents the reaction of propanoic acid (CH_3CH_2COOH) in water.

Solution

```
   H  H  O                              H  H  O
   |  |  ||                             |  |  ||     -          +
H-C-C-C-O-H  +  H2O   <=====>   H-C-C-C-O   +  H3O
   |  |                                 |  |
   H  H                                 H  H
```

Propanoic acid releases a proton and forms hydronium ion.

Try It Yourself #1

Write the reaction that represents the reaction of butanoic acid ($CH_3CH_2CH_2COOH$) in water.

Worked Example #2

Write the reaction that represents the reaction of pyruvate (shown below) in water.

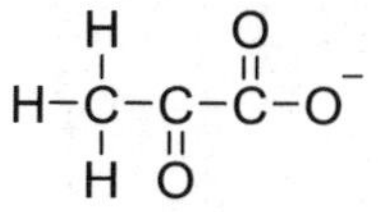

```
  H     O
  |     ||
H-C-C-C-O⁻
  |  ||
  H  O
```

pyruvate

Solution

```
  H     O                         H     O
  |     ||                        |     ||
H-C-C-C-O⁻  + H2O  ⇌  H-C-C-C-OH + HO⁻
  |  ||                           |  ||
  H  O                            H  O
```

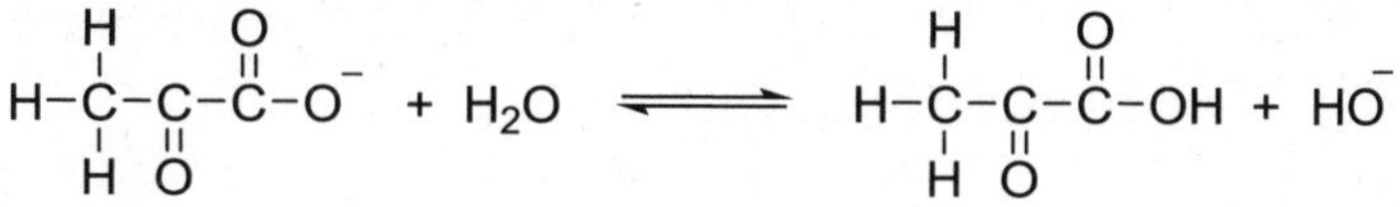

pyruvate

Pyruvate accepts a proton and forms hydroxide ion.

Try It Yourself #2

Write the reaction that represents the reaction of propanoate ion ($CH_3CH_2CH_2COO^-$) in water.

Worked Example #3

Label the conjugate acid-base pairs in the reaction that follows. Does water act as an acid or base in this reaction?

$$CH_3CH_2COO^- + H_3O^+ \rightleftharpoons CH_3CH_2COOH + H_2O$$

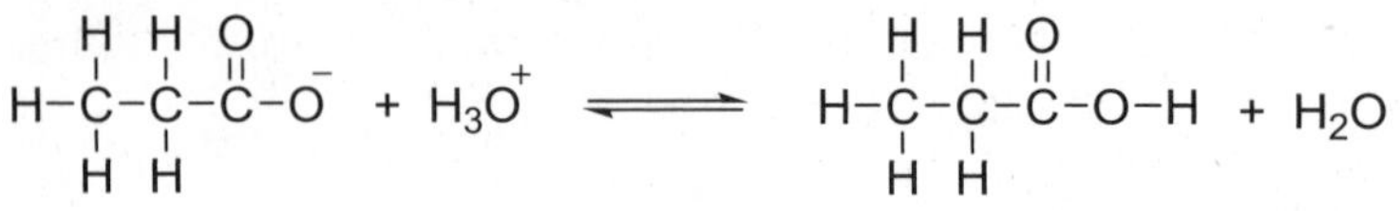

Solution

conjugate acid-base pair

$$CH_3CH_2COO^- + H_3O^+ \rightleftharpoons CH_3CH_2COOH + H_2O$$

conjugate acid-base pair

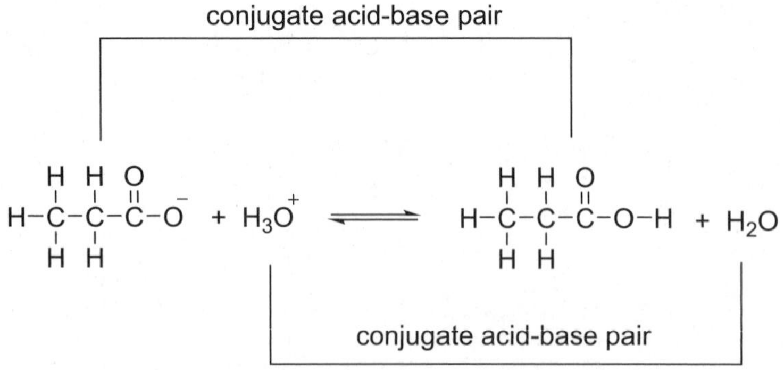

Water acts as an acid in this reaction, donating a proton to the base.

Try It Yourself #3

Label the conjugate acid-base pairs in the reaction below. Does water act as an acid or base in this reaction?

$$CH_3CH_2CH_2NH_3^+ + H_2O \rightleftharpoons CH_3CH_2CH_2NH_2 + H_3O^+$$

Does water accept or donate a proton? ______________

Water is a: ________________.

Does $CH_3CH_2CH_2NH_3^+$ accept or donate a proton? ________________

$CH_3CH_2CH_2NH_3^+$ is a: _________________________.

In a Nutshell: Strengths of Acids and Bases

An acid or base is classified as strong or weak depending on the extent to which it dissociates in water—in other words, the extent to which the forward reaction proceeds. A strong acid, HA, dissociates completely into its conjugate base (A^-) and hydronium ions (H_3O^+) in aqueous solution. Little if any undissociated acid, HA, exists in solution. The reaction has gone to completion and a single headed forward arrow is used when writing the chemical equation. Strong acids are also strong electrolytes because they form ions in solution. Strong acids include HCl, HNO_3, H_2SO_4, $HClO_4$, HBr, and HI.

A strong base completely dissociates in aqueous solution. Most strong bases are salts containing hydroxide ion and a group-1A or group-2A metal cation. Strong bases are also strong electrolytes because they form ions in solution. They consist of hydroxides that completely dissociate in water. Strong bases include KOH, $Ba(OH)_2$, NaOH, $Sr(OH)_2$, $Ca(OH)_2$, and LiOH.

Weak acids and bases differ from strong acids and bases in the extent to which they dissociate in water. Weak acids and weak bases undergo a reversible reaction in water that favors the reactants at equilibrium. Equilibrium arrows (two opposing half-headed arrows) are used to represent the equation for the reaction of a weak acid or a weak base with water. When a weak acid or a weak base dissolve in water, a mixture of both products and reactants is present at equilibrium. At equilibrium, the concentration of the reactants is greater than the concentration of products: The equilibrium lies to the left. A weak acid is a weak electrolyte because the concentration of ions is small. Organic compounds containing carboxylic acid functional groups, carbonic acid (H_2CO_3), phosphoric acid (H_3PO_4), the conjugate acids of amines, RNH_3^+, and water are weak acids.

Weak bases, like weak acids, react with water in a reversible reaction that produces an equilibrium mixture of reactants and products. The equilibrium lies to the left so the concentration of hydroxide ions is low. At equilibrium, an aqueous solution containing a weak base consists mainly of the un-ionized base. Weak bases are weak electrolytes because they form few ions in solution. Weak bases found in biochemistry include ammonia (NH_3), organic amines (RNH_2, R_2NH, and R_3N), a hydrogen carbonate ion (HCO_3^-), and the conjugate bases of carboxylic acids.

There many more weak acids than there are strong acids. There is an enormous range of acid strength among the weak acids. Similarly, there is a range of base strengths among the weak bases. The underlying reason for this range of strengths is a result of their chemical structure. An acid that produces a lower-energy, more stable conjugate base dissociates to a greater extent than an acid with a less stable conjugate base.

Amino acids, the building blocks of proteins, contain both an amine and a carboxylic acid functional group. In the cell, the amine group is in its conjugate acid form and the acid is in its conjugate base form. Therefore, amino acids are amphoteric molecules.

Worked Example #4

Indicate whether the following is a strong acid, strong base, weak acid, or weak base. For each, write the chemical equations that represent the dissociation in water.

a. $Ca(OH)_2$
b. CH_3NH_2
c. HBr

Tools: Use Tables 9-3 and 9-4

a. *$Ca(OH)_2$ is a strong base.*

$Ca(OH)_2 \longrightarrow Ca^{2+}\ (aq) + 2\ OH^-\ (aq)$

b. *CH_3NH_2 is a weak base. It is not on the lists of strong acids or bases.*

$CH_3NH_2 + H_2O \rightleftharpoons CH_3NH_3^+ + OH^-$

c. *HBr is a strong acid.*

$HBr + H_2O \longrightarrow H_3O^+ + Br^-$

Try It Yourself #4

Indicate whether the following is a strong acid, strong base, weak acid, or weak base. For each, write the chemical equations that represent the dissociation in water.

a. $CH_3CH_2CH_2COO^-$
b. $HClO_4$
c.
```
   H  H  H  H  O
   |  |  |  |  ||
H−C−C−C−C−C−OH
   |  |  |  |
   H  H  H  H
```

Tools: Use Tables 9-3 and 9-4

a. *The ion is:* ________________

b. The molecule is: ____________________

c. The molecule is: ________________________

In Chapter 4, you learned, a reaction at equilibrium can be disturbed by changing the temperature or changing the concentration of a reactant or product. Le Châtelier's principle predicts that, when a reaction at equilibrium is disturbed, the reaction responds by shifting in the direction that restores equilibrium: either the forward direction (a shift to the right) or the reverse direction (a shift to the left.) For example, if more reactant is added to a reversible reaction, the rate of the forward reaction will increase until a new equilibrium is established.

Worked Example #5

Consider the reversible reaction shown below, which occurs in ethylene glycol poisoning:

$$\underset{\text{oxalic acid}}{HO-\overset{\overset{O}{\|}}{C}-\overset{\overset{O}{\|}}{C}-OH} + \underset{\text{water}}{H_2O} \rightleftharpoons \underset{\text{oxalate ion}}{HO-\overset{\overset{O}{\|}}{C}-\overset{\overset{O}{\|}}{C}-O^-} + \underset{\text{hydronium ion}}{H_3O^+}$$

a. What substances are present at equilibrium?
b. At equilibrium, are the concentrations of oxalic acid and oxalate ion constant or changing?
c. How will the equilibrium shift if oxalic acid is removed from the solution?
d. How the equilibrium shift if more water is added to the solution?

a. Oxalic acid, water, oxalate ion, and hydronium ion are all present at equilibrium.
b. The concentrations of oxalic acid and oxalate ion are constant at equilibrium.
c. If oxalic acid is removed, the concentration of one of the reactants decreases; therefore, the equilibrium will shift to the left.
d. If more water is added, the concentration of one of the reactants increases; therefore, the equilibrium will shift to the right.

Try It Yourself #5

Phenylacetic acid builds up in the blood of people with phenylketonuria. In aqueous solution, phenylacetic acid undergoes the following reversible reaction:

$$C_6H_5\text{–}CH_2\text{–}C(=O)\text{–}OH + H_2O \rightleftharpoons C_6H_5\text{–}CH_2\text{–}C(=O)\text{–}O^- + H_3O^+$$

phenylacetic acid | water | phenylacetate ion | hydronium ion

a. What substances are present at equilibrium?

b. At equilibrium, are the concentrations of phenylacetic acid and phenylacetate ion constant or changing?

c. How will the equilibrium shift if phenylacetate ion is removed from the solution?

d. How the equilibrium shift if water is removed from the solution?

a. *The substances present at equilibrium are:* _________________________.

b. *The concentrations are:* ____________________________________.

c. *Phenylacetate ion is a reactant or a product?* ________________

The concentration of phenylacetate ____________*; therefore, the reaction shifts*

____________________.

d. *Water is a reactant or a product?* ________________

The concentration of water ____________*; therefore, the reaction shifts*

____________________.

Practice Problems for Acids and Bases

1. Write the reaction that represents the reaction of pentanoic acid, $CH_3CH_2CH_2CH_2COOH$, with water. Label the conjugate acid-base pairs in the reaction. Is pentanoic acid a strong acid, weak acid, weak base, or strong base?

2. Write the products for the following reaction between aniline and water. Label the conjugate acid-base pairs. Is aniline a strong acid, a weak acid, a weak base, or a strong base?

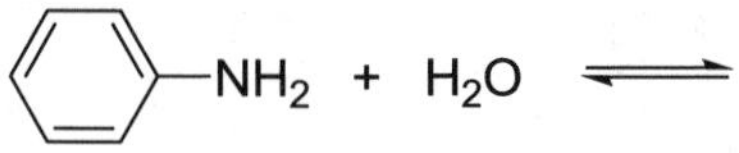

$$C_6H_5{-}NH_2 + H_2O \rightleftharpoons$$

aniline water

3. Dodecanoic acid or lauric acid is found in coconut oil. In aqueous solution, dodecanoic acid undergoes the following reversible reaction:

$$CH_3(CH_2)_{10}COOH + H_2O \rightleftharpoons CH_3(CH_2)_{10}COO^- + H_3O^+$$

dodecanoic acid water hydronium ion

a. How will the equilibrium shift if more water is added to the solution?

b. How will the equilibrium shift if dodecanoic acid is removed from the solution?

c. How will the equilibrium shift if more hydronium ion is added to the solution?

Extension Topic 9-1: The Acid Ionization Constant, K_a, and the Strength of a Weak Acid

The reaction of a weak acid in water is a reversible reaction, and in Chapter 4, you learned that the concentration of reactants and products is constant for a reversible reaction at equilibrium. Therefore, the concentration of the products divided by the concentration of reactants is constant, known as the acid dissociation constant, K_a. The value of K_a is a quantitative measure of the extent to which an acid dissociates.

For the reversible reaction of a weak acid, HA, in water, the chemical equation is

$$HA\ (aq) + H_2O\ (l) \rightleftharpoons H_3O^+\ (aq) + A^-\ (aq)$$

for which the acid dissociation constant, K_a, a unitless constant, is defined as

$$K_a = \frac{H_3O^+ \times A^-}{HA}$$

where brackets [ ] signify that the concentration is given in units of moles per liter, M. The equation shows that an acid with a larger K_a is a stronger acid than an acid with a smaller K_a.

To avoid the use of scientific notation, scientists calculate and report pK_a values. The pK_a of an acid is calculated by the following mathematical operation:

$$pK_a = -\log K_a$$

Since we are multiplying the log by −1, a "negative log," the larger the value of K_a, the smaller the value of pK_a, an inverse relationship.

Worked Example #E-1

What is the pK_a of succinic acid, which has a K_a of 6.46×10^{-5}?

$pK_a = -log[K_a] = -log\ (6.46 \times 10^{-5}) = 4.19$

Try It Yourself #E-1

What is the pK_a of adipic acid, which has a K_a of 3.80×10^{-5}?

Equation for calculating pK_a: ______________________

pK_a = ____________________

Worked Example #E-2

Some weak acids and their pK_a's are listed below. Which compound is the strongest acid? Which compound is the weakest acid?

Boric acid $pK_a = 9.14$

Glutaric acid $pK_a = 4.34$

Malonic acid $pK_a = 2.83$

Malonic acid is the strongest acid because it has the lower pK_a (larger K_a). Boric acid is the weakest acid because is has the higher pK_a (smaller K_a).

Try It Yourself #E-2

Some weak acids and their pK_a's are listed below. Which compound is the strongest acid? Which compound is the weakest acid?

phthalic acid $pK_a = 2.95$

maleic acid $pK_a = 1.93$

octanoic acid $pK_a = 4.90$

Strongest acid has (larger or smaller) pK_a: ____________________

Strongest acid: ____________________

Weakest acid has (larger or smaller) pK_a: ____________________

Weakest acid: ____________________

9.2 pH

The pH of a solution is a quantitative measure of the concentration of hydronium ions, H_3O^+, in aqueous solution. Aqueous solutions have a pH ranging from 0 to 14. A pH of less than 7 indicates an acidic solution, a pH greater than 7 indicates a basic solution, also known as an alkaline solution, and a pH equal to 7 indicates a neutral solution. Special dye-coated papers, known as pH indicator paper can be used to approximate the pH of a solution.

The pH in our cells (intracellular), extracellular fluids (outside the cell), our blood, and most solutions in our body are carefully regulated. Blood and intracellular fluids have a pH close to neutral: 7.35–7.45. A blood pH that drops below 7.35 causes a condition known as acidosis, and a blood pH above 7.45 causes a condition known as alkalosis.

In a Nutshell: Water

Water can act as an acid or a base, a property of amphoteric molecules. In pure water, a few water molecules react with other water molecules in a reaction known as the autoionization of water:

$$H_2O + H_2O \rightleftharpoons HO^- + H_3O^+$$

The autoionization of water produces an equal concentration of hydronium (H_3O^+) ions and hydroxide (OH^-) ions. In pure water, concentration of these ions is constant at 1.0×10^{-7} M (at 25 °C). The product of the hydronium ion concentration and the hydroxide ion concentration is equal to 1.0×10^{-14}, known as the ion-product constant for water, K_W. $K_W = [H_3O^+] \times [OH^-] = 1.0 \times 10^{-14}$. The brackets [ ] indicate that the units of concentration are given in moles per liter, M, molarity.

If we add acid to pure water, we disturb the equilibrium for the autoionization of water by increasing the amount of hydronium ions in solution. Le Châtelier's principle predicts a shift to the left will occur as a result of the added hydronium ions, which causes the hydroxide ion concentration to decrease proportionately, because, by definition, the ion-product constant remains constant at 1.0×10^{-14}. The quantitative relationship between the hydronium ion concentration and the hydroxide ion concentration are related through the ion-product constant in aqueous solution.

Worked Example #6

If an aqueous solution has a $[OH^-] = 4.8 \times 10^{-5}$, what is $[H_3O^+]$?

Solution

Use the equation for the ion-product constant:

$K_W = [H_3O^+] \times [OH^-]$

Set up the equation to solve for $[H_3O^+]$:

$$[H_3O^+] = \frac{K_W}{[HO^-]}$$

Substitute the values for K_w and $[OH^-]$ and solve:

$$[H_3O^+] = \frac{1 \times 10^{-14}}{4.8 \times 10^{-5}} = 2.1 \times 10^{-10}\ M$$

Try It Yourself #6

Complete the following table for an aqueous solution:

$[H_3O^+]$	$[OH^-]$
2.3×10^{-3}	
	1.7×10^{-10}
6.5×10^{-8}	

Use the equation for the ion-product constant.

Set up the equation to solve for $[H_3O^+]$ or $[OH^-]$.

Substitute the values for K_w and $[OH^-]$ or $[H_3O^+]$ and solve.

In a Nutshell: Calculating pH

The pH scale is a simple way to report the hydronium ion concentration without using scientific notation. The pH of a solution is defined as the negative logarithm of the hydronium ion concentration, $pH = -\log [H_3O^+]$. To convert pH to $[H_3O^+]$, use the formula $[H_3O^+] = 10^{-pH}$. Since pH represents a logarithmic scale, one pH unit represents a ten-fold change in hydronium ion concentration and hydroxide ion concentration—one order of magnitude. A $pH < 7$ represents an acidic solution, a $pH = 7$ represents a neutral solution, and a $pH > 7$ represents a basic solution.

Worked Example #7

What is the $[HO^-]$ of applesauce, which has a pH of 3.2? Is applesauce acidic or basic?

Applesauce is acidic because the pH is less than 7.

Calculate $[H_3O^+]$ from the pH value:

$[H_3O^+] = 10^{-pH} = 10^{-3.2}$

$[H_3O^+] = 6.3 \times 10^{-4}$ M

Calculate $[HO^-]$, using the equation for K_W:

$K_W = [H_3O^+] \times [OH^-]$

Set up the equation to solve for $[HO^-]$:

$$[HO^-] = \frac{K_W}{[H_3O^+]}$$

Substitute the values for K_w and $[H_3O^+]$ and solve:

$$[HO^-] = \frac{1 \times 10^{-14}}{6.3 \times 10^{-4}} = 1.6 \times 10^{-11} \text{ M}$$

Try It Yourself #7

What is the $[HO^-]$ of saliva, which has a pH of 6.8? Is this sample of saliva acidic, basic, or neutral?

Saliva is: ________________

Calculate $[H_3O^+]$ from the pH value:

Calculate $[HO^-]$, using the equation for K_W:

Set up the equation to solve for $[HO^-]$:

Substitute the values for K_w *and* $[H_3O^+]$ *and solve:*

Worked Example #8

What is the pH of a urine sample that has an $[HO^-]$ of 1.6×10^{-7} M? Is this sample of urine acidic, basic, or neutral?

Calculate $[H_3O^+]$*, using the equation for* K_W*:*

$K_W = [H_3O^+] \times [OH^-]$

Set up the equation to solve for $[H_3O^+]$*:*

$$[H_3O^+] = \frac{K_W}{[HO^-]}$$

Substitute the values for K_w *and* $[OH^-]$ *and solve:*

$$[H_3O^+] = \frac{1 \times 10^{-14}}{1.6 \times 10^{-7}} = 6.3 \times 10^{-8}\text{ M}$$

Calculate pH by substituting the value for $[H_3O^+]$ *into the equation for pH:*

$\text{pH} = -\log [H_3O^+] = -\log(6.3 \times 10^{-8}) = 7.2$

This sample of urine is basic, the pH is greater than 7.

Try It Yourself #8

What is the pH of lemon juice that has an $[HO^-]$ of 1.27×10^{-12} M? Is lemon juice acidic, basic, or neutral?

Calculate $[H_3O^+]$*, using the equation for* K_W*:*

Set up the equation to solve for $[H_3O^+]$*:*

Substitute the values for K_w *and* $[OH^-]$ *and solve:*

Calculate pH by substituting the value for [H_3O^+] into the equation for pH:

Lemon juice is: ______________________

In a Nutshell: Physiological pH

The pH of arterial blood ranges between 7.35 and 7.45 and is referred to as physiological pH. Many medical conditions can raise or lower the pH of the blood. Intracellular pH changes disrupt cell function by causing carboxylic acids and amines in biomolecules to change between their neutral and ionized forms. At physiological pH, carboxylic acids, amines, and phosphate esters are in their ionized forms. The charges on the ions in the cell enable the ions to stay within the cell and to bind to enzymes and chemically react as necessary.

The pH of a patient may be measured by taking a sample of blood and using a blood gas analyzer. The pH may also be measured testing the pH of urine by using either paper incorporating pH sensitive dyes or a pH probe.

Practice Problems for pH

1. Indicate whether the following solutions are acidic, basic, or neutral based on their pH.
   a. 12.4

   b. 1.2

   c. 7.0

   d. 6.9

e. 7.1

2. Complete the following table.

$[H_3O^+]$	$[HO^-]$	pH	Acidic, basic, or neutral
1×10^{-4}			
	1×10^{-2}		
1×10^{-7}			
	1×10^{-11}		

3. What is the pH of tomato juice that has a $[H_3O^+] = 3.98 \times 10^{-5}$? Calculate the $[HO^-]$ as well. Is tomato juice acidic, basic, or neutral?

4. What is the pH of a sodium bicarbonate (baking soda) solution that has a $[H_3O^+] = 6.31 \times 10^{-9}$? Calculate the $[HO^-]$ as well. Is baking soda acidic, basic, or neutral?

9.3 Acid-Base Neutralization Reactions

In a neutralization reaction, an acid is combined with a base. Neutralization reactions always go to completion—the forward reaction—regardless of the strength of the acid or the base. There are three common types of neutralization reactions that differ slightly in outcome depending on the type of base used: 1) an ionic compound containing hydroxide ion, OH^-, 2) an ionic compound containing hydrogen carbonate, HCO_3^-, or 3) ammonia, NH_3, or an organic amine (RNH_2, R_2NH, R_3N).

In a Nutshell: When Hydroxide Ion Is the Base

When an acid (H^+) and hydroxide ion (OH^-) are combined in an equal molar ratio, a solution of H_2O and a neutral salt (an ionic compound) is formed in the reaction, hence the term neutralization. The net reaction in a neutralization reaction between a hydroxide ion-containing salt and any acid involves only the hydroxide ion and the hydrogen ion:

$$OH^- \text{ (aq)} + H^+ \text{ (aq)} \longrightarrow H_2O \text{ (l)}$$

For acids that can donate more than one proton or salts with more than one hydroxide ion, coefficients are needed to balance the complete chemical equation. In these neutralization reactions, the cations and anions that form the neutral salt are considered spectator ions because they do not play a role in the neutralization reaction.

Worked Example #9

Aluminum hydroxide, $Al(OH)_3$, is one of the active ingredients in Maalox. Write the complete balanced equation and the net reaction for the reaction of aluminum hydroxide with stomach acid, HCl. What ions are spectator ions?

Complete balanced equation: $Al(OH)_3 \text{ (aq)} + 3\ HCl \text{ (aq)} \longrightarrow 3\ H_2O \text{ (l)} + AlCl_3 \text{ (aq)}$

base *acid* *water* *salt*

Net reaction: $OH^- \text{ (aq)} + H^+ \text{ (aq)} \longrightarrow H_2O \text{ (l)}$

The spectator ions are Al^{3+} *and* Cl^-

Since aluminum hydroxide yields 3 OH^- *per formula unit, three HCl molecules are required to neutralize one formula unit of* $Al(OH)_3$*. Therefore, the coefficient 3 must be placed before HCl (yielding 3* H^+*). Three water molecules are produced as well as the salt,* $AlCl_3$*.*

Try It Yourself #9

Write the balanced equation and the net reaction for the reaction between HBr and $Mg(OH)_2$. What are the spectator ions?

Unbalanced equation for the reaction of $Mg(OH)_2$ *and HBr:* ______________________

$Mg(OH)_2$ *produces how many* OH^- *per formula unit?* ______________________

Number of HBr molecules needed to neutral one formula unit of $Mg(OH)_2$*:* __________

Balanced equation for the reaction of $Mg(OH)_2$ and HBr: ______________________

Net reaction: ______________________________________

Spectator ions: ____________________________________

In a Nutshell: Bases that Contain Carbonate or Hydrogen Carbonate Ion

When an acid is combined with an ionic compound containing the carbonate ion (CO_3^{2-}), or the hydrogen carbonate ion (HCO_3^-), a neutral solution containing H_2O, a salt, and carbon dioxide (CO_2) gas are formed. The hydrogen carbonate ion and the carbonate ion react as bases by accepting one or two protons, respectively, from the acid to produce carbonic acid (H_2CO_3). Carbonic acid is an unstable compound that immediately decomposes to carbon dioxide and water:

$$H_2CO_3\ (aq) \rightleftharpoons H_2O\ (l) + CO_2(g)$$

The products of a neutralization reaction between an acid and an ionic compound containing the carbonate or hydrogen carbonate ion are water, a neutral salt, and carbon dioxide. Since carbon dioxide is a gas, we observe bubbling from this type of neutralization reaction.

Worked Example #10

Acid rain, which contains sulfuric acid, has caused the deterioration of limestone and marble statues in the United States and in Europe. Write the balanced equation for the reaction of sulfuric acid, H_2SO_4, and limestone, $CaCO_3$. Note that sulfuric acid loses two protons for every one molecule of H_2SO_h in a neutralization reaction. What ions are spectator ions?

Complete balanced equation: $CaCO_3\ (aq) + H_2SO_4\ (aq) \longrightarrow H_2O\ (l) + CO_2\ (g) + CaSO_4\ (aq)$

base acid water carbon dioxide salt

The spectator ions are Ca^{2+} *and* SO_4^{2-} *(sulfate).*

Try It Yourself #10

Write the balanced equation and the net reaction for the reaction between HCl and sodium hydrogen carbonate $NaHCO_3$. What are the spectator ions?

HCl produces how many protons? ____________________

Reaction of the proton with hydrogen carbonate: __________________________

Balanced equation: ____

Spectator ions: ______________________________________

In a Nutshell: Ammonia and Organic Amines as Bases

Ammonia and amines are common weak bases in biochemistry. These nitrogen-containing compounds with a lone pair of electrons on nitrogen react with acids in a neutralization reaction to form ammonium salts, RNH_3X^+. The conjugate acid, RNH_3^+, is a weak acid, which together with the X^- ion forms an ionic compound known as an ammonium salt.

This type of neutralization reaction is often used to convert a water insoluble amine into a water-soluble salt, especially for pharmaceutical applications. Amines are generally insoluble in water due to the hydrocarbon character of the rest of the molecule. However, reacting an amine with an acid forms the conjugate acid of the amine, RNH_3^+, a polyatomic cation, which, because of its charge, is soluble in water. The conjugate base formed in the reaction serves as the anion of the product salt.

Worked Example #11

Cymbalta, a drug used to treat depression in adults, is sold as the hydrochloride salt of the amine shown. Write the reaction of this molecule with HCl, showing the formation of the hydrochloride salt. Why is the hydrochloride salt more soluble than the amine?

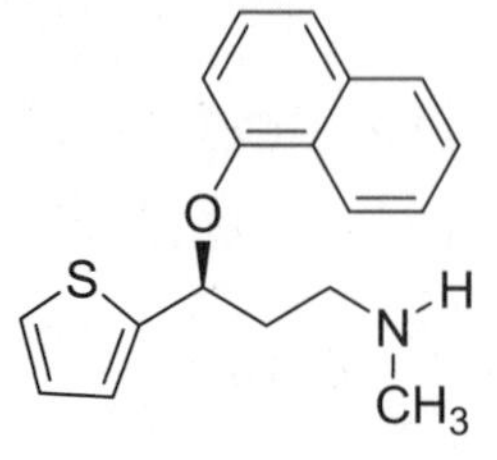

Cymbalta

Solution

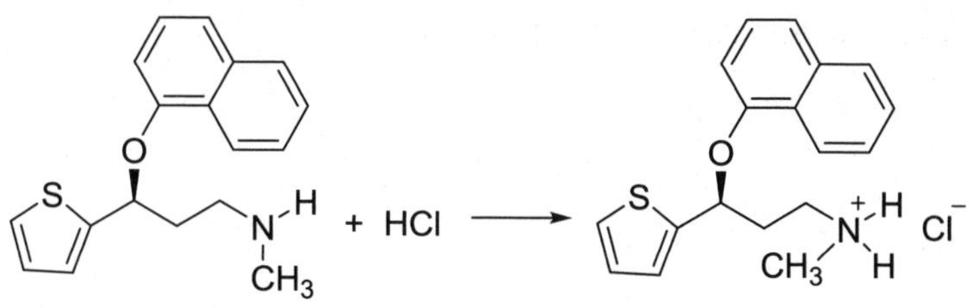

Cymbalta

The hydrochloride salt is more soluble than the amine because it has a charge on it.

Try It Yourself #11

Ultram, used to treat chronic pain, is sold as the hydrochloride salt of the amine shown. Write the reaction of this molecule with HCl, showing the formation of the hydrochloride salt.

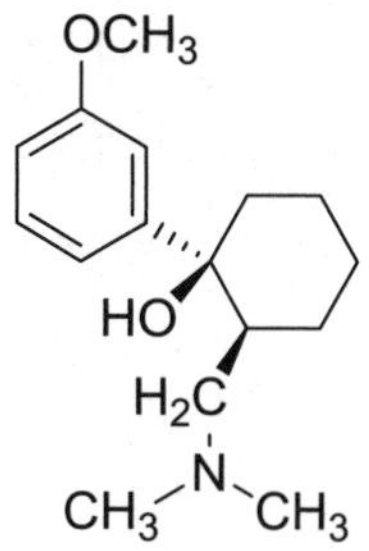

Ultram

Solution

Identity the amine. Identify the proton. Write the neutralization reaction that forms the salt.

Practice Problems for Acid-Base Neutralization Reactions

1. Write the balanced equation for the neutralization reaction between $HClO_4$ and $Ca(OH)_2$. Write the net reaction. What ions are spectator ions?

2. Write the balanced equation for the neutralization reaction between CH_3COOH and $CaCO_3$. Write the net reaction. What ions are spectator ions? Would you see bubbles in this neutralization reaction?

3. Promethazine hydrochloride, used to treat allergies, motion sickness, and nausea and vomiting, is sold as the hydrochloride salt of the amine shown. Write the reaction of this amine with HCl, showing the formation of the hydrochloride salt. Why is drug sold as the hydrochloride salt rather than the amine?

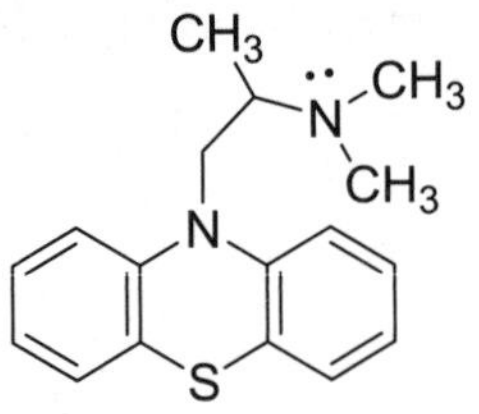

9.4 Buffers

A buffer is a solution that resists changes in pH by neutralizing any added acid (H^+) or base (OH^-). A buffer is a solution composed of a weak acid, HA, and its conjugate base, A^-, or a weak base, B, and its conjugate acid, BH^+. When an acid, H^+, is added to a buffer, such as a weak acid and its conjugate base, HA/A^-, the weak-base component of the buffer, A^-, neutralizes the added acid to form HA, producing more of the buffer. Thus, the concentration of H^+ and OH^- is not affected and the pH does not change. If a base, OH^-, is added to a buffer, the weak-acid component of the buffer, HA, neutralizes the added base to form water and A^-, producing more of the buffer, but not more OH^-. The pH does not change because the concentration of OH^- does not change. The amount of added acid or base that a buffer can neutralize is known as buffer capacity. The greater the concentration of the weak acid and it conjugate base of the buffer system, the greater its buffer capacity. The addition of protons or hydroxide ions in excess of the buffer capacity will cause the pH to change.

Worked Example #12

Consider the following buffer system, consisting of the weak acid, formic acid, HCOOH, and its conjugate base, $HCOO^-$.

$$HCOOH + H_2O \rightleftharpoons H_3O^+ + HCOO^-$$

a. How would this buffer respond if OH^- were added? Show the equation.
b. How would this buffer respond if H_3O^+ were added? Show the equation.

a. *The equilibrium would respond by shifting to the right:*

$$HCOOH + OH^- \longrightarrow HCOO^- + H_2O$$

b. *The equilibrium would respond by shifting to the left:*

$$HCOOH + H_2O \longleftarrow H_3O^+ + HCOO^-$$

Try It Yourself #12

Consider the following buffer system, consisting of the weak acid, boric acid (H_3BO_3), and its conjugate base, $H_2BO_3^-$.

$$H_3BO_3 + H_2O \rightleftharpoons H_3O^+ + H_2BO_3^-$$

a. How would this buffer respond if H_3O^+ were added? Show the equation.
b. How would this buffer respond if OH^- were added? Show the equation.

a. *The equilibrium shifts:* ______________________
   The equation: ____________________________
b. *The equilibrium shifts:* ________________________
   The equation: ______________________________

In a Nutshell: Acid-Base Homeostasis

Maintenance of proper pH in the body is known as acid-base homeostasis, where homeostasis refers to a balanced system. Acid-base homeostasis is maintained by buffers and by regulation of breathing rate as well as by secretions by the kidneys. The primary buffers in intracellular fluids are the phosphate buffer, the hemoglobin buffer system in red blood cells, and protein buffers. The primary buffer in the blood is the carbonic acid (H_2CO_3)-hydrogen carbonate ion (HCO_3^-) buffer.

Most of the carbon dioxide in solution reacts with water to form carbonic acid in a reversible reaction. The kidneys release hydrogen carbonate ion (HCO_3^-) into the blood and together with carbonic acid form the carbonic acid-hydrogen carbonate buffer system. The body responds immediately to acid-base imbalances with the carbonic acid-hydrogen carbonate buffer and by regulating breathing rate. Increasing respiration removes carbon dioxide via the lungs and decreases the concentration of carbonic acid formed. This decrease in carbonic acid concentration causes the buffer reaction to shift, consuming H^+ and preventing acidosis. Eventually, the kidneys release more hydrogen carbonate ions into the blood, which consumes H^+ and raises the pH of the blood.

A decreased respiration rate causes the partial pressure of carbon dioxide to increase, increasing the concentration of carbon dioxide dissolved in the blood. This increase in carbon dioxide concentration causes an increase in the formation of carbonic acid and causes a shift in the buffer system, generating more protons and preventing alkalosis. The kidneys excrete more hydrogen carbonate ions in the urine, which produces more H^+ and lowers the pH of the blood.

When blood pH falls below its normal range, acidosis occurs. Acidosis can occur when breathing is too weak or too slow, preventing carbon dioxide from being removed from the lungs rapidly enough. The excess carbon dioxide that builds up in the blood, causing more carbonic acid to be formed and causing an increase in H^+ concentration. Metabolic acidosis is caused by a metabolic disorder, such as kidney failure, diabetes, or starvation.
When blood pH rises above its normal range, alkalosis occurs. Respiratory alkalosis occurs when someone hyperventilates and removes carbon dioxide from the lungs faster than it is produced. The concentration of H^+ drops and the pH of the blood rises. Metabolic alkalosis can be caused by excessive vomiting, ingestion of excessive amounts of antacids, and some adrenal-gland diseases.

Practice Problems for Buffers

1. Indicate which of the following solutions are buffers.

   a. H_2CO_3 (carbonic acid, a weak acid) and HCO_3Na

   b. HBr (strong acid) and NaBr

   c. NaCl and NaOH

   d. CH_3CO_2H and CH_3CO_2Na

2. Consider the following buffer system, consisting of the weak acid, benzoic acid, C_6H_5COOH, and its conjugate base, $C_6H_5COO^-$.

   $C_6H_5COOH + H_2O \rightleftharpoons H_3O^+ + C_6H_5COO^-$

a. How would this buffer respond if H_3O^+ were added? Show the equation.

b. How would this buffer respond if OH^- were added? Show the equation.

Chapter 9 Quiz

1. Write the chemical equation for the reaction of HI in water. Label the conjugate acid-base pairs. Is HI a strong acid, weak acid, strong base, or weak base?

2. Write the chemical equation for the reaction of $CH_3CH_2NH_2$ in water. Label the conjugate acid-base pairs. Is $CH_3CH_2NH_2$ a strong acid, weak acid, strong base, or weak base?

3. The reaction of pyruvic acid in aqueous solution is shown below.

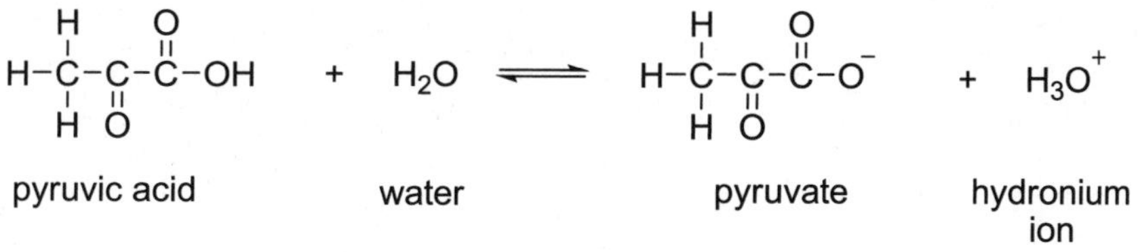

```
  H     O                                 H     O
  |     ||                                |     ||               +
H-C-C-C-OH      +    H2O   <=====>    H-C-C-C-O⁻      +    H3O
  |  ||                                   |  ||
  H  O                                    H  O

pyruvic acid         water                pyruvate           hydronium
                                                                 ion
```

a. What substances are present at equilibrium?

b. How will the equilibrium shift if pyruvate is added to the solution?

c. How will the equilibrium shift if hydronium ion is removed from the solution?

d. How will the equilibrium shift is pyruvic acid is added to the solution?

4. Write the balanced equation for the neutralization reaction between HNO_3 and $Al(OH)_3$.

5. Effexor, used to treat depression is sold as the hydrochloride salt. Write the equation for the neutralization reaction between the amine shown below and HCl.

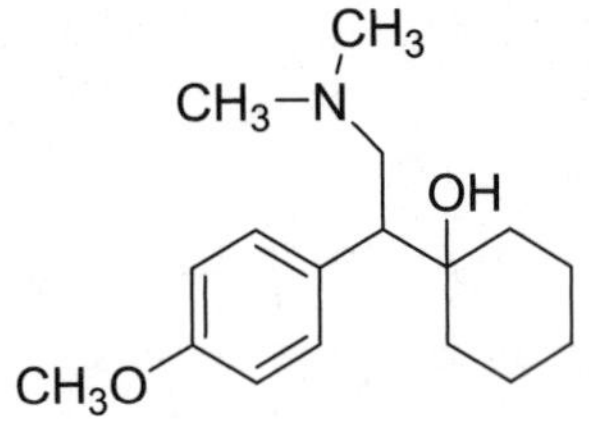

CH_3
CH_3-N
OH
CH_3O

6. Complete the following table for an aqueous solution:

$[H_3O^+]$	$[OH^-]$
4.8×10^{-5}	
	9.1×10^{-2}
7.3×10^{-9}	

7. Are the following solutions acidic, basic, or neutral based on their pH.

a. 7.0

b. 1.1

c. 14

d. 10.6

e. 2.3

8. Complete the following table:

$[H_3O^+]$	$[HO^-]$	pH	Acidic, basic, or neutral
1×10^{-7}			
	1×10^{-11}		
1×10^{-12}			

9. Ketchup has a $[H_3O^+] = 1.29 \times 10^{-4}$. What is the pH of ketchup? Is ketchup neutral, acidic, or basic? Calculate the $[OH^-]$ for ketchup.

10. Consider the following buffer system of phosphoric acid, H_3PO_4, and its conjugate base, dihydrogen phosphate, $H_2PO_4^-$.

$$H_3PO_4 + H_2O \rightleftharpoons H_3O^+ + H_2PO_4^-$$

a. How would this buffer react if H_3O^+ were added? Show the equation.

b. How would this buffer react if OH^- were added? Show the equation.

Chapter 9

Answers to Additional Exercises

33 Ethylene glycol is metabolized into oxalic acid and glyoxylic acid. Oxalic acid contains a carboxylic acid functional group rather than an alcohol functional group.

35

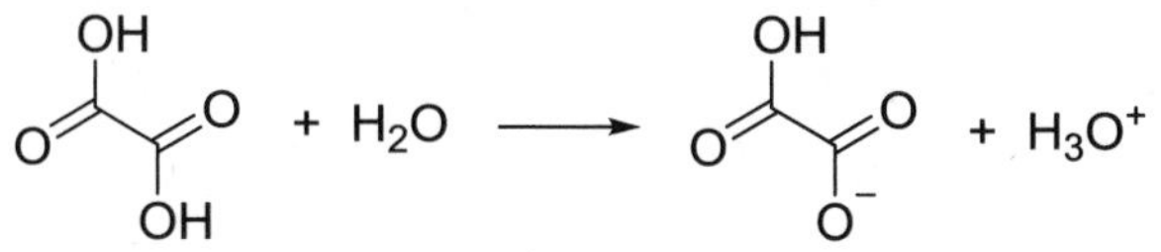

$$HO{-}C(=O){-}C(=O){-}OH + H_2O \longrightarrow HO{-}C(=O){-}C(=O){-}O^- + H_3O^+$$

37 Veterinarians can use drugs, such as other alcohols that counteract the effects of oxalic and glyoxlic acids. In humans, hemodialysis can be performed to eliminate the unmetabolized ethylene glycol and the toxic acids from the blood.

39 Acids have a sour taste, can neutralize bases, and turn blue litmus paper red. Bases have a bitter taste, slippery feel, neutralize acids, and turn red litmus paper blue.

41 The hydronium ion is formed when the proton is released in aqueous solutions. The Lewis dot structure is $H{-}\ddot{O}^+(H){-}H$.

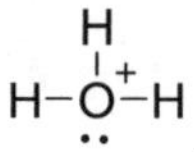

43 donates, accepts

45

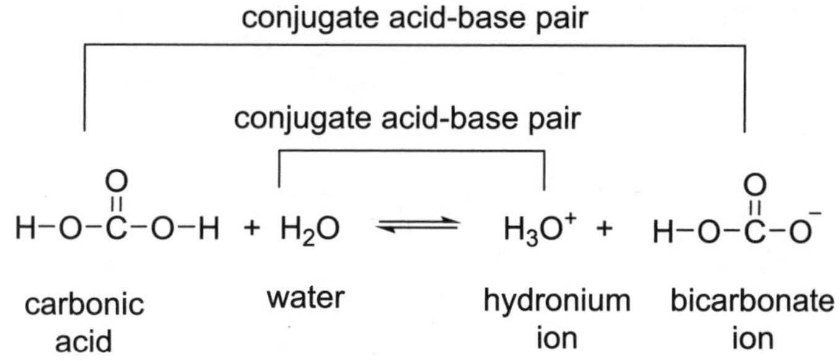

conjugate acid-base pair

conjugate acid-base pair

$$H{-}O{-}C(=O){-}O{-}H + H_2O \rightleftharpoons H_3O^+ + H{-}O{-}C(=O){-}O^-$$

carbonic acid | water | hydronium ion | bicarbonate ion

Water acts as a base.

47

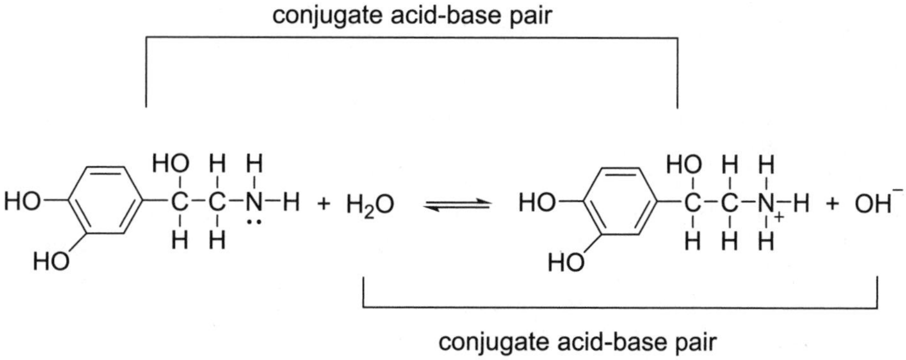

conjugate acid-base pair

$$(HO)_2C_6H_3{-}CH(OH){-}CH_2{-}\ddot{N}H_2 + H_2O \rightleftharpoons (HO)_2C_6H_3{-}CH(OH){-}CH_2{-}NH_3^+ + OH^-$$

conjugate acid-base pair

Water acts as an acid.

49 $Ca(OH)_2$ (s) + H_2O (l) → Ca^{2+} (aq) + 2 OH^-(aq). Calcium hydroxide is a base because it produces hydroxide ions.

51 (b), (d), and (e) are bases.

53 a. H_3O^+ b. H_2O c. $CH_3CH_2NH_2CH_3^+$ d. $CH_3CH_2CH_2CH_2COOH$ e. HCO_3H

55 If a concentrated solution of a strong acid is spilled on the skin, it will cause a serious chemical burn. HCl is present in the body.

57

conjugate acid-base pair

$$HI + H_2O \longrightarrow I^- + H_3O^+$$

Hydrogen iodide is a strong acid because it completely dissociates into hydronium ion and iodide in water.

59 $Ba(OH)_2(s) + H_2O(l) \rightarrow Ba^{2+}(aq) + 2\ OH^-(aq)$

It is considered a dissolution rather than a reaction because the ionic bonds in barium hydroxide are replaced by ion dipole interactions that occur when the ions are surrounded by water.

61 Weak acids do not completely dissociate in water; they undergo a reversible reaction. At equilibrium, the weak acid, its conjugate base, and hydronium ion are present. Strong acids completely dissociate, only the conjugate base and hydronium ion are present in solution.

63 a. dissociation of a strong base b. dissociation of a strong acid c. dissociation of a weak base

65 a. $HCl + H_2O \rightarrow H_3O^+ + Cl^-$

The acid is a strong acid. The reaction is not reversible.

HCl, acid; H_2O, base; H_3O^+, conjugate acid; Cl^-, conjugate base

b. $HNO_3 + H_2O \rightarrow H_3O^+ + NO_3^-$

The acid is strong acid. The reaction is not reversible.

HNO_3, acid; H_2O, base; H_3O^+, conjugate acid; NO_3^-, conjugate base

c.

$$\underset{\text{acid}}{H{-}\overset{\overset{\displaystyle O}{\|}}{C}{-}O{-}H} + \underset{\text{base}}{H_2O} \rightleftharpoons \underset{\text{conjugate acid}}{H_3O^+} + \underset{\text{conjugate base}}{H{-}\overset{\overset{\displaystyle O}{\|}}{C}{-}O^-}$$

The acid is a weak acid. The reaction is reversible.

d. $H_2CO_3 + H_2O \rightleftharpoons H_3O^+ + HCO_3^-$

The acid is a weak acid. The reaction is reversible.

H_2CO_3, acid; H_2O, base; H_3O^+, conjugate acid; HCO_3^-, conjugate base

67 a.

```
      O H O
      ‖ | ‖
H-O-C-C-C-O-H
        |
        H
```

b.

```
      O H O                                    O H O
      ‖ | ‖                       +            ‖ | ‖
H-O-C-C-C-O-H  +  H2O  ⇌  H3O  +  H-O-C-C-C-O⁻
        |                                        |
        H                                        H
```

69 a. Propanoic acid, water, propanoate, and hydronium ion are present at equilibrium.

b. At equilibrium, the reactants are favored because propanoic acid is a weak acid. Weak acids do not fully dissociate at equilibrium.

c. The concentrations of propanoic acid and propanoate are constant at equilibrium.

d. The two opposing arrows indicate that both the forward and reverse reactions occur simultaneously.

e.

conjugate acid-base pair

conjugate acid-base pair

```
  H H O                         H H O
  | | ‖                         | | ‖             +
H-C-C-C-OH  +  H2O  ⇌  H-C-C-C-O⁻  +  H3O
  | |                           | |
  H H                           H H
```

propanoic acid — water — propanoate — hydronium ion

71

conjugate acid-ase pair

conjugate acid-ase pair

```
   Cl O                           Cl O
   |  ‖                  +        |  ‖
Cl-C-C-O-H  + H2O  ⇌  H3O  +  Cl-C-C-O⁻
   |                              |
   Cl                             Cl
```

trichloroacetic acid — trichloroacetate

a. Trichloroacetic acid, water, trichloroacetate, and hydronium ion are present at equilibrium.

b. The concentration of trichloroacetic acid and trichloroacetate are constant at equilibrium.

c. The reaction will shift toward the reactants.

d. The reaction will shift toward the left, toward the reactants.

73

$$(CH_3)_3N + H_2O \rightleftharpoons HO^- + (CH_3)_3NH^+$$

a. The reaction will shift to the left. Le Châtelier's principle predicts that the reaction will shift to the left to consume the added hydroxide until equilibrium is reached.

b. The reaction will shift to the left. Le Châtelier's principle predicts that the reaction will shift to the left to consume the excess products until equilibrium is reached.

75 pH indicator paper can be used to approximate the pH of a solution. pH indicator paper changes color according to the concentration of the hydronium ions in solution. Red litmus paper will turn blue in the presence of a base.

77 a. Basic, pH > 7

b. Acidic, pH < 7

c. Basic, pH > 7

d. Acidic, pH < 7

e. Basic, pH > 7

79 When a condition causes excessive amounts of acid or base to enter the blood, the buffer capacity is exceeded. Acidosis occurs when the blood pH falls below its normal range. Alkalosis occurs when blood pH rises above its normal range.

81

$[H_3O^+]$	$[OH^-]$	Is the solution acidic, neutral, or basic?
1.0×10^{-3}	**1.0×10^{-11}**	**Acidic pH = 3**
1.0×10^{-12}	1.0×10^{-2}	**Basic pH = 12**
1.0×10^{-7}	**1.0×10^{-7}**	**Neutral pH = 7**
1.0×10^{-5}	**1.0×10^{-9}**	**Acidic pH = 5**
1.0×10^{-9}	1.0×10^{-5}	**Basic pH = 9**

83 Calculate pH by substituting the value for $[H_3O^+]$ into the equation for pH:

$pH = -\log [H_3O^+] = -\log (3.2 \times 10^{-4}) = 3.5$

Apple juice is acidic, pH < 7.

Calculate $[OH^-]$, using the equation for K_W.

$K_W = [H_3O^+] \times [OH^-]$

Set up the equation to solve for $[OH^-]$:

$$OH^- = \frac{1 \times 10^{-14}}{H_3O^+}$$

Substitute the values for K_w and $[H_3O^+]$ and solve:

$$OH^- = \frac{1 \times 10^{-14}}{3.2 \times 10^{-4}} = 3.1 \times 10^{-11} \text{ M}$$

85 Calculate pH by substituting the value for $[H_3O^+]$ into the equation for pH:

$pH = -\log [H_3O^+] = -\log (3.2 \times 10^{-7}) = 6.5$

Milk is acidic, pH < 7.

Calculate $[OH^-]$, using the equation for K_W.

$K_W = [H_3O^+] \times [OH^-]$

Set up the equation to solve for $[OH^-]$:

$$[HO^-] = \frac{K_W}{[H_3O^+]}$$

Substitute the values for K_w and $[H_3O^+]$ and solve:

$$OH^- = \frac{1 \times 10^{-14}}{3.2 \times 10^{-7}} = 3.1 \times 10^{-8} \text{ M}$$

87 Physiological pH is the pH of arterial blood. It ranges between 7.35 and 7.45. It is slightly basic compared with the pH of water (pH = 7.0).

89 At physiological pH, amines are in their ionized form. The charge on the ion keeps it inside the cell. It also enables them to bind to enzymes and chemically react.

91 Arterial blood pH is measured with a blood gas analyzer. The pH of urine is measured with dipsticks containing dyes made specifically for testing urine.

93

a. $HBr (aq) + NaOH (aq) \rightarrow H_2O (l) + NaBr (aq)$
b. $3\ HBr (aq) + Al(OH)_3 (aq) \rightarrow 3\ H_2O (l) + AlBr_3 (aq)$. Since aluminum hydroxide yields 3 OH^- per formula unit, three HBr molecules are required to neutralize one formula unit of $Al(OH)_3$. Therefore, the coefficient 3 must be placed before HBr (yielding 3 H^+). Three water molecules are produced as well as the salt $AlBr_3$.
c. $2\ HBr (aq) + Sr(OH)_2 (aq) \rightarrow 2\ H_2O (l) + SrBr_2 (aq)$. Since strontium hydroxide yields 2 OH^- per formula unit, two HBr molecules are required to neutralize one formula unit of $Sr(OH)_2$. Therefore, the coefficient 2 must be placed before HBr (yielding 2 H^+). Two water molecules are produced as well as the salt $SrBr_2$

95

a. $HNO_3 (aq) + KOH (aq) \rightarrow H_2O (l) + KNO_3 (aq)$
b. $2\ HNO_3 (aq) + Mg(OH)_2 (aq) \rightarrow 2\ H_2O (l) + Mg(NO_3)_2 (aq)$. Since magnesium hydroxide yields 2 OH^- per formula unit, two HNO_3 molecules are required to neutralize one formula unit of $Mg(OH)_2$. Therefore, the coefficient 2 must be

placed before HNO_3 (yielding 2 H^+). Two water molecules are produced as well as the salt $Mg(NO_3)_2$

c. $3\ HNO_3\ (aq) + Al(OH)_3\ (aq) \rightarrow 3\ H_2O\ (l) + Al(NO_3)_3\ (aq)$. Since aluminum hydroxide yields 3 OH^- per formula unit, three HNO_3 molecules are required to neutralize one formula unit of $Al(OH)_3$. Therefore, the coefficient 3 must be placed before HNO_3 (yielding 3 H^+). Three water molecules are produced as well as the salt $Al(NO_3)_3$.

97 Antacid tablets supply the base for the neutralization reaction with stomach acid, HCl.

99

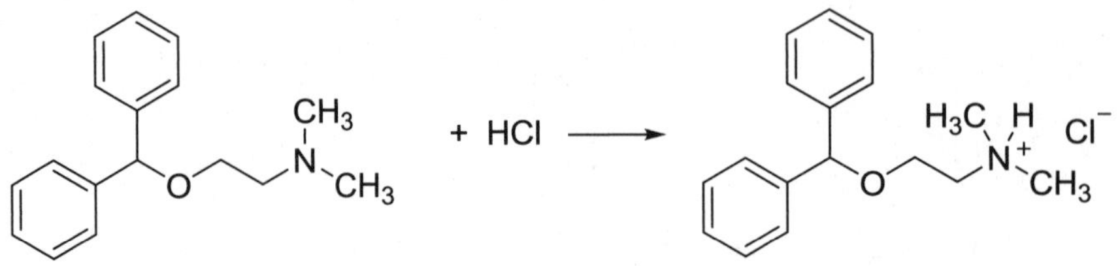

CH_3 N CH_3 O + HCl → O H_3C H N^+ CH_3 Cl^-

The salt is charged and therefore soluble in water.

101 A buffer is a solution that resists changes in pH by neutralizing any added acid or base.

103 When a base, OH^-, is added to a buffer, the weak acid component of the buffer, HA, neutralizes the added base, OH^-, to form water and A^-, producing more of the buffer, but not more OH^-.

105 The amount of added acid or base that a buffer can neutralize is known as the buffer capacity. The greater the concentration is of the weak acid and the conjugate base of the buffer system, the greater is its buffer capacity.

107 When a small amount of base enters the bloodstream, carbonic acid reacts with the OH^- ion to form bicarbonate ion and water. These products do not increase the concentration of H_3O^+ and therefore do not change pH.

109 a. The concentration of carbon dioxide decrease when a person hyperventilates. Carbon dioxide is removed faster than it is produced.

b. As the concentration of carbon dioxide decreases, the H_3O^+ concentration decreases and the pH rises.

111 When sodium hydrogen carbonate is removed from the body, there is an excess of carbonic acid in the blood. Therefore, the pH will decrease. The patient should be given the hydrogen carbonate portion to regulate the pH.

113 In type I diabetes, the pancreas has stopped producing insulin, so diabetics need to replace it.

115 Protein drugs cannot be taken orally because the acidic environment of the gastrointestinal tract will degrade the drugs.

117 A protein does not keep its original shape or its cellular function in the stomach.

119 Enteric coatings are stable in acidic environment; therefore, the enteric coating controls where in the digestive tract the medication is absorbed.

E9.7 $pK_a = -\log [K_a] = -\log (7.59 \times 10^{-3}) = 2.12$

E9.9 Phosphoric acid is the strongest acid of the three; it has the smallest pK_a. Citric acid is the weakest acid of the three; it has the largest pK_a.

E9.11 The strongest acid is oxalic acid. The weakest acid is caffeic acid.

Chapter 10

The Reactions of Organic Functional Groups in Biochemistry

Chapter Summary

In this chapter, you have seen that biochemistry is organic chemistry. You discovered that functional groups react in the same characteristic way under a given set of conditions. You learned about four types of organic reactions seen in biochemistry. These reactions are oxidation-reduction reactions, group transfer reactions (esterification and amidation), hydrolysis reactions, and hydration and dehydration reactions. These reactions will be seen again as you learn about the roles of biomolecules in metabolism.

To understand organic reactions, it is important that you be able to readily identify the common functional groups. Generally, it is only one functional group in a molecule that undergoes a change in a particular chemical reaction; the other functional groups in the molecule are not involved. The benefit of leaning a few functional groups and their corresponding chemical properties is that you can then predict the outcome of a chemical reaction for any molecule based on the functional groups that it contains.

Almost all biochemical reactions are catalyzed by a specific enzyme whose function is to increase the rate of the reaction. In the cell, there are only six basic reaction types. These six reaction types consist of oxidation-reduction reactions, group transfer reactions, hydrolysis reactions, hydration and dehydration reactions, isomerization reactions, and condensation reactions and ATP/ADP hydrolysis. Associated with each of the six basic reaction types is a major class of enzymes that catalyzes this type of reaction. Enzymes are classified according to the type of reaction they catalyze.

10.1 Oxidation-Reduction Reactions

Oxidation-reduction reactions are some of the most important reactions in nature because they are involved with energy transfer. Combustion reactions are a type of oxidation-reduction reaction. All combustion reactions require oxygen, O_2, and complete combustion converts an organic compound containing C, H, and sometimes O into carbon dioxide (CO_2), water (H_2O), and energy. The cell uses cellular respiration to perform combustion reactions. In the cell, glucose undergoes combustion through a series of separate oxidation-reduction reactions to produce carbon dioxide, water, and energy. As with all combustion reactions, cellular respiration requires oxygen, O_2, and this is why we breathe.

In a Nutshell: Chemical Definitions of Oxidation and Reduction

An oxidation-reduction reaction (often abbreviated redox) is characterized by the transfer of electrons from one reactant to the other reactant. The reactant that loses electrons undergoes oxidation; the reactant that gains electrons undergoes reduction. By definition, where there is oxidation, there must also be reduction.

Generally in inorganic compounds, when a metal cation is formed from a metal, the metal has undergone oxidation. The reaction of two electrons and two protons (H^+) to form H_2 is a reduction reaction. When a nonmetal anion is formed from a nonmetal, the nonmetal has undergone reduction.

In a Nutshell: Oxidation-Reduction Reactions of Organic Compounds

Redox reactions involving organic compounds also involve the transfer of electrons. In organic reactions, it is easier to recognize oxidation and reduction reactions by looking for a change in the number of hydrogen atoms or oxygen atoms in the reactants compared to products. If the functional group change involves an increase in the number of oxygen atoms and/or a decrease in the number of hydrogen atoms, then it has lost electrons and therefore has been oxidized. A functional group that does the reverse has a decrease in the number of oxygen atoms and/or an increase in the number of hydrogen atoms has been reduced.

In an oxidation-reduction reaction, the reactant that causes the substrate to be oxidized is referred to as the *oxidizing agent,* which itself is being reduced in the process. In general, the abbreviation [O] is placed above the reaction arrow to indicate an oxidizing agent is present as a reactant. The reactant that causes the substrate to be reduced is referred to as the *reducing agent,* which itself is being oxidized in the process. In general, the abbreviation [H] is placed above the reaction arrow to indicate a reducing agent is present as a reactant.

In a Nutshell: Common Oxidation and Reduction Reactions of Functional Groups

Each functional group has a characteristic chemical behavior that is independent of the structure of the rest of the molecule. By learning how a given functional group behaves in the presence of an oxidizing or reducing agent, we can predict the outcome for an oxidation or reduction reaction of any molecule that contains this functional group.

Primary alcohols are oxidized to aldehydes. Aldehydes can be oxidized to carboxylic acids. Secondary alcohols can be oxidized to ketones. In these reactions, the number of hydrogen atoms decreases and/or the number of oxygen atoms increases (oxidation). Tertiary alcohols do not react with oxidizing agents.

In the presence of a reducing agent, a carboxylic acid can be reduced to an aldehyde, and an aldehyde can be reduced to a primary alcohol. Ketones can be reduced to secondary alcohols.

An oxidizing agent is required to convert an alkane into an alkene and to convert an alkene into an alkyne. The reverse of each of these reactions is a reduction. Alkynes can be reduced to alkenes, and alkenes can be reduced to alkanes.

The reduction of an alkene to an alkane is the most common reduction reaction of hydrocarbons. In the laboratory, this reaction is typically carried out with hydrogen gas (H_2) as the reducing agent in the presence of a catalyst (Pd, Pt, Ni, etc.), in a reaction known as catalytic hydrogenation. Catalytic hydrogenation is used by the food industry to prepare "partially hydrogenated" fats. These fats have a longer shelf-life and a more solid spreadable consistency. Unfortunately, during the catalytic hydrogenation process, some double bonds are not reduced, but instead isomerized from *cis* to *trans*. The fats containing *trans* double bonds are less healthy than even saturated fats.

Worked Example #1

Determine whether the following reactions represent an oxidation or a reduction of the organic substance shown. Explain your reasoning by showing the hydrogen and/or oxygen atoms that have been gained or lost from the reactant.

a.

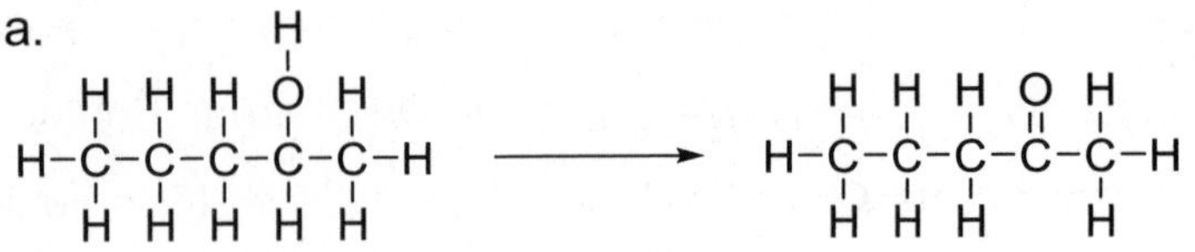

```
            H
            |
    H  H  H  O  H                       H  H  H  O  H
    |  |  |  |  |                       |  |  |  ||  |
H - C - C - C - C - C - H   ------>   H - C - C - C - C - C - H
    |  |  |  |  |                       |  |  |     |
    H  H  H  H  H                       H  H  H     H
```

b.

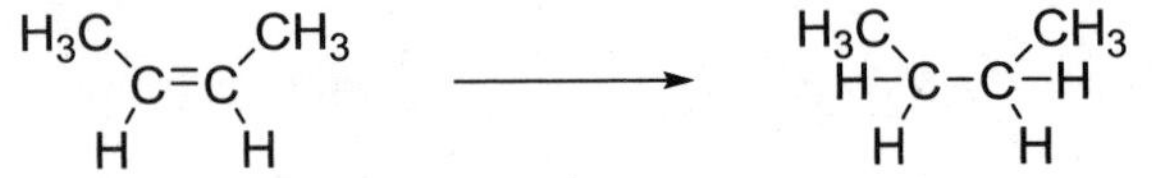

```
H3C       CH3                     H3C      CH3
   \     /                           \    /
    C = C          ------>          H-C - C-H
   /     \                           /     \
  H       H                         H       H
```

Solutions

a.

```
            (H)
             |
    H  H  H  O  H                       H  H  H  O  H
    |  |  |  |  |                       |  |  |  ||  |
H - C - C - C - C - C - H   ------>   H - C - C - C - C - C - H
    |  |  |  |  |                       |  |  |     |
    H  H  H (H) H                       H  H  H     H
```

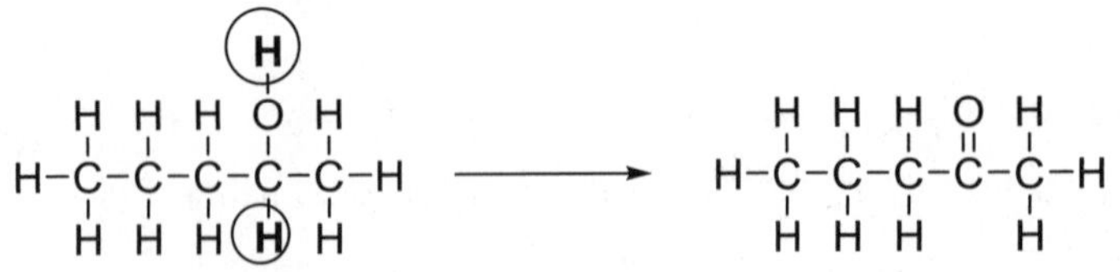

alcohol ketone

The functional group changes from an alcohol to a ketone. There is a decrease in the number of hydrogen atoms and no change in the number of oxygen atoms. Therefore, the reactant has undergone an oxidation.

b.

```
H3C       CH3                     H3C      CH3
   \     /                           \    /
    C = C          ------>          H-C - C-H
   /     \                           /     \
  H       H                        (H)     (H)
```

alkene alkane

The functional group changes from an alkene to an alkane. There is an increase in the number of hydrogen atoms and no change in the number of oxygen atoms. Therefore, the reactant has undergone a reduction.

Try It Yourself #1

Determine whether the following reactions represent an oxidation or a reduction of the organic substance shown. Explain your reasoning by showing the hydrogen and/or oxygen atoms that have been gained or lost from the reactant.

a.

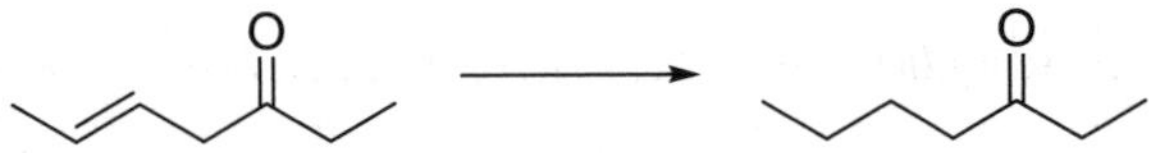

O ⟶ O

b.

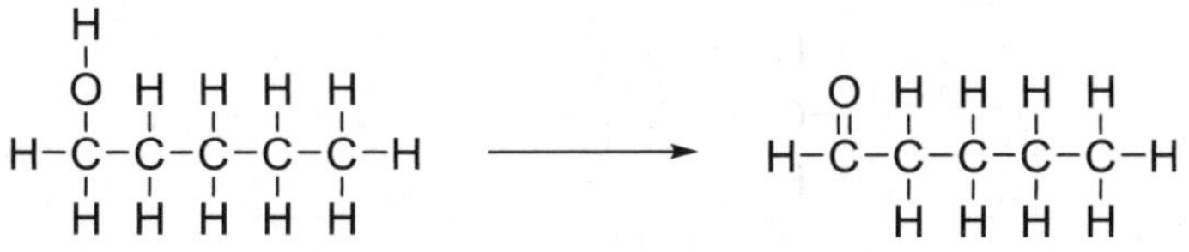

```
  H
  |
  O  H  H  H  H                 O  H  H  H  H
  |  |  |  |  |                 ‖  |  |  |  |
H-C--C--C--C--C-H   ------>   H-C--C--C--C--C-H
  |  |  |  |  |                    |  |  |  |
  H  H  H  H  H                    H  H  H  H
```

a. *The functional group changes from ________________ to ____________________.*

The number of hydrogen atoms: ___________________________

The number of oxygen atoms: ___________________________

The hydrogen atoms or oxygen atoms that have changed:

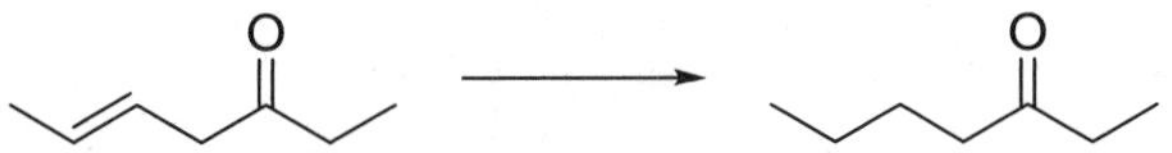

```
O        O
```

The reaction undergoes ______________________.

b. *The functional group changes from ________________ to ____________________.*

The number of hydrogen atoms: ___________________________

The number of oxygen atoms: ___________________________

The hydrogen atoms or oxygen atoms that have changed:

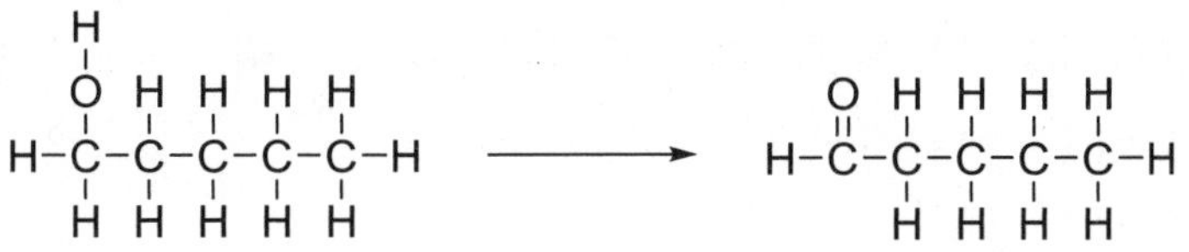

```
  H
  |
  O  H  H  H  H                 O  H  H  H  H
  |  |  |  |  |                 ‖  |  |  |  |
H-C--C--C--C--C-H   ------>   H-C--C--C--C--C-H
  |  |  |  |  |                    |  |  |  |
  H  H  H  H  H                    H  H  H  H
```

The reaction undergoes: ______________________.

Worked Example #2

Predict the product formed in the following reaction.

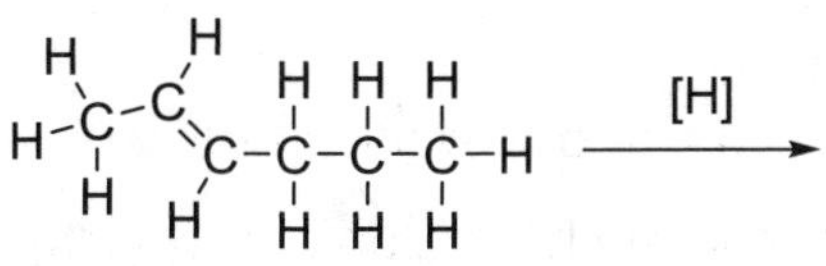

```
H     H
 \   /   H  H  H
H-C-C    |  |  |          [H]
  |  \\C-C--C--C-H     -------->
  H  /   |  |  |
    H    H  H  H
```

Solution

The [H] indicates that the reaction is a reduction reaction. The functional group is an alkene, which undergoes reduction.

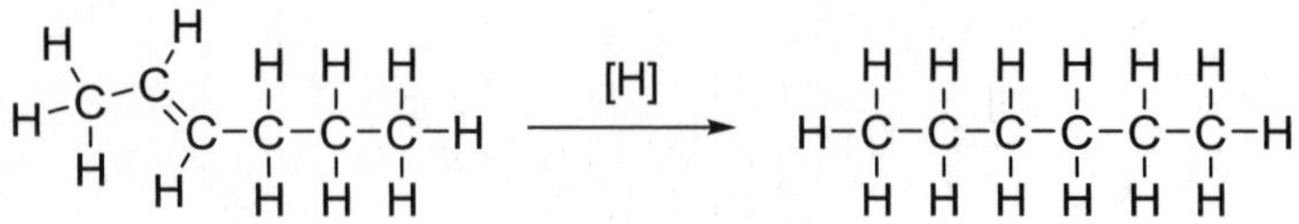

```
H    H
 \  /    H H H               [H]          H H H H H H
H-C-C    | | |                            | | | | | |
  |  \\C-C-C-C-H      ------------->    H-C-C-C-C-C-C-H
  H  /   | | |                            | | | | | |
    H    H H H                            H H H H H H
```

Try It Yourself #2

Predict the product formed in the following reaction.

```
  H O
  | ||          [O]
H-C-C-H     ---------->
  |
  H
```

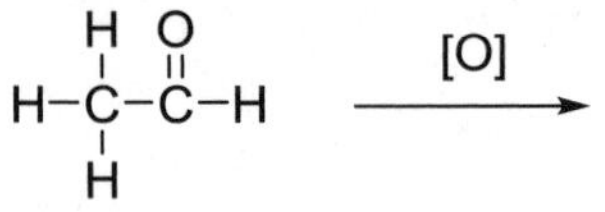

[O] indicates that the reaction is: ____________________.

The product is:

Worked Example #3

Predict the structure of the product formed in each of the following reactions. Write the name of the functional group affected in the reactant and the product.

a.

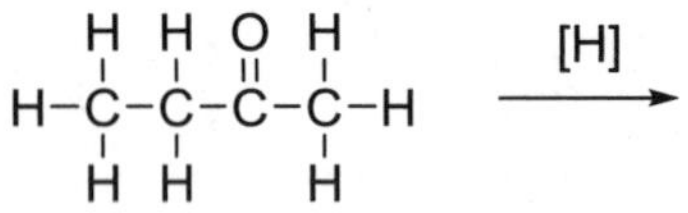

```
  H H O H          [H]
  | | || |
H-C-C-C-C-H    ---------->
  | |   |
  H H   H
```

b.

```
                H H H H O        [O]
                | | | | ||
$H_2O$ +      H-C-C-C-C-C-H  ---------->
                | | | |
                H H H H
```

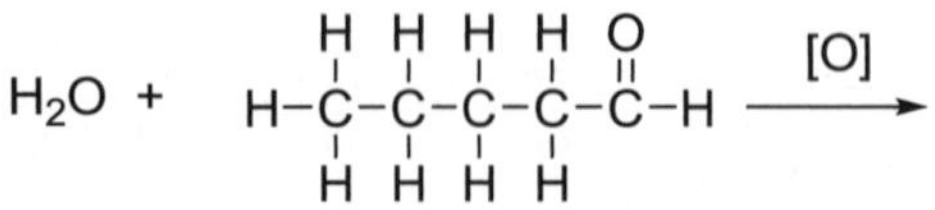

a. *The reactant is a ketone. The [H] above the arrow indicates the ketone is reduced. The reduction of a ketone produces a secondary alcohol.*

```
                                                             H
                                                             |
                  H  H  O  H               [H]         H  H  O  H
                  |  |  ||  |                          |  |  |  |
2H+ + 2e +  H–C–C–C–C–H          ———→     H–C–C–C–C–H
                  |  |     |                           |  |  |  |
                  H  H     H                           H  H  H  H

                  ketone                               2° alcohol
```

b. *The reactant is an aldehyde. The [O] above the arrow signifies that the aldehyde undergoes oxidation. The oxidation of an aldehyde produces a carboxylic acid.*

```
              H  H  H  H  O          [O]          H  H  H  H  O
              |  |  |  |  ||                       |  |  |  |  ||
H2O +     H–C–C–C–C–C–H   ———→   H–C–C–C–C–C–O–H  + 2H+ + 2e
              |  |  |  |                           |  |  |  |
              H  H  H  H                           H  H  H  H

              aldehyde                             carboxylic acid
```

Try It Yourself #3

Predict the structure of the product formed in each of the following reactions. Write the name of the functional group affected in the reactant and the product.

```
a.       H  H  H  O          [H]
         |  |  |  ||
      H–C–C–C–C–H         ———→
         |  |  |
         H  H  H
```

```
b.                H
                  |
         H  H  H  O  H  H        [O]
         |  |  |  |  |  |
      H–C–C–C–C–C–C–H        ———→
         |  |  |  |  |  |
         H  H  H  H  H  H
```

a. *The reactant is:* ____________.

The [H] signifies: ________________.

The product is: ________________.

The structure of the product is:

b. *The reactant is:* ____________.

The [O] signifies: ________________.

The product is: ________________.

The structure of the product is:

A biological cell uses coenzymes to transfer electrons in oxidation-reduction reactions. The enzymes that catalyze oxidation-reduction reactions, *oxidoreductases,* require a coenzyme to achieve their catalytic activity. While the enzyme is chemically unchanged in the reaction, the coenzyme *does* undergo a chemical change in the reaction. The coenzymes for *oxidoreductases* essentially shuttle electrons between molecules in a biological cell. A common coenzyme seen in oxidation-reduction reactions involving carbon-oxygen bonds is nicotinamide adenine dinucleotide, NAD^+/NADH. NAD^+ is the oxidized form of the coenzyme and NADH is the reduced form of the coenzyme. In its NADH form, the coenzyme has two more electrons than in its NAD^+ form. The convention when writing a biochemical pathway is to show the structure of the substrate and the resulting product on either side of the main reaction arrow (the straight arrow). The abbreviations for the coenzyme are written on the reactant and product sides of a curved arrow that intersects the main reaction arrow.

A common coenzyme seen in the oxidation-reduction reactions involving the metabolism of hydrocarbon functional groups is flavin adenine dinucleotide, abbreviated FAD/$FADH_2$. FAD is an oxidizing agent that is reduced to $FADH_2$ upon receiving two electrons in the form of a hydride ion, H^-, and a proton, H^+. Conversely, $FADH_2$ is a reducing agent that can deliver two electrons in the form of a hydride ion, H^-, and a proton, H^+, when oxidized to FAD.

Worked Example #4

Answer the following questions about the reaction shown below.

$$\begin{array}{c} COO^- \\ | \\ H-C-OH \\ | \\ H-C-H \\ | \\ COO^- \end{array} \xrightarrow{NAD^+ \quad NADH} \begin{array}{c} COO^- \\ | \\ C=O \\ | \\ H-C-H \\ | \\ COO^- \end{array}$$

malate — oxaloacetate

a. Is malate oxidized or reduced in this reaction? How can you tell?

b. What is the coenzyme and is it undergoing an oxidation or reduction?

c. Is NAD^+ an oxidizing agent or a reducing agent?

d. What functional group changes in the reaction?

Solutions

a. *Malate is oxidized because the product, oxaloacetate, has fewer hydrogen atoms than the substrate.*

b. *The coenzyme is NAD^+. NAD^+ is reduced to NADH as indicated by the added H in the formula.*

c. *NAD^+ is an oxidizing agent because it causes the substrate to get oxidized.*

d. *The secondary alcohol is converted into a ketone.*

Try It Yourself #4

Answer the following questions about the reaction shown below.

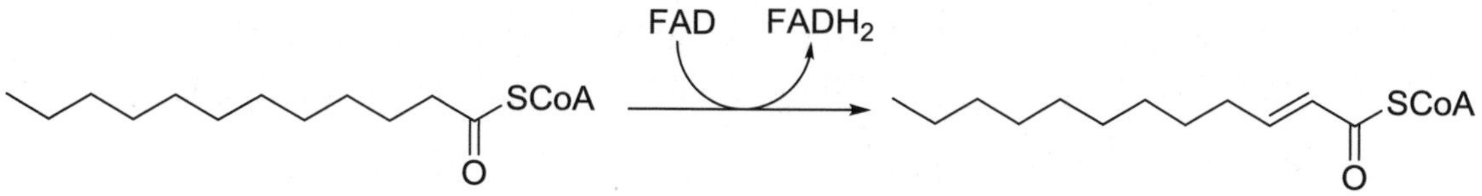

FAD $FADH_2$

SCoA → SCoA

a. Is substrate oxidized or reduced in this reaction? How can you tell?

b. What is the coenzyme and is it undergoing an oxidation or reduction?

c. Is FAD an oxidizing agent or a reducing agent?

d. What functional group changes in the reaction?

Solutions

a. *The number of hydrogen atoms:* ______________________________

The number of oxygen atoms: ______________________________

The substrate is (oxidized or reduced): ______________________________.

b. *Coenzyme:* ______________________________

Does it (gain or lose) hydrogen atoms? ______________________________

The coenzyme is: ______________________________.

c. *The substrate is (oxidized or reduced):* ______________________________.

FAD is (an oxidizing agent or reducing agent): ______________________________.

d. *The functional group that changed:* ______________________________

Some oxidation reactions produce free radicals, which can be damaging to cells and particularly the DNA in cells. Free radicals are molecules, atoms, or ions that contain an odd number of valence electrons. Radicals are unstable and reactive because they contain an atom that does not have an octet. To attain an octet, free radicals remove electrons from other molecules—a reduction of the free radical and an oxidation of the other molecule—in a type of chain reaction that can ultimately damage the cell.

Antioxidants are substances that are believed to prevent the destructive oxidation of substances in the cell. Antioxidants work by stopping the chain reaction initiated by the free radicals or preventing the formation of the radical species in the first place. Antioxidants work because they are reducing agents, so they themselves are oxidized in the process.

Practice Problems for Oxidation-Reduction Reactions

1. Determine whether the following reactions represent an oxidation or a reduction of the organic substance shown. Explain your reasoning by showing the hydrogen and/or oxygen atoms that have been gained or lost from the reactant.

a.

```
              H
              |
   H  H  H  O  H  H              H  H  H  O  H  H
   |  |  |  |  |  |              |  |  |  ||  |  |
H–C–C–C–C–C–C–H   ——→   H–C–C–C–C–C–C–H
   |  |  |  |  |  |              |  |  |     |  |
   H  |  H  H  H  H              H  |  H     H  H
      |                             |
   H–C–H                         H–C–H
      |                             |
      H                             H
```

b.

```
   H  H  H  O                     H  H  H  O
   |  |  |  ||                    |  |  |  ||
H–C–C–C–C–O–H   ——→   H–C–C–C–C–H
   |  |  |                        |  |  |
   H  H  H                        H  H  H
```

2. Predict the structure of the product formed in each of the following reactions.

a. $\text{FAD} \xrightarrow{[H]}$

b. $\text{NADH} \xrightarrow{[O]}$

c.

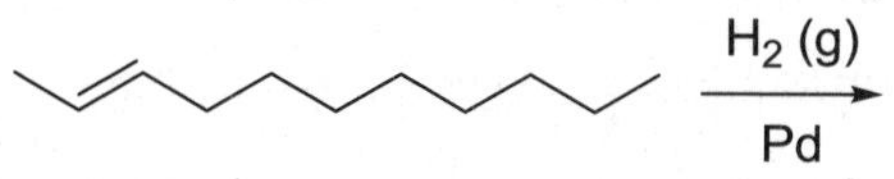

$\xrightarrow[\text{Pd}]{H_2\text{ (g)}}$

3. Predict the structure of the product formed in each of the following reactions. Write the name of the functional group affected in the reactant and the product.

a.

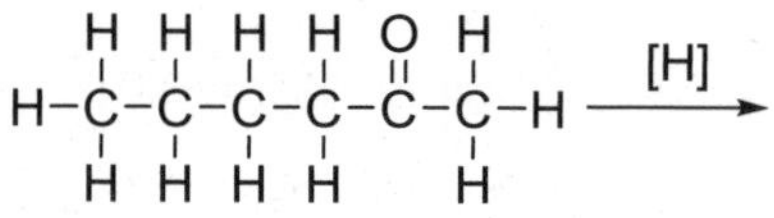

```
   H  H  H  H  O  H
   |  |  |  |  ‖  |          [H]
H–C––C––C––C––C––C–H  ——————>
   |  |  |  |     |
   H  H  H  H     H
```

b.

```
            H
            |
   H  H  H  O
   |  |  |  |          [O]
H–C––C––C––C–H   ——————>
   |  |  |  |
   H  H  H  H
```

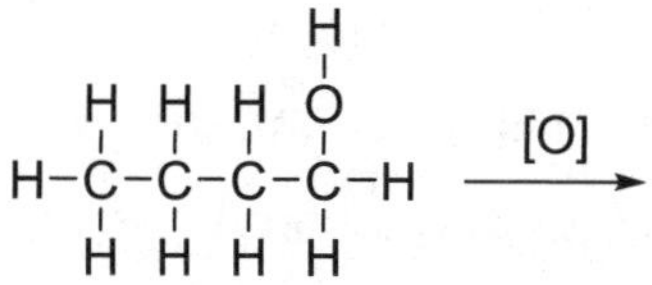

10.2 Group Transfer Reactions: Esterification and Amidation Reactions-

Carboxylic acids and their derivatives—esters, thioesters, and amides—undergo many reactions. Biochemists refer to this broad class of reactions as acyl group transfer reactions because the group transferred is an acyl group, a carbonyl group, and its attached R group.

An acyl transfer reaction is so named because it appears as if the acyl group on the carboxylic acid moves from the —OH of the carboxylic acid to the heteroatom of an alcohol (HOR), thiol (HSR), or amine (HNR_2) to produce an ester, thioester, or amide. A molecule of water is produced in these reactions as well.

In a Nutshell: Esters and Thioesters

An ester and water are formed in the reaction of an alcohol with a carboxylic acid in the presence of a catalyst. In an analogous manner, thioesters are formed by the reaction of a thiol with a carboxylic acid. A thiol is a functional group analogous to an alcohol, with a sulfur instead of oxygen RSH. A thioester is a functional group analogous to an ester with sulfur instead of oxygen attached to the carbonyl group.

To predict the chemical structure of the ester or thioester formed in the reaction between a carboxylic acid and an alcohol or a thiol, follow these guidelines: 1) Write the carboxylic acid with its OH group pointing to the right and place the alcohol or thiol next to it on the right with its H—O or H—S group pointing to the left. 2) Remove the elements of water, OH and H, by removing the OH from the carboxylic acid and the H from the alcohol or thiol, and then form a single bond between the atoms from which the H and OH were removed.

The fragrance of fruits and perfumes is often due to esters. Esters are readily prepared in the laboratory in esterification reactions by heating together a carboxylic acid and an alcohol in the presence of an acid catalyst (H^{+}). Different scents can be created by varying the structure of the carboxylic acid and the alcohol.

Fats are esters derived from three fatty acids and glycerol. Our cells store fat in the form of glycerol esters. The esterification of three fatty acids (the same or different) and one glycerol molecule produces a fat.

Worked Example #5

Write the products formed in the following esterification reactions.

a.
```
    H H H           H H O
    | | |           | | ||        catalyst
  H-C-C-C-O-H  +  H-C-C-C-O-H  ----------->
    | | |           | |
    H H H           H H
```

b.
```
    H H H           H O
    | | |           | ||        catalyst
  H-C-C-C-S-H  +  H-C-C-O-H  ----------->
    | | |           |
    H H H           H
```

a. *Rewrite the structures so that the carboxylic acid OH is pointing to the right and the OH group on the alcohol is on the left.*

```
  H H O              H H H
  | | ||             | | |
H-C-C-C-O-H    H-O-C-C-C-H
  | |                | | |
  H H                H H H
```

Remove the OH group on the carboxylic acid and the H on the oxygen atom of the alcohol to form two incomplete structures and a water molecule:

```
  H H O            H H H
  | | ||           | | |
H-C-C-C          O-C-C-C-H    +    H2O
  | |              | | |
  H H              H H H
```

Form a carbon-oxygen single bond between the carbonyl carbon and the OR group to produce an ester:

```
  H H O   H H H
  | | ||  | | |
H-C-C-C-O-C-C-C-H
  | |     | | |
  H H     H H H
```

b. *Rewrite the structures so that the carboxylic acid OH is pointing to the right and the SH group on the thiol is on the left.*

```
  H O           H H H
  | ||          | | |
H-C-C-O-H  H-S-C-C-C-H
  |             | | |
  H             H H H
```

Remove the OH group on the carboxylic acid and the H on the sulfur atom of the thiol to form two incomplete structures and a water molecule:

```
  H O      H H H
  | ||     | | |
H-C-C    S-C-C-C-H  +  H2O
  |        | | |
  H        H H H
```

Form a carbon-sulfur single bond between the carbonyl carbon and the SR group to produce a thioester:

```
  H O   H H H
  | ||  | | |
H-C-C-S-C-C-C-H
  |     | | |
  H     H H H
```

Try It Yourself #5

Write the products formed in the following esterification reactions.

a.

```
   H H                 H O
   | |                 | ||                catalyst
H-C-C-O-H  +  H-C-C-O-H  ──────────→
   | |                 |
   H H                 H
```

b.

```
   H H H O                        H H H
   | | | ||                       | | |          catalyst
H-C-C-C-C-O-H     +     H-S-C-C-C-H   ──────────→
   | | |                          | | |
   H H H                          H H H
```

a. *Rewrite the structures with the OH group on the carboxylic acid and the hydroxyl group on the alcohol pointing the correct way.*

Remove the OH group and the H atom.

The two partial structures are:

The other product is: ______________________________.

Connect the two resulting partial structures.

The structure of the product is:

b. *Rewrite the structures with the OH group on the carboxylic acid and the SH group on the thiol pointing the correct way.*

Remove the OH group and the H atom.

The two partial structures are:

The other product is: ______________________________.

Connect the two resulting partial structures.

The structure of the product is:

In a Nutshell: Amidation Reactions

An amidation reaction is the formation of an amide from a carboxylic acid and a primary or secondary amine, in the presence of a catalyst. Amidation reactions are used to build proteins from amino acids. To predict the structure of the amide formed when a carboxylic acid reacts with an amine in the presence of a catalyst, follow these guidelines: 1) Write the carboxylic acid with its OH group pointing to the right and place the amine next to it on the right with its NH group pointing to the left. 2) Remove the elements of water, OH and H, by removing the OH from the carboxylic acid and the H from the amine, and then form a single bond between the atoms from which the H and OH were removed. These steps are analogous to the formation of an ester and a thioester. Note: A tertiary amine does not contain an N—H bond; therefore, it cannot undergo an amidation reaction.

Worked Example #6

Write the products formed in the following amidation reaction.

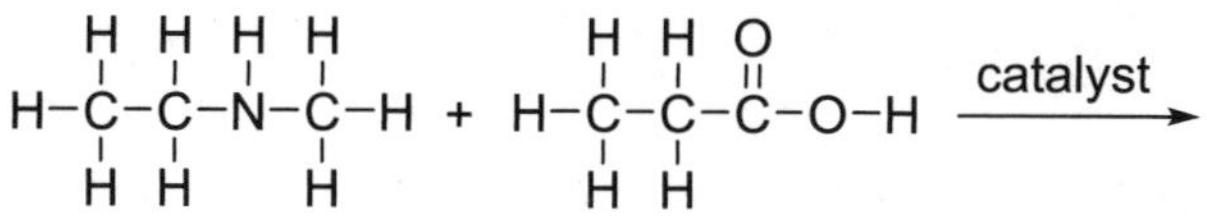

```
  H H H H        H H O
  | | | |        | | ||        catalyst
H-C-C-N-C-H  + H-C-C-C-O-H   ---------->
  | |   |        | |
  H H   H        H H
```

Rewrite the structures so that the carboxylic acid OH is pointing to the right and the NH group on the amine is on the left.

```
                          H
                          |
 H  H  O              H-C-H
 |  |  ||                 |
H-C-C-C-O-H      H-N
 |  |                     |
 H  H                 H-C-H
                          |
                      H-C-H
                          |
                          H
```

Remove the OH group on the carboxylic acid and the H on the nitrogen atom of the amine to form two incomplete structures and a water molecule:

```
                       H
                       |
  H  H  O          H-C-H
  |  |  ||             |
H-C-C-C                N          + H2O
  |  |                 |
  H  H             H-C-H
                       |
                   H-C-H
                       |
                       H
```

Form a carbon-nitrogen single bond between the carbonyl carbon and the NR_2 group to produce an amine:

```
                 H
                 |
  H  H  O  H-C-H
  |  |  ||       |
H-C-C-C----N
  |  |           |
  H  H       H-C-H
                 |
             H-C-H
                 |
                 H
```

Try It Yourself #6

Write the products formed in the following amidation reaction.

```
  H  H  H              H  H  H  O
  |  |  |              |  |  |  ||        catalyst
H-C-C-N-H    +    H-C-C-C-C-O-H  ---------->
  |  |                 |  |  |
  H  H                 H  H  H
```

Rewrite the structures with the OH group on the carboxylic acid and the NH group on the amine pointing the correct way.

Remove the OH group and the H atom.

The two partial structures are:

The other product is: ______________________________.

Connect the two resulting partial structures.

The structure of the product is:

Practice Problems for Group Transfer Reactions: Esterification and Amidation Reactions

1. Predict the product formed in the following esterification reactions.

a.

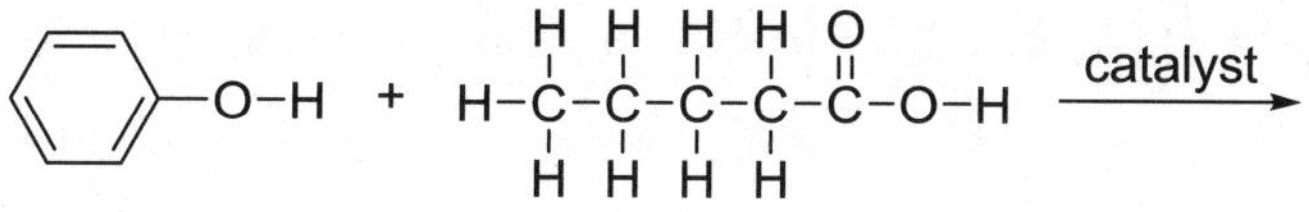

```
                H  H  H  H  O
                |  |  |  |  ||               catalyst
(benzene)-O-H + H-C--C--C--C--C-O-H      ------------>
                   |  |  |  |
                   H  H  H  H
```

b.

```
   H  H  H  H  O
   |  |  |  |  ||                        catalyst
H-C--C--C--C--C-O-H   +   H-SCoA     ------------>
   |  |  |  |
   H  H  H  H
```

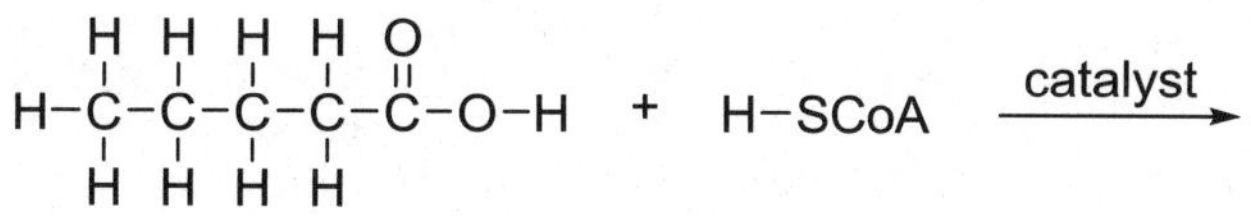

2. Predict the product formed in the following amidation reaction.

```
  H H H O             H H H
  | | | ||            | | |     catalyst
H-C-C-C-C-O-H  +  H-N-C-C-H   ----------->
  | | |               | |
  H H H               H H
```

10.3 Hydrolysis Reactions

Hydrolysis reactions are some of the most common reactions in both the laboratory and in biochemistry. In this section, we describe the hydrolysis of a carboxylic acid derivative into a carboxylic acid into an alcohol, a thiol, or an amine. The hydrolysis reactions are essential the opposite of the acyl transfer reactions described in Section 10.2. Water is the agent that breaks the bond between the carbonyl carbon and the O, S, or N atom of these carboxylic acid derivatives. Two products are formed in the hydrolysis of a carboxylic acid derivative and one of these is a carboxylic acid or its conjugate base. The other product formed depends on the functional group of the carboxylic acid derivative. If it is an ester, an alcohol is formed; if a thioester, a thiol is formed; if an amide, an amine (or the conjugate acid of an amine) is formed. In the case of hydrolysis of an amide, the carboxylic acid and the amine products formed subsequently undergo an acid-base neutralization reaction to form the conjugate base of the acid and the conjugate acid of the amine.

To predict the structure of the products formed in a hydrolysis reaction of an ester, thioester, or amide, follow these guidelines: 1) Break the single bond between the carbonyl carbon and the heteroatom (O, S, or N) of an ester, thioester or amide, to produce two partial structures: an acyl group fragment and a heteroatom-R fragment. 2) Add an —OH group to the carbonyl carbon of the acyl group fragment, forming a carboxylic acid. The new bond is a C—O bond. Water supplies the OH group. 3) Add a hydrogen atom, H—, to the heteroatom R' fragment to form an alcohol, thiol, or amine. The new bond is a heteroatom—H bond. Water supplies the H atom. 4) In the case of an amide, the carboxylic acid and the amine produced react in an acid-base reaction to produce the conjugate base of the carboxylic acid and the conjugate acid of the amine.

In a Nutshell: Applications of Hydrolysis Reactions

A common hydrolysis reaction seen in biochemistry is the digestion of proteins. The digestion of proteins is aided by the acidic environment, where amide bonds joining amino acids are hydrolyzed into their individual amino acid components, with the help of hydrolase enzyme. The carboxylic acids formed are found in their conjugate base form ($RCOO^-$), and the amines, in their conjugate acid form (RNH_3^+).

Soaps have been prepared for centuries by the hydrolysis of vegetable oils. Vegetable oils are triglycerides, which upon heating in the presence of a strong base (OH^-), produce glycerol and three fatty acids in their conjugate base form. The hydrolysis of an ester in aqueous base is known as saponification.

Hydrolysis is the first step in the metabolism of fats, which occurs in the small intestine. The reaction is similar to saponification, except it occurs in a mildly basic environment and with the help of lipase enzymes.

Worked Example #7

Write the structure of the products formed in the following hydrolysis reactions.

a. hydrolysis of an ester

```
              H H     O H H H H          acid
              | |     ‖ | | | |        catalyst
H2O  +  H-C-C-O-C-C-C-C-C-H    ———————→
              | |       | | | |
              H H       H H H H
```

b. hydrolysis of a thioester

```
              H H O     H          acid
              | | ‖     |        catalyst
H2O  +  H-C-C-C-S-C-H    ———————→
              | |       |
              H H       H
```

c. hydrolysis of an amide

```
                  H
                  |
                H-C-H
                  |
              H   |    O H H
              |   |    ‖ | |          catalyst
H2O  +  H-C-C-N-C-C-C-H   ─────────→
              |   |        | |
              H H H        H H
```

Solutions

a. Break the single bond between the carbonyl carbon and the O atom of the ester to produce two partial structures: an acyl group and an O—R fragment.

```
  H H               O H H H H
  | |               ‖ | | | |
H-C-C-O             C-C-C-C-C-H
  | |                 | | | |
  H H                 H H H H
```

Add an —OH group to the carbonyl carbon of the acyl group fragment to form a carboxylic acid. The new bond is a C—O bond. Water supplied the OH group.

```
      O H H H H
      ‖ | | | |
  H-O-C-C-C-C-C-H
        | | | |
        H H H H
```

Add a hydrogen atom, H—, to the oxygen of the heteroatom-R' fragment to form an alcohol. The new bond is an O—H bond. Water supplies the H atom.

```
  H H
  | |
H-C-C-O-H
  | |
  H H
```

b. Break the single bond between the carbonyl carbon and the S atom of the thioester, to produce two partial structures: an acyl group and an S—R fragment.

```
  H H O          H
  | | ‖          |
H-C-C-C        S-C-H
  | |            |
  H H            H
```

Add an —OH group to the carbonyl carbon of the acyl group fragment to form a carboxylic acid. The new bond is a C—O bond. Water supplied the OH group.

```
  H H O
  | | ‖
H-C-C-C-O-H
  | |
  H H
```

Add a hydrogen atom, H—, to the sulfur of the heteroatom-R' fragment to form an alcohol. The new bond is an S—H bond. Water supplies the H atom.

```
       H
       |
  H-S-C-H
       |
       H
```

c. *Break the single bond between the carbonyl carbon and the N atom of the amide, to produce two partial structures: an acyl group and an N—R fragment.*

```
       H
       |
    H-C-H
       |
    H  |             O  H  H
    |  |             ||  |  |
 H-C--C-N            C-C-C-H
    |  |  |             |  |
    H  H  H             H  H
```

Add an —OH group to the carbonyl carbon of the acyl group fragment to form a carboxylic acid. The new bond is a C—O bond. Water supplied the OH group.

```
        O  H  H
        ||  |  |
  H-O-C-C-C-H
           |  |
           H  H
```

Add a hydrogen atom, H—, to the nitrogen of the heteroatom-R' fragment to form an alcohol. The new bond is an N—H bond. Water supplies the H atom.

```
       H
       |
    H-C-H
       |
    H  |
    |  |
 H-C--C-N-H
    |  |  |
    H  H  H
```

Remember that the hydrolysis of the amide initially produces an amine and carboxylic acid. These products undergo a subsequent neutralization reaction to form the conjugate acid of the amine and the conjugate base of the carboxylic acid. Therefore, the following amine salt is the result:

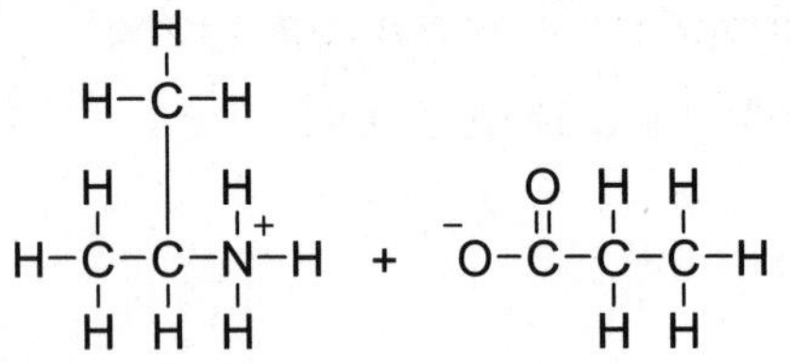

```
      H
      |
    H-C-H
      |
   H  |  H
   |  |  |+                 O  H  H
 H-C--C--N-H     +      -   ||  |  |
   |  |  |                 O-C--C--C-H
   H  H  H                      |  |
                                H  H
```

conjugate acid of amine   conjugate base of acid

Try It Yourself #7

Write the structure of the products formed in the following hydrolysis reactions.

a.

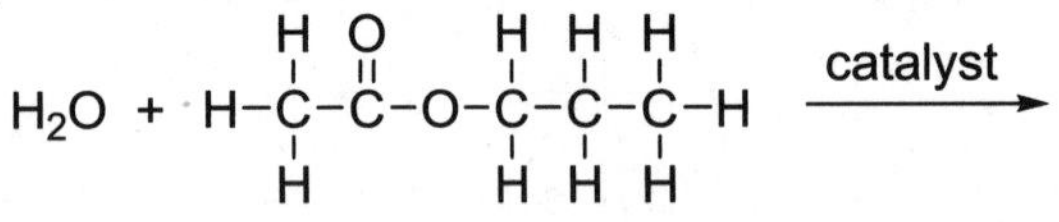

```
             H  O     H  H  H
             |  ||    |  |  |              catalyst
H2O  +  H-C--C--O--C--C--C-H          ----------->
             |        |  |  |
             H        H  H  H
```

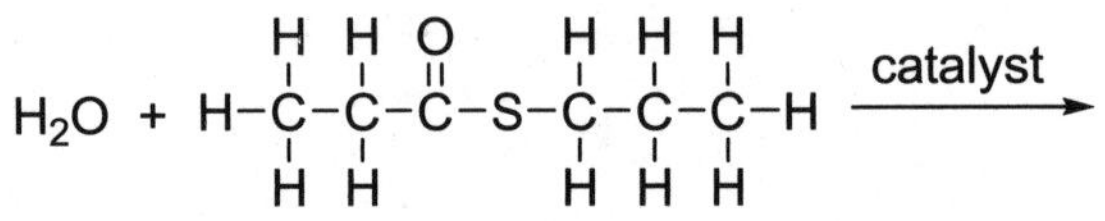

b.

```
             H  H  O     H  H  H
             |  |  ||    |  |  |              catalyst
H2O  +  H-C--C--C--S--C--C--C-H          ----------->
             |  |        |  |  |
             H  H        H  H  H
```

c.

```
             H  H  H     O  H  H
             |  |  |     ||  |  |              catalyst
H2O  +  H-C--C--C--N--C--C--C-H          ----------->
             |  |  |  |      |  |
             H  H  H  H      H  H
```

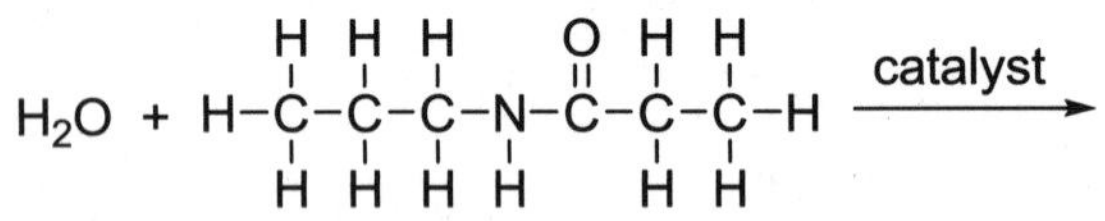

Solutions

a. *Break the single bond between the carbonyl carbon and the heteroatom of the carboxylic acid derivative to produce two partial structures—an acyl group and the heteroatom-R fragment:*

Add an —OH group to the carbonyl carbon of the acyl group fragment:

Add a hydrogen atom, H-, to the oxygen of the heteroatom-R' fragment:

If the reaction is the hydrolysis of an amide, remember the acid-base reaction between to the products.
The products are:

b. Break the single bond between the carbonyl carbon and the heteroatom of the carboxylic acid derivative to produce two partial structures—an acyl group and the heteroatom-R fragment:

Add an —OH group to the carbonyl carbon of the acyl group fragment:

Add a hydrogen atom, H-, to the oxygen of the heteroatom-R' fragment:

If the reaction is the hydrolysis of an amide, remember the acid-base reaction between to the products.
The products are:

c. *Break the single bond between the carbonyl carbon and the heteroatom of the carboxylic acid derivative to produce two partial structures—an acyl group and the heteroatom-R fragment:*

Add an —OH group to the carbonyl carbon of the acyl group fragment:

Add a hydrogen atom, H-, to the oxygen of the heteroatom-R' fragment:

If the reaction is the hydrolysis of an amide, remember the acid-base reaction between to the products.
The products are:

Practice Problems for Hydrolysis Reactions

1. Predict the products formed in the following hydrolysis reactions.

a.
```
     H  H  O       H  H
     |  |  ||      |  |                      catalyst
  H--C--C--C--O--C--C--H    +  H2O   ----------->
     |  |          |  |
     H  H          H  H
```

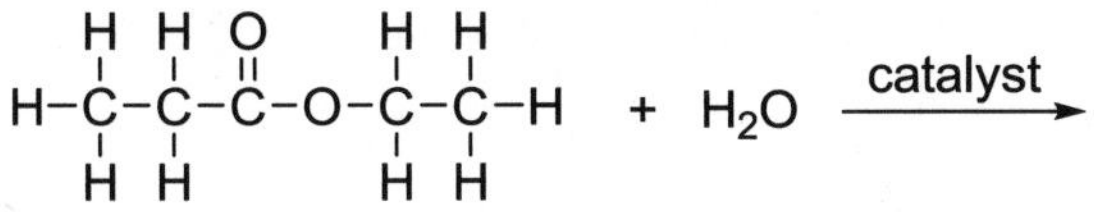

b.
```
        H
        |
     H--C--H
     H  |      O  H  H  H
     |  |      ||  |  |  |                      catalyst
  H--C--C--S--C--C--C--C--H    +  H2O   ----------->
     |  |          |  |  |
     H  H          H  H  H
```

c.
```
           H
           |
        H--C--H
        H  |      O  H  H
        |  |      ||  |  |                     catalyst
     H--C--C--N--C--C--C--H  +  H2O   ----------->
        |  |  |       |  |
        H  H  H       H  H
```

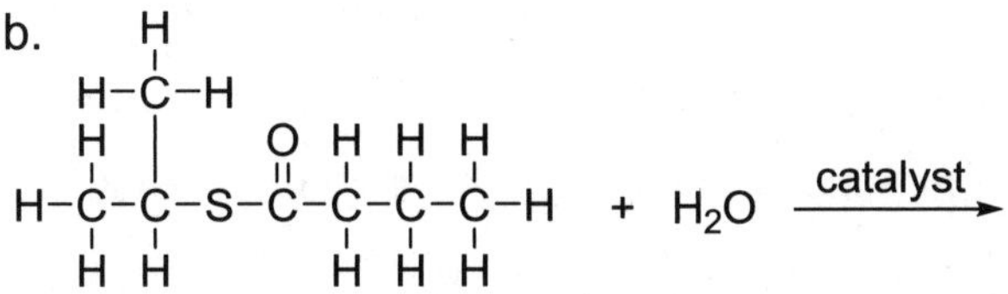

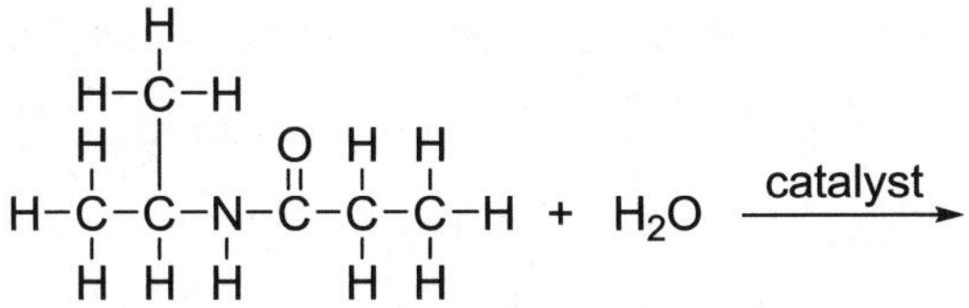

2. Predict the product(s) formed in the following reactions.

a.
```
     H  H  H  O      H  H
     |  |  |  ||     |  |                          catalyst
  H--C--C--C--C--N--C--C--H  +  H2O   ----------->
     |  |  |      |  |  |
     H  H  H      H  H  H
```

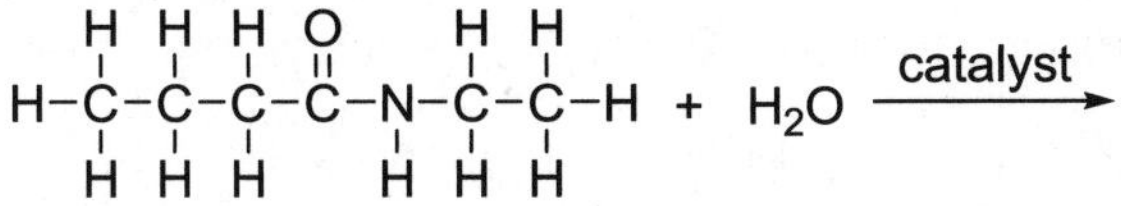

b.

$$H_2O + \begin{array}{ccccccccccccc} & & H & & H & & H & & O & & & & H & \\ & & | & & | & & | & & \| & & & & | & \\ H & - & C & - & C & - & C & - & C & - & O & - & C & - & H \\ & & | & & | & & | & & & & & & | & \\ & & H & & H & & H & & & & & & H & \end{array} \xrightarrow{\text{catalyst}}$$

10.4 Hydration and Dehydration Reactions

Hydration reactions are another class of reactions that use water is a reactant. In a hydration reaction, the elements of water, H and OH, add to an alkene to produce an alcohol. The reverse of a hydration reaction, elimination of the elements of water, H and OH, from and alcohol to form an alkene is known as a dehydration reaction. Hydration and dehydration reactions are catalyzed by the class of enzymes known as lyases.

In a Nutshell: Hydration

In a hydration reaction water reacts with an alkene to produce an alcohol. The H and OH atoms from a molecule of water form bonds to each of the carbon atoms of the alkene and the carbon-carbon double bond becomes a carbon-carbon single bond. In the case of an unsymmetrical alkene, the OH from a molecule of water adds to the double-bond carbon with fewer hydrogen atoms and the H atom from a molecule of water adds to the double-bond carbon with more hydrogen atoms. When a reaction forms one structural isomer over another structural isomer, it is known as a regioselective reaction. Hydration of an alkene is a regioselective reaction, known as Markovnikov's rule.

In biochemical applications of the hydration reaction, a carbonyl group is usually adjacent to the carbon-carbon double bond, which directs the outcome of the reaction. The H atom adds to the double-bond carbon atom closer to the carbonyl groups, known as the α-carbon; and the OH group adds to the double-bond carbon farther from the carbonyl group, known as the β-carbon. The carbonyl group itself is unchanged in the biochemical reaction. A biochemical hydration reaction is seen in the biochemical pathway that occurs in muscle and liver cells when fatty acids are metabolized for energy.

In a Nutshell: Dehydration

The loss of a water molecule from an alcohol to produce an alkene is known as a dehydration reaction, essentially the reverse of a hydration reaction. In a dehydration reaction, —OH and —H are eliminated from the substrate to form a molecule of water, H_2O, and a carbon-carbon double bond is formed in the substrate between the carbon atoms that formerly had the —OH group and —H atom. If the substrate in a dehydration reaction can lead to two possible alkene products (structural isomers), the alkene with fewer hydrogen atoms on the double bond is formed preferentially over the alkene with more hydrogen atoms, an observation known as Zaitsev's rule.

In biochemical dehydration reactions, the hydroxyl group is usually located β to a carbonyl group. Dehydration then occurs with elimination of an α hydrogen and the β hydroxyl group forming water and an α-β unsaturated carbonyl compound. The carbonyl group is itself unchanged, but necessary for the biochemical dehydration.

Worked Example #8

Predict the product formed in the following hydration reaction:

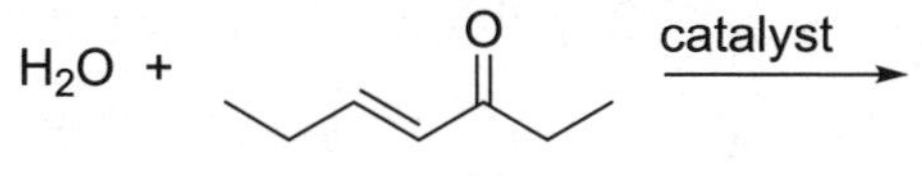

H_2O + O catalyst →

Solution

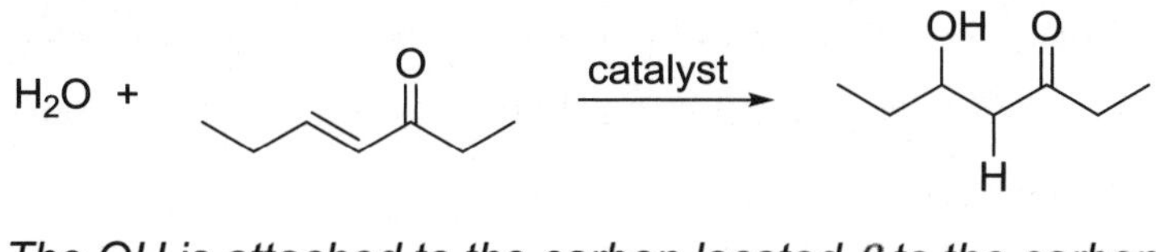

H_2O + O catalyst → OH O H

The OH is attached to the carbon located β to the carbonyl group.

Try It Yourself #8

Predict the product formed in the following hydration reaction:

```
        H  H  H  H  O  H
        |  |  |  |  ‖  |      catalyst
H2O + H-C--C--C=C--C--C-H   ---------->
        |  |           |
        H  H           H
```

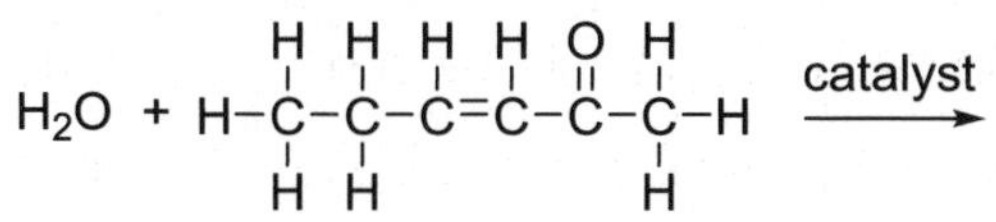

Identify the double bond.

Which carbon atom is β to the carbonyl group?

Structure of the product:

Worked Example #9

Predict the product formed in the following dehydration reaction:

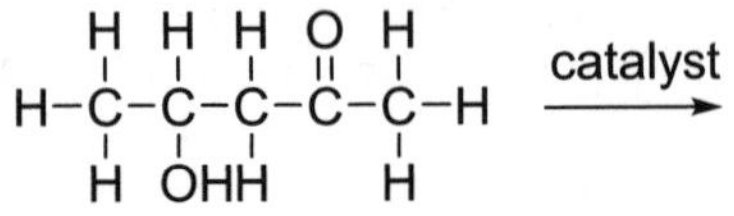

```
   H  H  H  O  H
   |  |  |  ||  |        catalyst
H-C-C-C-C-C-H      ------->
   |  |  |       |
   H  OHH       H
```

Solution

```
   H  H  H  O  H                        H  H       O  H
   |  |  |  ||  |        catalyst         |  |       ||  |
H-C-C-C-C-C-H      ------->    H-C-C=C-C-C-H  +  H2O
   |  |  |       |                        |      |       |
   H  OHH       H                        H      H       H
```

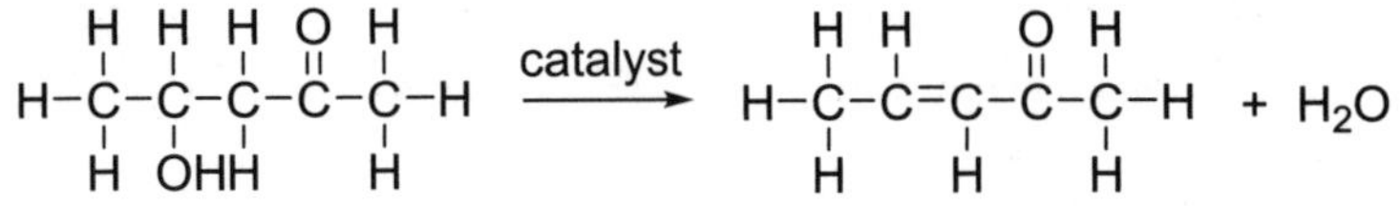

Remember to remove the β OH group and the α H atom when determining where to place the double bond.

Try It Yourself #9

Predict the product formed in the following dehydration reaction:

```
   H  H  H  O  H  H  H
   |  |  |  ||  |  |  |        catalyst
H-C-C-C-C-C-C-C-H      ------->
   |  |  |       |  |  |
   H  OHH       H  H  H
```

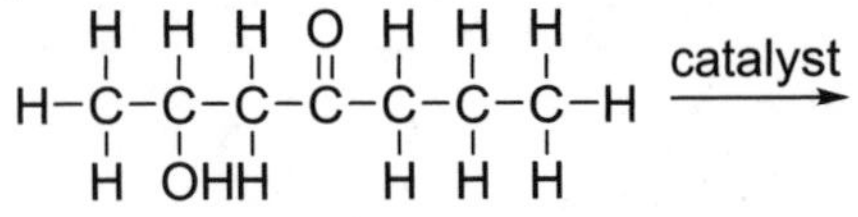

Identify the OH group.

Which H atom should be removed?

Structure of the product:

Practice Problems for Hydration and Dehydration Reactions

1. Predict the product formed in the following dehydration reaction.

   a. OH O H catalyst

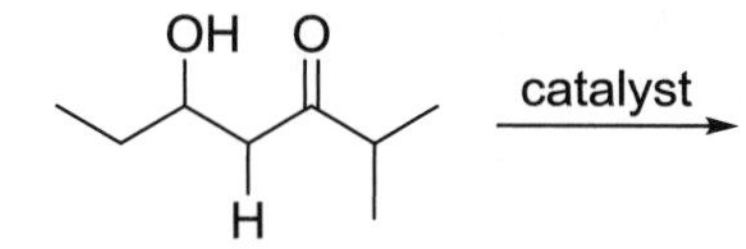

   b. H H H O H H H–C–C–C–C–C–C–H H H H H OH catalyst

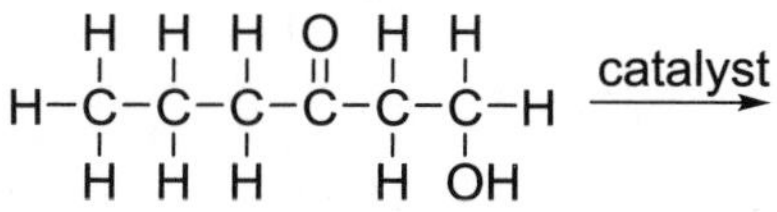

2. Predict the product formed in the following hydration reaction.

   a. O SCoA + H_2O catalyst

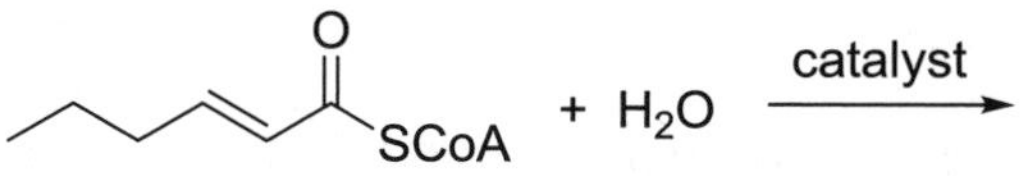

   b. O CH_3 + H_2O catalyst

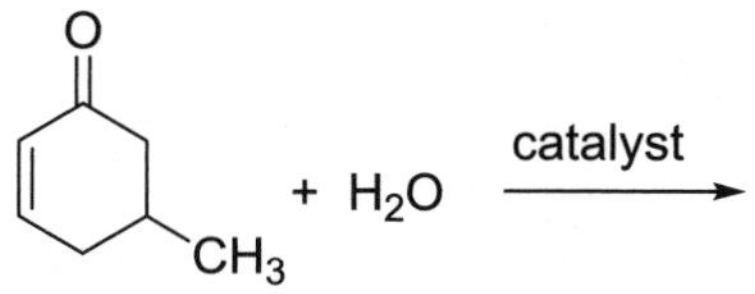

3. For the reactions below, indicate whether they are hydration or dehydration reactions and add a curved arrow showing water to the equation.

a.

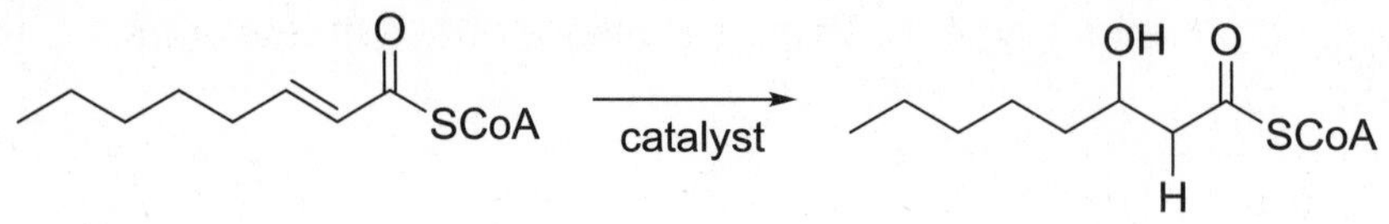

O

SCoA

catalyst

OH O

SCoA

H

b.

H H O H H H
H–C–C–C–C–C–C–H
OH H H H H

H H O H H H
H–C=C–C–C–C–C–H
H H H

Chapter 10 Quiz

1. Predict the product for the following oxidation-reduction reactions.

a. $NAD^+ \xrightarrow{[H]}$

b. 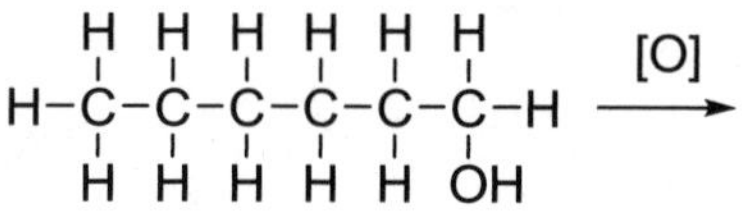 $CH_3-CH_2-CH_2-CH_2-CH_2-CH_2OH \xrightarrow{[O]}$

c. 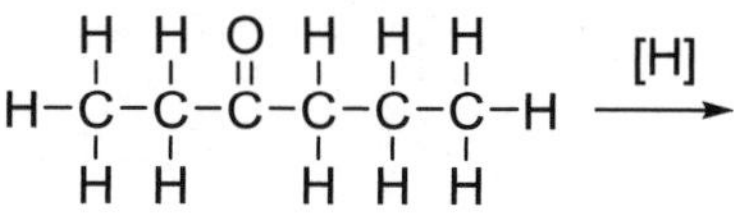 $CH_3-CH_2-C(=O)-CH_2-CH_2-CH_3 \xrightarrow{[H]}$

2. For the oxidation-reduction reactions below, determine whether the organic compound has undergone an oxidation or a reduction by placing an “H” or “O” in the brackets above the arrow.

a. 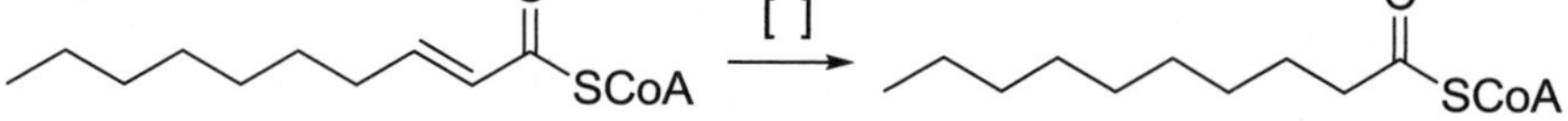 O, SCoA —[ ]→ O, SCoA

b.  OH —[ ]→ O

c.

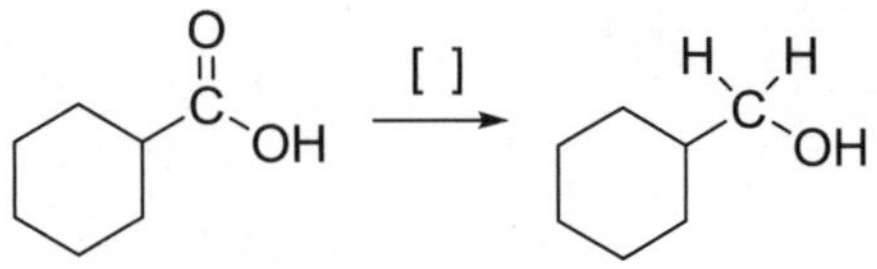

```
    O                        H  H
    ‖          [ ]            \ /
    C          ——→             C
 ⬡/   \OH                  ⬡/   \OH
```

3. Predict the products for the following dehydration or hydration reactions.

a.

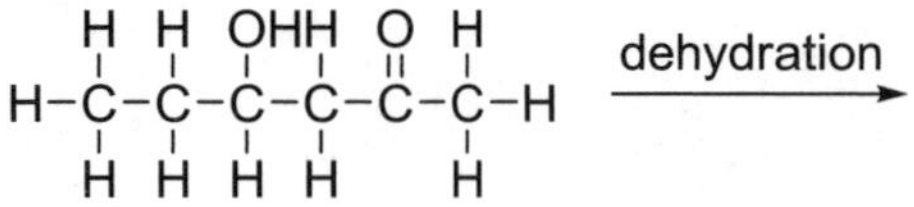

```
     H  H  OHH  O  H
     |  |  |  |  ‖  |            dehydration
   H-C-C-C-C-C-C-H           ——————————→
     |  |  |  |     |
     H  H  H  H     H
```

b.

```
     H  O        H
     |  ‖        |
   H-C-C-C=C-C-H  +  H2O  ——————→
     |     |  |  |
     H     H  H  H
```

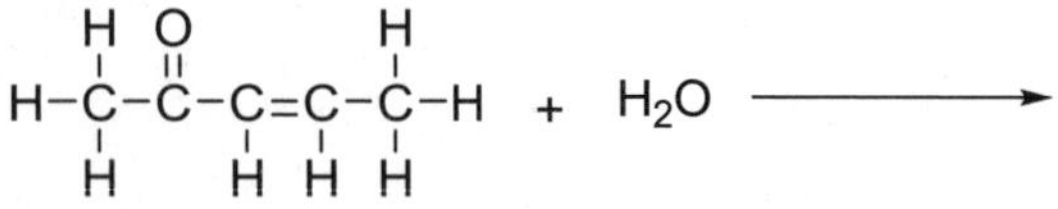

4. Determine whether the reactions shown below are hydration or dehydration reactions and add a curved arrow showing water to the equation.

a.

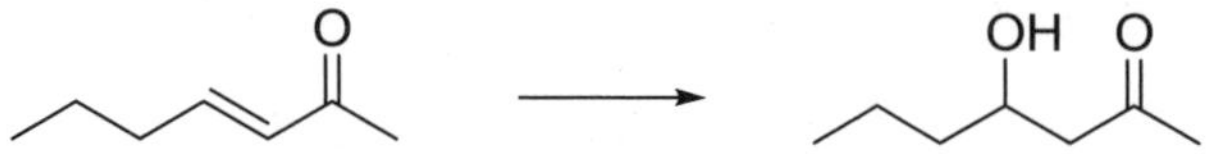

```
          O                 OH  O
          ‖       ——→        |   ‖
```

b.

```
     H  O        H                H  O  H  H  H
     |  ‖        |                |  ‖  |  |  |
   H-C-C-C=C-C-H    ——————→    H-C-C-C-C-C-H
     |     |  |  |                |     |  |  |
     H     H  H  H                H     H  OHH
```

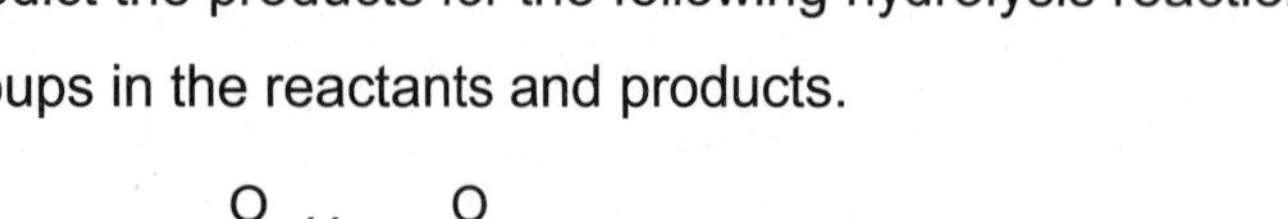

5. Predict the products for the following hydrolysis reactions. Identify the functional groups in the reactants and products.

a.

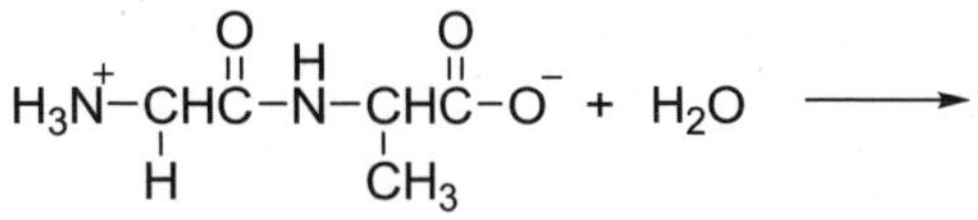

```
          O         O
   +      ‖  H      ‖   -
   H3N-CHC-N-CHC-O   +  H2O  ——————→
        |      |
        H      CH3
```

b.

```
  H H H   O H
  | | |   ‖ |
H-C-C-C-S-C-C-H + H2O ——→
  | | |     |
  H H H     H
```

c.

```
       O H H
       ‖ | |
(C6H5)-O-C-C-C-H + H2O ——→
         | |
         H H
```

6. Write the products formed in the following esterification reactions.

a.

```
  H H O           H H OHH
  | | ‖           | | | |     catalyst
H-C-C-C-OH  +  H-C-C-C-C-H  ————————→
  | |           | | | |
  H H           H H H H
```

b.

```
                H H H O
                | | | ‖      catalyst
HSCoA  +     H-C-C-C-C-OH  ————————→
                | | |
                H H H
```

7. Write the products formed in the following amidation reaction.

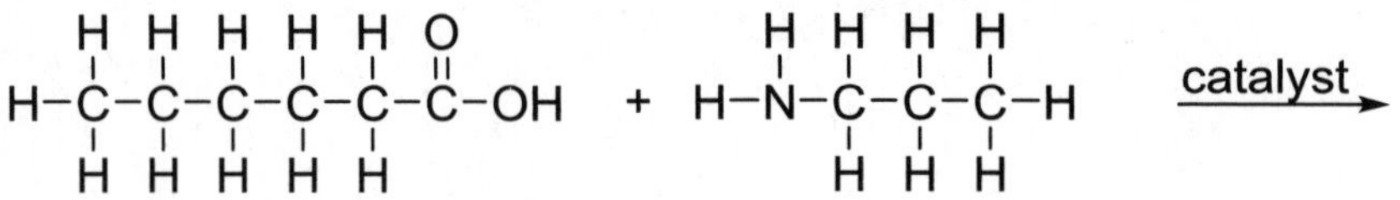

```
    H   H   H   H   H   O              H   H   H   H
    |   |   |   |   |   ||             |   |   |   |
H - C - C - C - C - C - C - OH  +  H - N - C - C - C - H      catalyst
    |   |   |   |   |                      |   |   |        ──────────►
    H   H   H   H   H                      H   H   H
```

8. Identify the following reactions as oxidation-reduction reactions, hydration or dehydration reactions, hydrolysis reactions, or group transfer-esterification or amidation.

a.

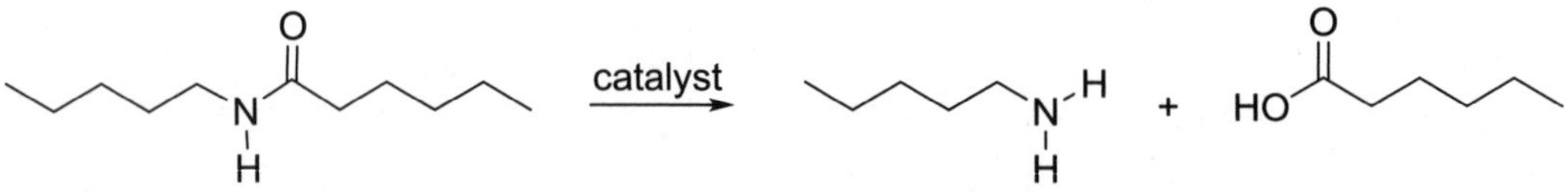

O N H —catalyst→ N H H + HO O

b.

O —catalyst→ OH

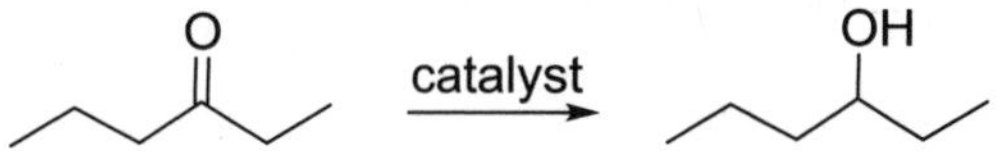

9. Identify the following reactions as oxidation-reduction reactions, hydration or dehydration reactions, hydrolysis reactions, or group transfer-esterification or amidation.

a.

O ——→ O

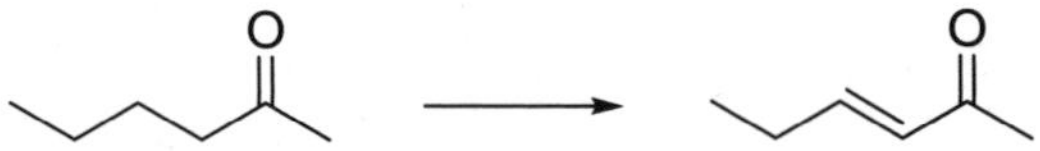

b.

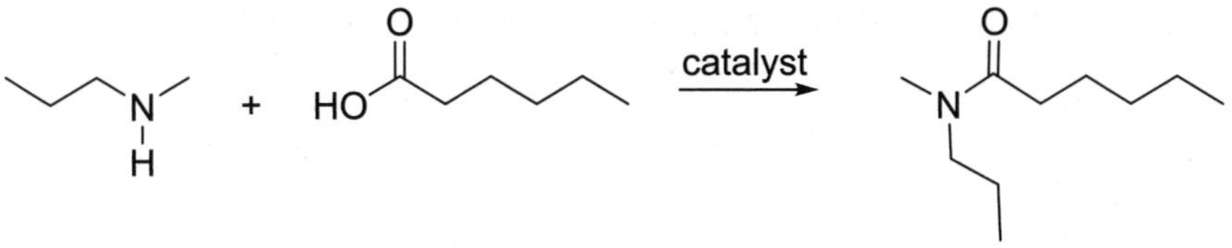

N
H
+ HO
O
catalyst
N
O

10. Identify the following reactions as oxidation-reduction reactions, hydration or dehydration reactions, hydrolysis reactions, or group transfer-esterification or amidation.

a. OH
+ HO
O
catalyst
O
O

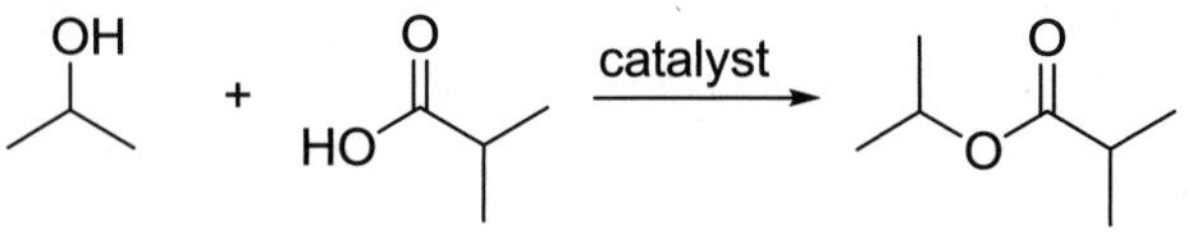

b.

H_2O +
O
OH O

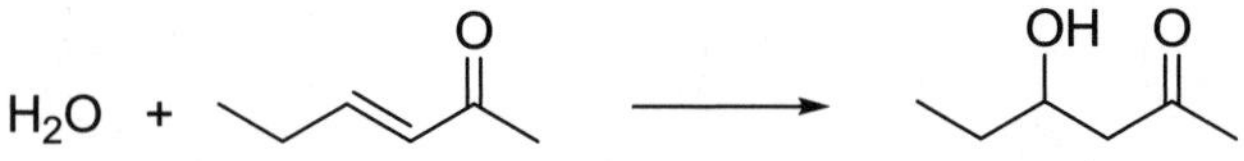

Chapter 10

Answers to Additional Exercises

25 Vitamins allow cells to perform the metabolic reactions necessary for living.

27 Riboflavin is essential to the metabolism of carbohydrates, proteins, and fats.

29 If a newborn has jaundice, the newborn will need light therapy to treat it. The light therapy degrades riboflavin.

31 Water soluble vitamins are readily eliminated through the urine and need to be replenished on a daily basis. Most natural foods are rich in vitamins; there are also supplements that contain vitamins.

33 The functional groups are the reactive sites in a molecule.

35 a. $R-\overset{O}{\overset{\|}{C}}-R$ b. $-\overset{O}{\overset{\|}{C}}-O-R$ c. $-\overset{O}{\overset{\|}{C}}-S-R$ d. $-\ddot{N}-$ (with a bond down from N)

37 A substrate is a particular reactant for a reaction catalyzed by an enzyme.

39 When an organic compound undergoes combustion, carbon dioxide and water are formed as products.

41 The reactant that undergoes loss of electrons in an oxidation-reduction reaction is said to undergo *oxidation*.

43 a. Oxidation

$$Zn: \longrightarrow Zn^{2+} + 2\,e^-$$

Reduction

$$2\,H^+ + 2\,e^- \longrightarrow H-H$$

Cl^- is a spectator ion.

b. Oxidation

$$Na\cdot \longrightarrow Na^+ + e^-$$

Reduction

$$:\ddot{\underset{..}{Cl}}-\ddot{\underset{..}{Cl}}: + 2\,e^- \longrightarrow 2\,:\ddot{\underset{..}{Cl}}:^-$$

45 An organic reactant that loses hydrogen atoms has undergone oxidation.

47 An organic reactant that loses oxygen atoms has undergone reduction.

49 a.

$$HO-\overset{O}{\overset{\|}{C}}-CH_2-CH_2-CH_3 \xrightarrow{[H]} HO-CH_2-CH_2-CH_2-CH_3$$

carboxylic acid → primary alcohol

The reaction represents a reduction because there is a decrease in the number of oxygen atoms in the product.

b.

```
   H  H  H  H  H
   |  |  |  |  |
H–C–C–C–C–C–H   +   8 O2   --[O]-->   5 CO2  +  6 H2O
   |  |  |  |  |
   H  H  H  H  H
```

The combustion reaction is an oxidation reaction because the carbon atoms have a decrease in the number of hydrogen atoms attached and an increase in the number of oxygen atoms attached.

c.

[H]

The reaction represents a reduction because there is an increase in the number of hydrogen atoms in the product.

d.

```
     H  H  H  H  H                  O  H  H  H  H  H
     |  |  |  |  |                  ‖  |  |  |  |  |
HO–C–C–C–C–C–H   --[O]-->   H–C–C–C–C–C–C–H
     |  |  |  |  |                     |  |  |  |  |
     H  H  H  H  H                     H  H  H  H  H
```

The reaction represents an oxidation because there is a decrease in the number of hydrogen atoms in the product.

51

HO, O $\xrightarrow[Pt]{H_2(g)}$ HO, O

This reaction is a reduction reaction. The product has an increase in the number of hydrogen atoms.

53 An unsaturated fat is more likely to be a liquid.

55 a.

```
   H  H  H  O                 H  H  H  O
   |  |  |  ‖                 |  |  |  ‖
H–C–C–C–C–H   --[O]-->   H–C–C–C–C–OH
   |  |  |                    |  |  |
   H  H  H                    H  H  H
```

b.

O, H $\xrightarrow{[O]}$ O, OH

c.

O, OH $\xrightarrow{[H]}$ O, H or OH, H

d. [H]; OH; H; or; OH; H

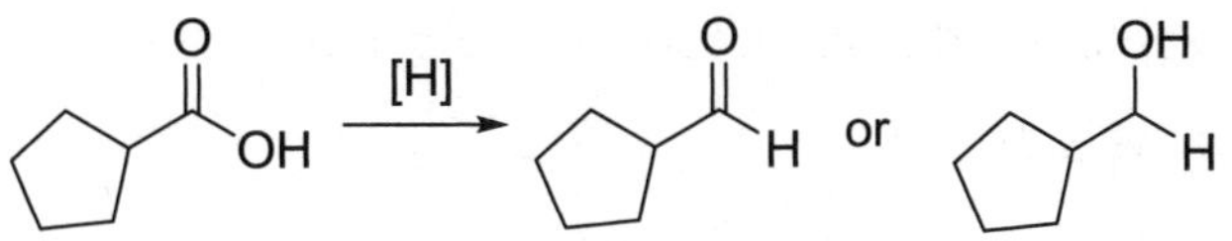

57 An enzyme is chemically unchanged in the reaction; a coenzyme does undergo a chemical change in the reaction.

59 a. $C_3H_8 + 5\ O_2 \rightarrow 3\ CO_2 + 4\ H_2O$

b. $H_3C{-}CH{=}CH{-}CH_3 + H_2 \xrightarrow{\text{catalyst}} H_3C{-}CH_2{-}CH_2{-}CH_3$

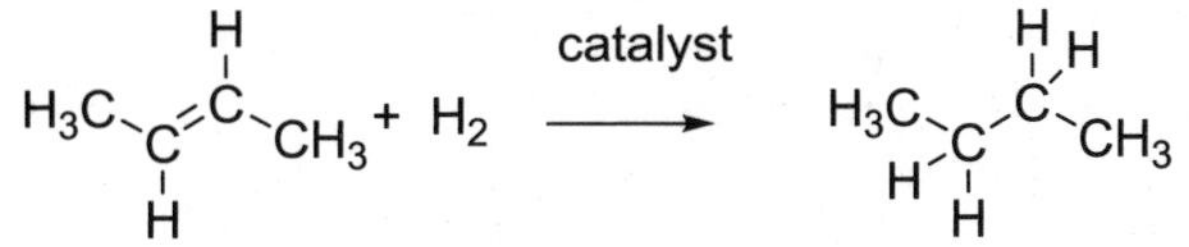

c. $CH_3{-}CH_2{-}CH(OH){-}CH_3 \xrightarrow{NAD^+ \rightarrow NADH + H^+} CH_3{-}CH_2{-}C(=O){-}CH_3$

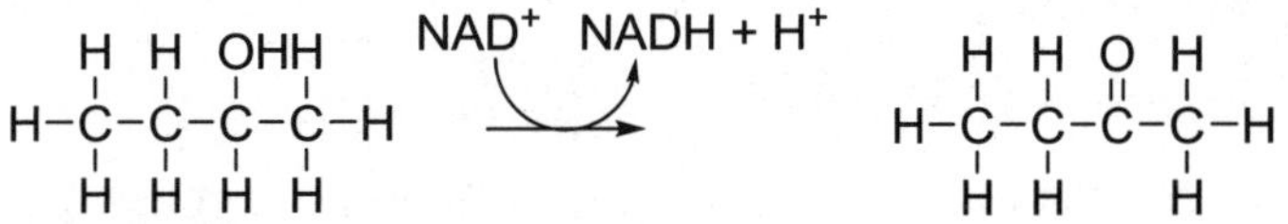

d. $FADH_2 + NAD^+ \rightarrow FAD + NADH + H^+ + 2\ e^-$

61 a. $H_3C{-}CH(CH_3){-}CH(OH){-}CH_3 \xrightarrow{[O]} H_3C{-}CH(CH_3){-}C(=O){-}CH_3$

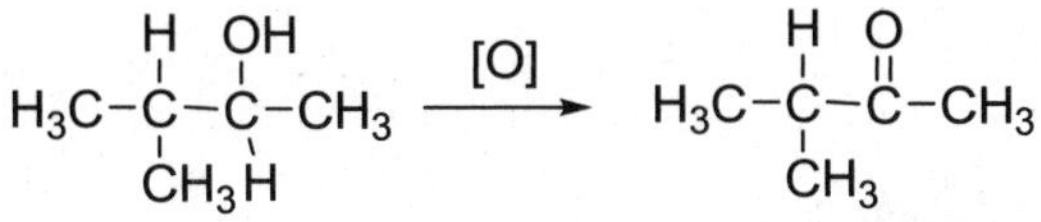

b. S−CoA [O] S−CoA

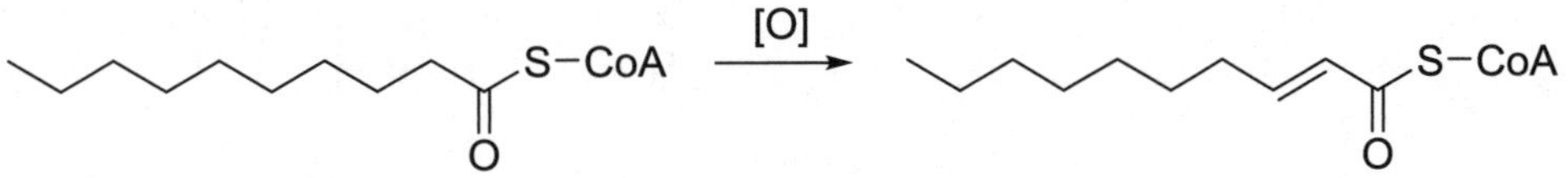

c. $FADH_2$

d. $C_6H_5{-}C(=O){-}H \xrightarrow{[O]} C_6H_5{-}C(=O){-}OH$

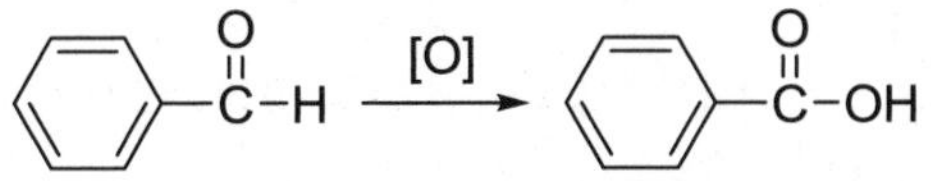

63 d. False. A free radical does not contain an octet of electrons.

65 *Transferase* enzymes are the major class of enzymes that catalyze acyl transfer reactions in biochemistry.

67 a. $H_3C{-}OH + (CH_3)_2CH{-}COOH \xrightarrow{\text{catalyst}} (CH_3)_2CH{-}C(=O){-}O{-}CH_3 + H_2O$

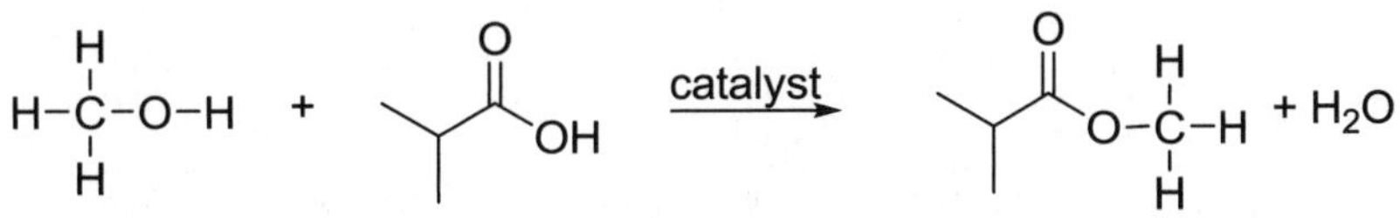

b. $(CH_3)_2CH{-}COOH + (CH_3)_2CH{-}OH \xrightarrow{\text{catalyst}} (CH_3)_2CH{-}C(=O){-}O{-}CH(CH_3)_2 + H_2O$

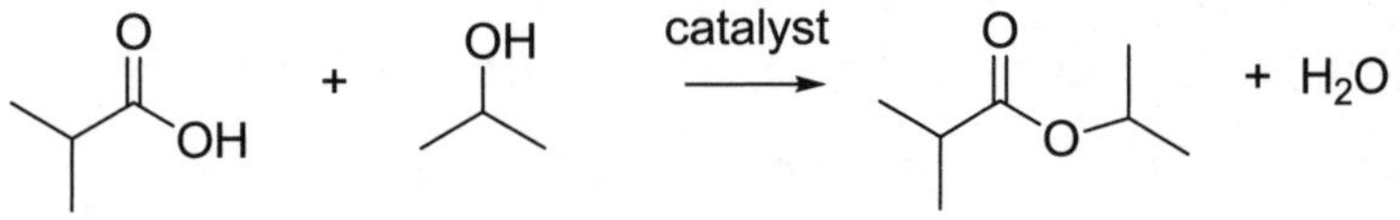

c.

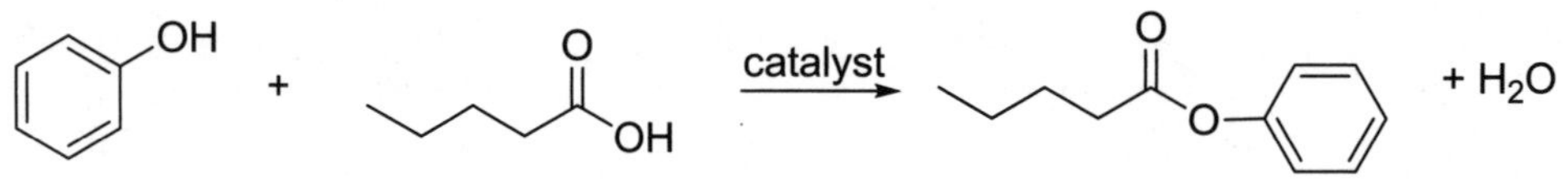

OH + O OH catalyst O O + H_2O

d.

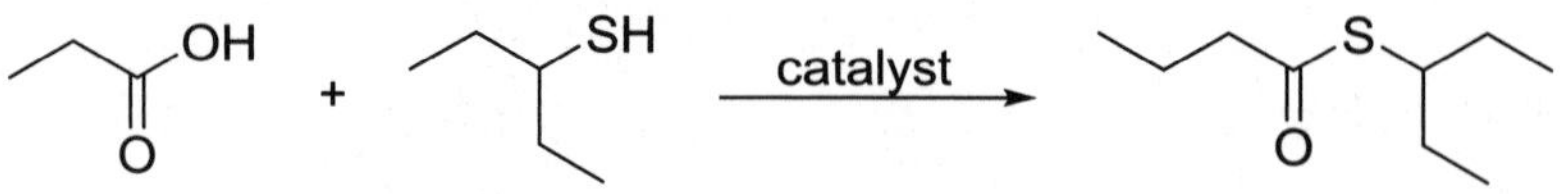

OH O + SH catalyst S O

69

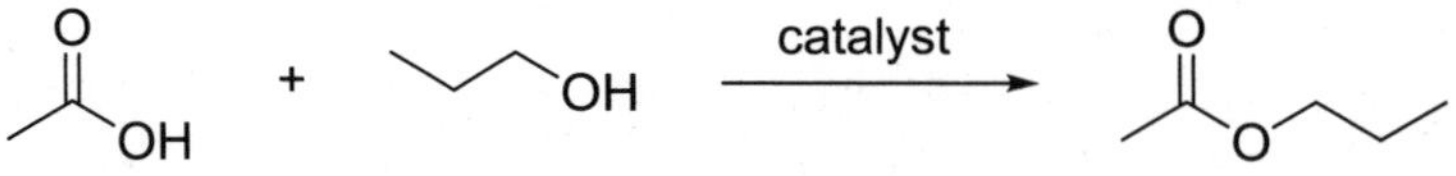

O OH + OH catalyst O O

acetic acid propan-1-ol

The ester smells like pear.

71 Three fatty acid molecules react with one glycerol molecule to produce a fat molecule. An esterification reaction produces the fat molecule.

73 a.

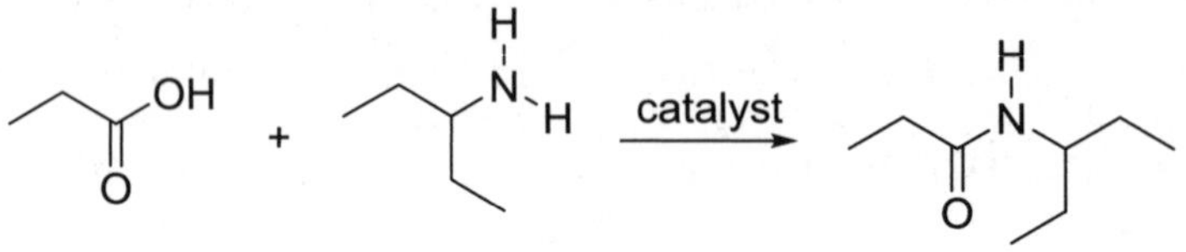

OH O + H N H catalyst H N O

b.

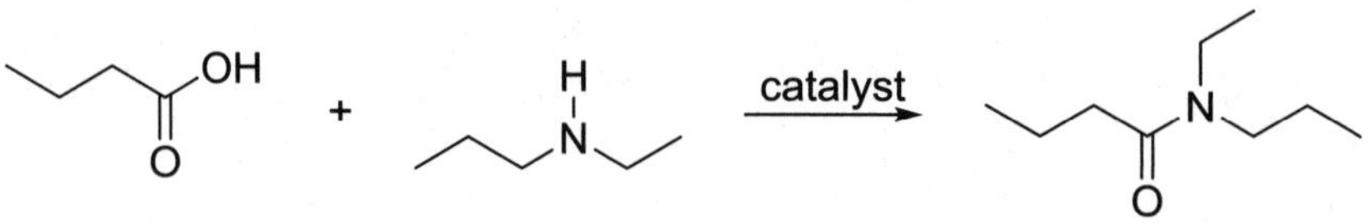

OH O + H N catalyst N O

c.

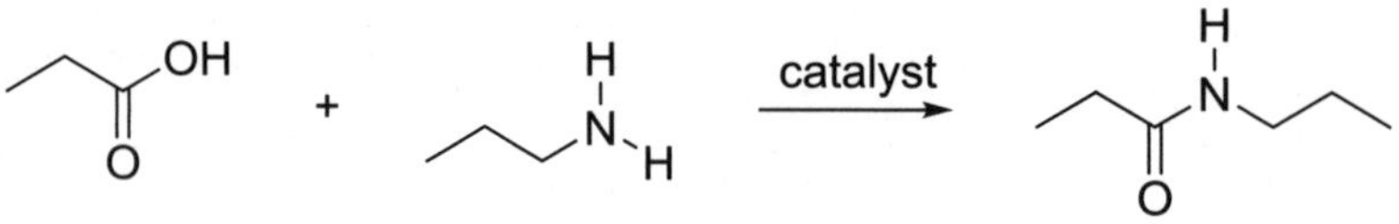

OH O + H N H catalyst H N O

75

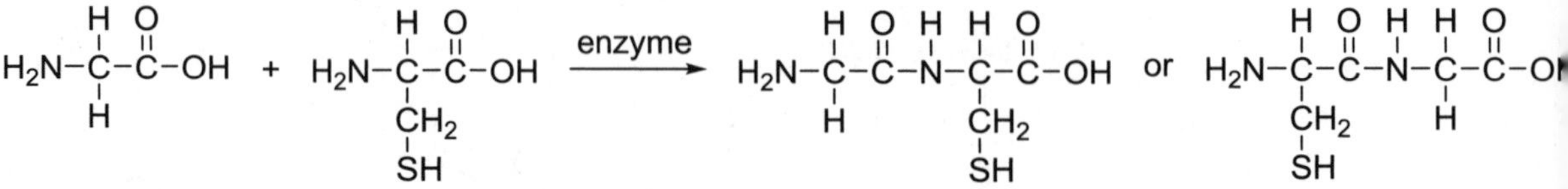

H_2N–CH$_2$–C(=O)–OH + H_2N–CH(CH$_2$SH)–C(=O)–OH —enzyme→ H_2N–CH$_2$–C(=O)–NH–CH(CH$_2$SH)–C(=O)–OH or H_2N–CH(CH$_2$SH)–C(=O)–NH–CH$_2$–C(=O)–OH

Glycine Cysteine

77 Two hydrolysis reactions that occur during digestion are the metabolism of proteins into amino acids and of fats into fatty acids and glycerol.

79 a.

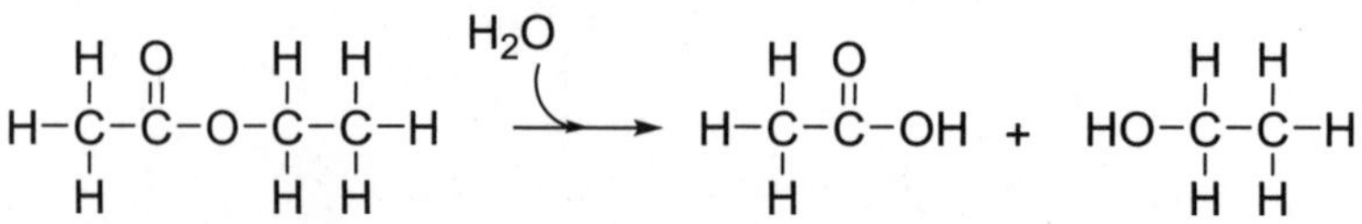

H–CH$_2$–C(=O)–O–CH$_2$–CH$_2$–H —H_2O→ H–CH$_2$–C(=O)–OH + HO–CH$_2$–CH$_2$–H

b.

```
    H CH3  O H H        H2O          H CH3             O H H
    |  |   ‖ | |         (           |  |              ‖ | |
  H-C--C-S-C-C-C-H      ─────►     H-C--C-SH  +  HO-C-C-C-H
    |  |     | |                     |  |              | |
    H  H     H H                     H  H              H H
```

c.

```
    H H    O H         H2O        H H H            _ O H
    | |    ‖ |          (         | | |  +           ‖ |
  H-C-C-N-C-C-H       ─────►    H-C-C-N-H  +      O-C-C-H
    | | |    |       catalyst     | | |              |
    H H H    H                    H H H              H
```

81 a. The acyl group being transferred contains six carbon atoms.

b. The bond between the carbon atom and the sulfur atom in the thioester is broken.

c. The acyl group is being transferred from a sulfur atom to an oxygen atom.

d. The reaction is classified as a hydrolysis reaction because water is a reactant in the reaction.

83 The acidic environment of the stomach and hydrolase enzymes aid the digestion of proteins. Hydrolysis of amide bonds occurs during the digestion of proteins, which produces individual amino acids.

85

```
   H H O H H   O H H O  _    pepsin      H H O  _       H H O  _          H H O  _
 + | | ‖ | |   ‖ | | ‖                 + | | ‖        + | | ‖           + | | ‖
 H-N-C-C-N-C---C-N-C-C-O    ───────►   H-N-C-C-O   +  H-N-C-C-O      +  H-N-C-C-O
   | |     |       |                     | |            | |               | |
   H H     HC-CH3  HC-OH                 H H            H HC-CH3          H HC-OH
           |       |                                      |                 |
           CH2     CH3                                    CH2               CH3
           |                                              |
           CH3                                            CH3
```

87 The digestion of fats occurs in the small intestine. *Lipase* enzymes catalyze the hydrolysis of fats.

89 Lyase enzymes catalyze hydration and dehydration reactions.

91 a. hydration

O H_2O → O, OH

b. dehydration

OH O → H_2O → O

c. hydration

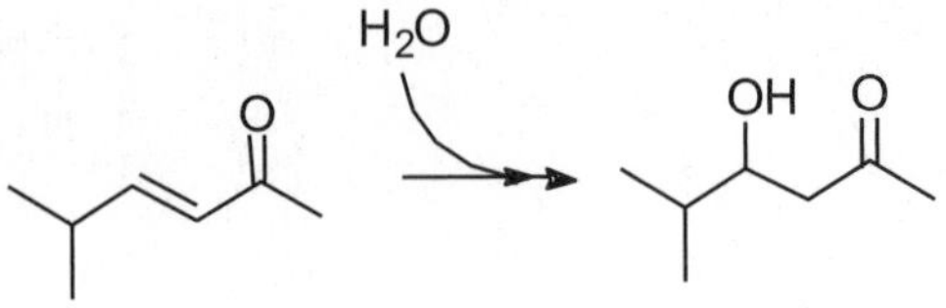

H_2O

OH O

93 a.

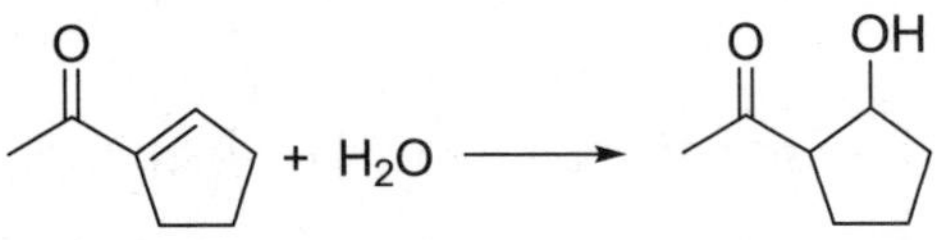

O OH

+ H_2O

b.

COO^- H_2O COO^-

OH

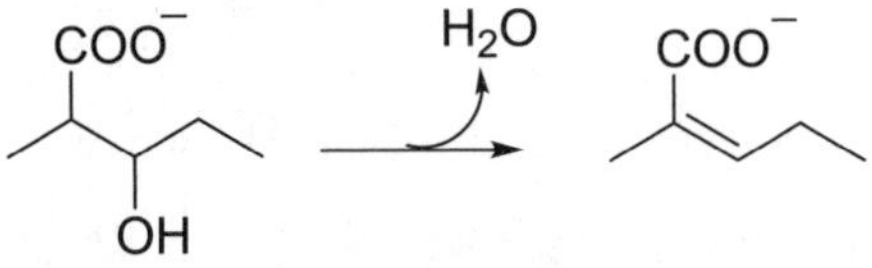

95 a. hydrolysis b. dehydration reaction

97 PKU is caused by a deficiency in the enzyme *phenylalanine hydroxylase (PAH).*

99 BH_4, tetrahydrobiopterin, is the coenzyme required by PAH.

101 The treatment for PKU is managed through a strict diet that eliminates foods that include phenylalanine, such as beef, chicken, fish, and dairy products. They must also take tyrosine supplements because they cannot synthesize this amino acid.

Chapter 11

Carbohydrates: Structure and Function

Chapter Summary

In this chapter, you learned about carbohydrates and how they provide fuel for the cell. You first studied the structure of simple carbohydrates and then learned how the simple carbohydrates form more complex carbohydrates. You observed the role that complex carbohydrates play in living things. Finally, you learned that carbohydrates serve an important function as cell markers.

11.1 An Overview of Carbohydrates and Their Function

Carbohydrates, also known as sugars, are the most abundant of the biomolecules. The primary function of carbohydrates in our diet is to supply our cells with energy over the short term, making possible for our cells to perform functions necessary for living. Half of the earth's carbon exists in the form of carbohydrates. Plants produce carbohydrates and oxygen from carbon dioxide and water during an anabolic process that requires energy, provided by sunlight. This process is called photosynthesis.

When we consume starch, a complex carbohydrate, it is metabolized into many glucose molecules, a process that begins in the mouth. Digestion continues in the small intestine; glucose then enters the blood stream. Given the signal from the hormone insulin, glucose is transported through the cell membrane and into the cell, where it can be metabolized further into carbon dioxide and water. This process, known as cellular respiration, produces energy for the cell. The energy from sunlight is stored in plants in the form of starch through photosynthesis. The starch in the plant is converted into energy by our cells through cellular respiration.

While the primary function for carbohydrates is as a fuel for cells, they also serve many other important functions. Carbohydrates provide the molecular building blocks for synthesis of more complex molecules such as RNA and DNA. Carbohydrates are also important in cellular recognition.

When the structure of carbohydrates were first determined, scientists noted that their composition fit the molecular formula $(CH_2O)_n$, where n is an integer 1, 2, 3,.... So it was first hypothesized that carbohydrates might be hydrates of carbon and they were given the name carbohydrates. The names of most carbohydrates end in *-ose*.

Carbohydrates are subdivided into four categories depending on the number of monosaccharide units produced when the carbohydrate is hydrolyzed. Monosaccharides are carbohydrates that cannot be hydrolyzed into simpler carbohydrates; hence, they are also known as simple sugars. Disaccharides are hydrolyzed into two monosaccharides. Oligosaccharides are hydrolyzed into three or more monosaccharides, and polysaccharides are hydrolyzed into many monosaccharides. Polysaccharides are biological polymers of monosaccharides. A polymer, in general, is a large molecule composed of the same or similar repeating smaller units—monomers. Monosaccharides are the monomer units of polysaccharides.

11.2 The Structure of the Monosaccharides

Monossacharides are defined as carbohydrates that cannot be hydrolyzed into simpler carbohydrates; thus, they are also known as simple sugars. Monosaccharides are crystalline, colorless solids with a sweet taste.

In a Nutshell: The Chemical Structure of Monosaccharides

There are three structural features that identify a monosaccharide: 1) the number of carbon atoms in the chain, 2) the type of carbonyl-containing functional group (aldehyde or ketone), and 3) the stereochemical configurations of all the centers of chirality. The common monosaccharides are composed of a straight chain of three to six carbon atoms, with one carbonyl group that is either an aldehyde at C(1) or a ketone at C(2), and a hydroxyl group at all other carbon atoms.

Most monosaccharides are chiral because they contain one or more centers of chirality. A center of chirality is a tetrahedral carbon atom with four different atoms or groups of atoms bonded to it. In a monosaccharide, the centers of chirality each have one hydroxyl group (—OH) and one hydrogen atom (—H). It is common practice to draw monosaccharides as Fischer projections. In a Fischer projection, the carbon chain of the monosaccharide is written vertically (from top to bottom) with the carbonyl group at the top. Each hydroxyl group and hydrogen atom on a center of chirality appears as a horizontal bond, either to the left or to the right, intersecting the vertical carbon chain. In a Fischer projection, a center of chirality occurs whenever a horizontal line intersects a vertical line. In the Fischer projection of a monosaccharide, the two possible configurations at centers of chirality are OH group left/H atom right or vice versa. The last carbon atom in the chain is not a center of chirality because it contains two identical atoms-two hydrogen atoms. Although a Fischer projection is drawn as a flat structure, it is understood that the carbon atoms with four bonds are tetrahedral in shape, the horizontal bonds are understood to be projecting toward the viewer (wedges) and the vertical bonds are understood to be projecting away from the viewer (dashes).

Nonsuperimposable mirror image stereoisomers are known as stereoisomers and the centers of chirality have the opposite configuration. Most monosaccharides have multiple centers of chirality. The maximum number of stereoisomers can be calculated using the formula 2^n, where n is the number of centers of chirality. Diastereomers have a different configuration at one or more centers of chirality, but not at every center of chirality. Enantiomers have a different configuration at every center of chirality. Monosaccharides that are enantiomers are identified by the prefixes D- and L- or (+) or (−), while diastereomers are given different names altogether.

Since nature only produces D-sugars, it is useful to know how to look at a Fischer projection and determine whether or not it is a D-sugar. A D-sugar has the —OH group on the center of chirality farthest from the carbonyl group on the right. Conversely, in an L-sugar the OH group on the center of chirality farthest from the carbonyl group is on the left.

There is an analogous series of D- and L-sugars containing a ketone carbonyl group. The most common ketone containing monosaccharide is D-fructose. The carbonyl carbon is not a center of chirality because it is not tetrahedral.

Worked Example #1

Fischer projections of three different monosaccharides are shown below:

Gulose:
CHO
HO—H
HO—H
H—OH
HO—H
CH_2OH

Threose:
CHO
HO—H
H—OH
CH_2OH

Arabinose:
CHO
HO—H
H—OH
H—OH
CH_2OH

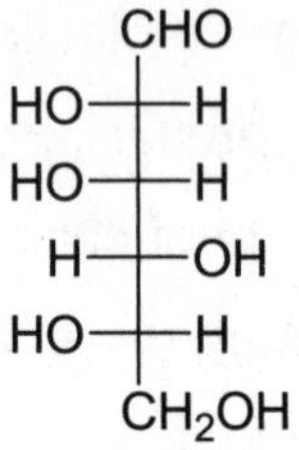

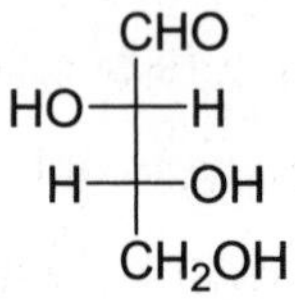

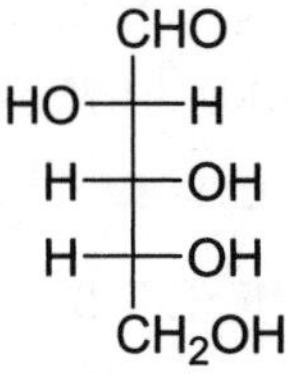

Gulose Threose Arabinose

a. Are these structures D-sugars or L-sugars? How can you tell?
b. Write the structure of the enantiomer of gulose shown. What is the name of the enantiomer?

a. L-Gulose—the OH group on the crosshair farthest from the CHO group—points to the left. D-Threose—the OH group on the crosshair farthest from the CHO group—points to the right. D-Arabinose—the OH group on the crosshair farthest from the CHO group—points to the right.

b.

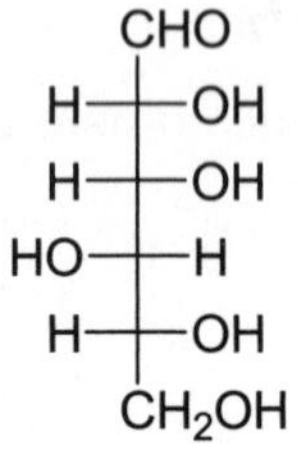

CHO
H—OH
H—OH
HO—H
H—OH
CH_2OH

D-Gulose

Try It Yourself #1

Fischer projections of three different monosaccharides are shown below:

Allose:
CHO
H—OH
H—OH
H—OH
H—OH
CH_2OH

Xylose:
CHO
H—OH
HO—H
H—OH
CH_2OH

Erythrose:
CHO
H—OH
H—OH
CH_2OH

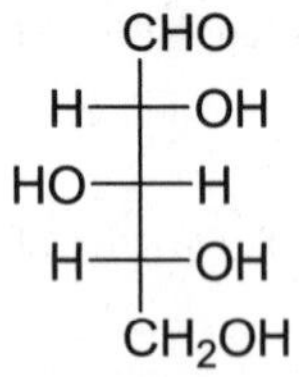

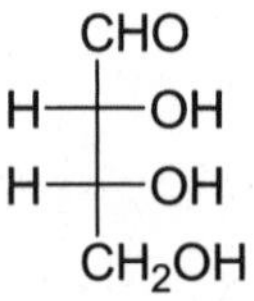

Allose Xylose Erythrose

a. Are these structures D-sugars or L-sugars? How can you tell?
b. Write the structure of the enantiomer of erythrose shown. What is the name of the enantiomer?

Solutions

a. *Allose—the OH group on the crosshair farthest from the aldehyde—points to:*

_____________________. It is a(n) _______-sugar.

Xylose—the OH group on the crosshair farthest from the aldehyde—points to:

_____________________. It is a(n) _______-sugar.

Erythrose—the OH group on the crosshair farthest from the aldehyde—points to:

_____________________. It is a(n) _______-sugar.

b. CHO

H—OH

H—OH

CH_2OH

Erythrose

Erythrose enantiomer

The name of the enantiomer is: __________________________________.

Worked Example #2

The structures of D-idose, L-idose, and D-altrose are shown below:

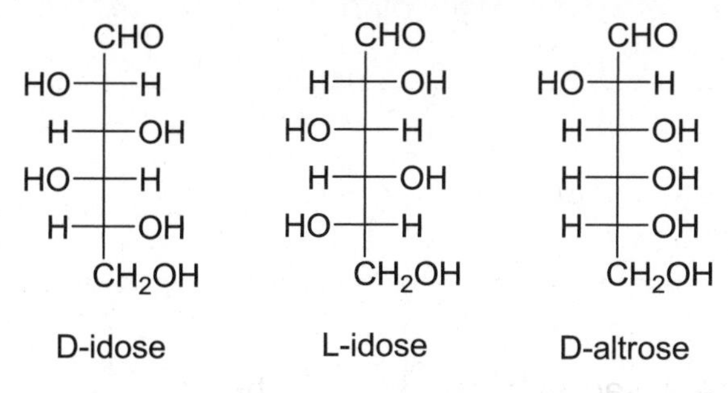

D-idose L-idose D-altrose

a. Are D-idose and L-idose enantiomers or diastereomers?

b. Are D-idose and D-altose enantiomers or diastereomers?

a. *D-idose and L-idose are nonsuperimposable mirror images of each other; therefore, they are enantiomers.*

b. *D-idose and D-altose are diastereomers. Only one center of chirality is different between the two structures.*

Try It Yourself #2

The structure of D-ribose and D-lyxose are shown below.

D-ribose	D-lyxose
CHO	CHO
H–C–OH	HO–C–H
H–C–OH	HO–C–H
H–C–OH	H–C–OH
CH_2OH	CH_2OH

D-ribose D-lyxose

a. Write the structure for L-lyxose.

b. Are D-lyxose and L-lyxose enantiomers or diastereomers?

c. Are D-ribose and D-lyxose enantiomers or diastereomers?

a.

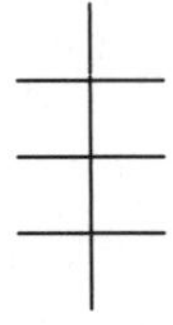

L-lyxose

b. Are D-lyxose and L-lyxose mirror images of each other? ____________

They are: ____.

c. Are D-ribose and D-lyxose mirror images of each other? ____________

They are: ____.

In a Nutshell: Ring Forms of Monosaccharides

Monosaccharides with five or more carbons spontaneously undergo a ring forming reaction in aqueous solution. A new carbon-oxygen single bond forms between the oxygen atom on the last center of chirality and the carbonyl carbon. This new bond creates a five- or six-membered ring with a new hydroxyl group on what was formerly the carbonyl carbon. A six-membered ring containing five carbon atoms and one oxygen atom is called a pyranose ring. The other common ring is a five-membered ring containing four carbon atoms and one oxygen atom is called a furanose ring.

In forming a pyranose or furanose ring, a new center of chirality is created at what was formerly the carbonyl carbon. The new center of chirality, only present in the ring form is known as the anomeric center. The ring from with the OH group at the anomeric center below the ring is known as the α-anomer. The ring form with the OH group at the anomeric center above the ring is known as the β-anomer.

The two ring forms of a monosaccharide are rapidly interconverting via the open chain form of the monosaccharide, creating an equilibrium mixture of all three forms: the open-chain form, the α-anomer, and the β-anomer. Equilibrium favors both ring forms over the open-chain form. The open-chain form is usually depicted as a Fischer projection, and the ring forms are depicted as Haworth projection.

Haworth projections are generally reserved for carbohydrates where stereochemical configuration is the distinguishing feature. Fischer projections are readily translated into Haworth structures. Convert a Fischer projection into a Haworth structure by turning the Fischer projection 90° in the clockwise direction: Groups that appear on the right side in a Fischer projection appear below the ring in a Haworth projection, and groups that appear on the left side in a Fischer projection appear above the ring in a Haworth projection. The following conventions are used when writing Haworth structures: 1) Always write a furanose ring with the ring oxygen at the top, and pyranose ring with the ring oxygen at the top right. 2) The center of chirality farthest form the carbonyl is the carbon atom with a bond to the ring oxygen and to the CH_2OH group, which always appears above the ring and at the top left in a D-sugar. 3) The ring is written so that the anomeric center is positioned at the far right with the —OH group above or below the ring depending on whether it is the β- or α-anomer, respectively.

Worked Example #3

Two forms of mannose are shown below.

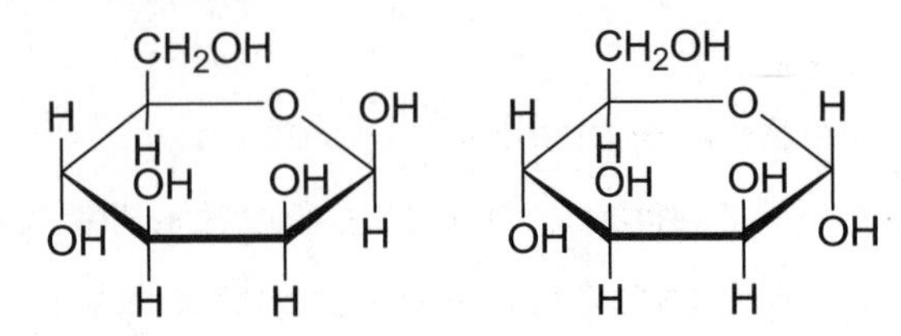

a. Are these sugars D-sugars or L-sugars? How can you tell?

b. Name each from of mannose: ___-___ mannose and ____-____ mannose.

c. Do the Haworth projections show these monosaccharides to be pyranoses or furanoses? Explain.

d. Are these two forms of mannose enantiomers or diastereomers?

a. *These sugars are D-sugars because the C6 CH_2OH group lies above the ring.*

b. *In the structure on the left the OH group attached to the anomeric center is above the ring; therefore, it is β-D-mannose. In the structure on the right the OH group attached to the anomeric center lies below the ring; therefore, it is α-D-mannose.*

c. *These sugars are pyranoses. They are represented by a six-membered ring with an oxygen atom in it.*

d. *These two forms of mannose are diastereomers. Only the positions of the H and OH groups on the anomeric have changed.*

Try It Yourself #3

Two forms of allose are shown below.

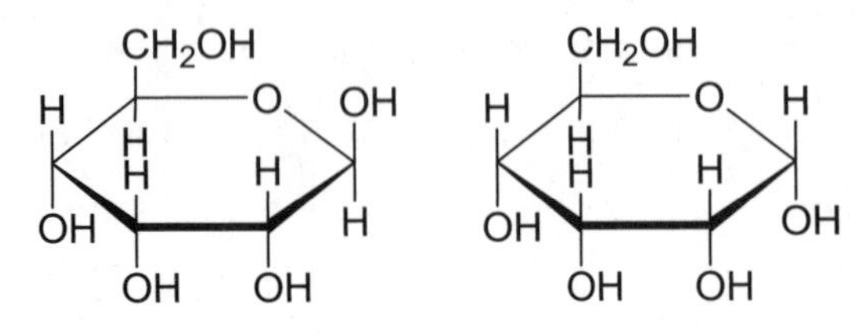

a. Are these sugars D-sugars or L-sugars? How can you tell?

b. Name each form of allose: ___-___ allose and ____-____ allose.

c. Do the Haworth projections show these monosaccharides to be pyranoses or furanoses? Explain.

d. Are these two forms of mannose enantiomers or diastereomers?

Solutions

a. *The position of the group on C6 is ________ the ring.*

These sugars are: ____________________.

b. *In the sugar on the left, the ____ on the anomeric center lies __________ the ring. The sugar on the left is ___-___ allose. In the sugar on the right, the ____ on the anomeric center lies __________ the ring. The sugar on the right is ___-___ allose.*

c. *The structures shown are ____-membered rings with an oxygen atom. They are: _________________.*

d. *Are the structures mirror images? __________*

They are: ______________________________.

Practice Problems for the Structure of The Common Monosaccharides

1. The Fischer projections for three monosaccharides are shown below.

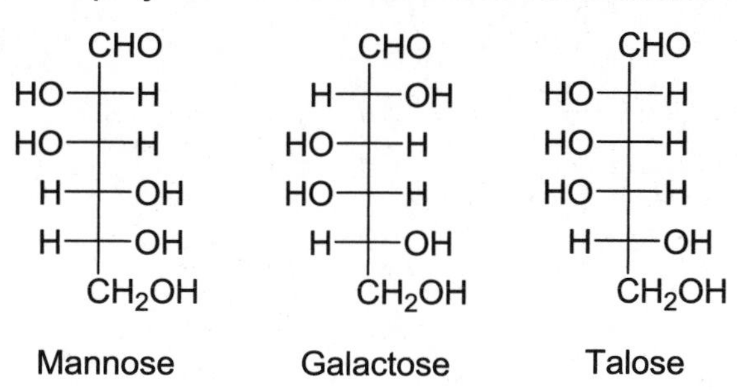

a. Are these sugars D-sugars or L-sugars? How can you tell?

b. Write the structure for L-mannose.

c. Are L-mannose and D-mannose enantiomers or diastereomers?

d. Are D-galatose and D-talose enantiomers or diastereomers?

2. The Haworth projection of ribose is shown below.

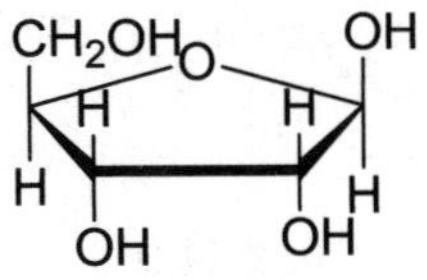

ribose

a. Does the Haworth projection show this sugar to be a pyranose or a furanose?

b. Is the structure shown a D- or L-sugar? Explain.

c. Which anomer of ribose is shown?

3. The Haworth projection of altrose is shown below.

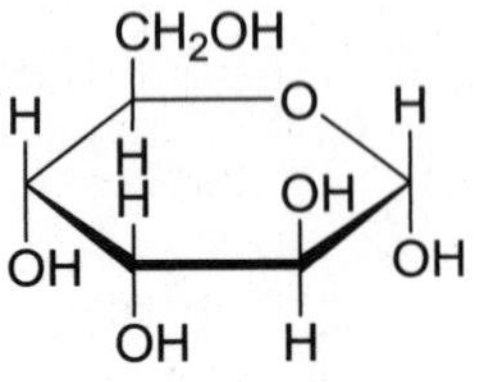

altrose

a. Does the Haworth projection show this sugar to be a pyranose or a furanose?

b. Is the structure shown a D- or L-sugar? Explain.

c. Which anomer of altrose is shown?

11.3 The Structure of Complex Carbohydrates

In a Nutshell: Disaccharides

Disaccharides are defined as carbohydrates that upon hydrolysis (reaction with water) yield two monosaccharides: either identical or different monosaccharides. The two monosaccharide components of a disaccharide are joined by a covalent bond between the anomeric center of one monosaccharide and the oxygen atoms of one of the hydroxyl groups on the other monosaccharide. Three structural features identify the disaccharide: 1) the identity of the two monosaccharide components; 2) the configuration of the anomeric center(s), α- or β-, and 3) which hydroxyl group on the second monosaccharide is involved in the covalent bond joining the two monosaccharides. The C—O bond joining the two monosaccharide components is known as the glycosidic bond or glycosidic linkage.

The glycosidic bond is always formed between the anomeric center of one monosaccharide and one of the hydroxyl groups on the second monosaccharide. Thus, instead of an —OH group at the anomeric center, a disaccharide has an —OR group, at the anomeric center of the first monosaccharide, where R is the second monosaccharide. The anomeric center at a glycosidic linkage is a stable functional group, so it does not open and close in aqueous solution, and therefore cannot change configuration form the α-anomer to the β-anomer. A catalyst or an enzyme is required to hydrolyze a glycosidic bond. The enzymes that facilitate the hydrolysis of a glycosidic linkage are known as *glycosidases*. The most common glycosidic linkages are α(1→4) and β(1→4).

Worked Example #4

Mannobiose is a disaccharide composed of only mannose in an α(1→4) linkage.

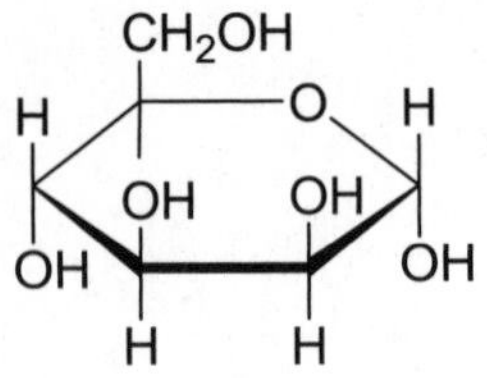

CH_2OH

mannose

a. Write the structure of mannobiose.

b. Label the anomeric carbons in your structure.

c. What monosaccharide(s) are produced upon hydrolysis of mannobiose?

a. and b.

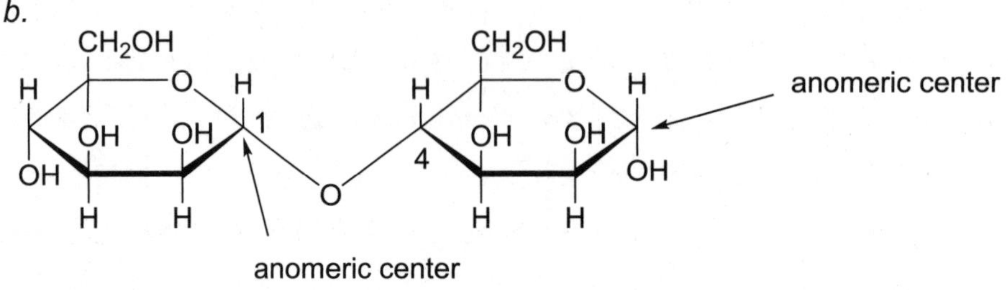

anomeric center

anomeric center

c. *The monosaccharide mannose is produced upon hydrolysis of mannobiose.*

Try It Yourself #4

The structure of isomaltose is shown below.

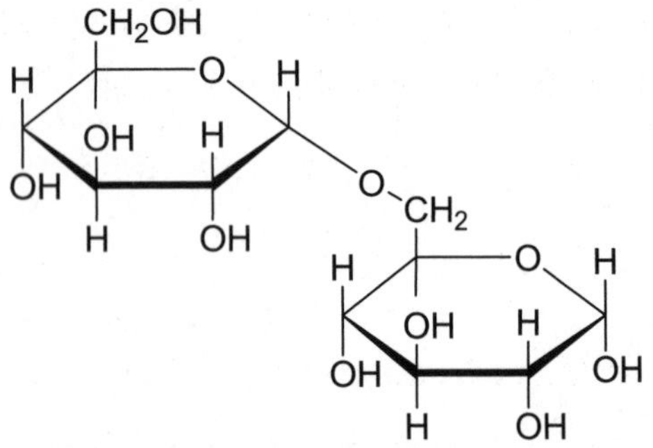

a. What are the monosaccharide components of isomaltose?

b. What type of linkage exists between the monosaccharides in isomaltose?
c. Why is isomaltose classified as a disaccharide?

a. *The monosaccharide components are:* ________________________.

b. *The linkage is:* ____.

c. *Isomaltose is a dissacharide because:* ________________________.

In a Nutshell: Polysaccharides

Polysaccharides are biological polymers, where a polymer is any large molecule with the same or similar repeating structural units. The repeating unit is known as the monomer. The most abundant polysaccharides in plants or animals are polymers of the monomer D-glucose, which differ in whether their glycosidic linkages are $\alpha(1\rightarrow4)$ or $\beta(1\rightarrow4)$. The important polysaccharides include: amylopectin, cellulose, and glycogen.

Plants synthesize starch and cellulose from glucose. Plants store energy in the form of starch, a mixture of 20% amylose and 80% amylopectin. When we consume starch, we hydrolyze it into glucose during digestion. Glucose is metabolized into carbon dioxide, water, and energy. In the presence of excess blood sugar, glucose is converted into glycogen, the polymeric storage form of glucose in mammals. Glycogen, amylose, and amylopectin are all polysaccharides composed of glucose monomers connected by $\alpha(1\rightarrow4)$ glycosidic linkages.

Amylose, the minor constituent of starch, is an unbranched polysaccharide; each glucose monomer is connected like a chain to the next monomer in the same type of linkage: $\alpha(1\rightarrow4)$. The overall molecule adopts helical shape, due to extensive hydrogen bonding between the hydroxyl groups.

Amylopectin, the major component of starch, is similar to amylase except that it is a branched polymer. The branches are created by $\alpha(1\rightarrow6)$ glycosidic bonds every 25–30 glucose units. The overall shape of amylopectin is spherical.

In humans and animals, glucose is stored as glycogen in liver and muscle cells. Glycogen serves as energy storage in animals. Glycogen is similar to amylopectin, except that it is even more highly branched: $\alpha(1\rightarrow6)$ glycosidic linkages approximately every 10 glucose units.

Cellulose, the main component of wood, paper, and cotton, is another polysaccharide composed of glucose monomers. Cellulose contains unbranched $\beta(1\rightarrow4)$ glycosidic linkages. These distinctive linkages give cellulose a different overall shape. Cellulose has a flat, sheet-like appearance, which provides structural rigidity for the plant.

In the laboratory, cellulose, starch, and glycogen can all be hydrolyzed into D-glucose in the presence of an acid. In the body, starch taken in through the diet is hydrolyzed in the mouth and small intestine by enzymes that specifically hydrolyze $\alpha(1\rightarrow4)$ linkages. These enzymes are known as *α-glycosidases*. Since humans do not have enzymes that hydrolyze $\beta(1\rightarrow4)$ linkages, cellulose cannot be digested. Some animals, however, can digest cellulose because bacteria that live in their digestive tracts produce *β-glycosidases.* These animals include horses, cows, giraffes, and other ruminants.

Worked Example #5

For each of the statements below, indicate the polysaccharides for which the statement is true. (More than one selection may be chosen.)

amylose amylopectin glycogen cellulose

a. ____ Provides structure in plants
b. ____ Is hydrolyzed by the enzyme *β-glycosidase*
c. ____ is helical in shape
d. ____ Is branched in appearance
e. ____ Has $\alpha(1\rightarrow4)$ linkages

Solutions

a. cellulose
b. cellulose
c. amylose

d. amylopectin and glycogen

e. amylose, amylopectin, and glycogen

Try It Yourself #5

For each of the statements below, indicate the polysaccharides for which the statement is true. (More than one selection may be chosen.)

amylose amylopectin glycogen cellulose

a. ____ Serves as a glucose storage molecule in animals and humans

b. ____ Contains $\alpha(1\rightarrow6)$ linkages

c. ____ Has a layered sheet like appearance

d. ____ Can be digested by ruminants (horses, cows, and giraffes)

Solutions

a. _______________

b. _______________

c. _______________

d. _______________

Practice Problems for The Structure of Complex Carbohydrates

1. Nigerose is a disaccharide that is a product of the hydrolysis of the polysaccharides in black mold. It contains two glucose monosaccharides joined in an $\alpha(1\rightarrow3)$ linkage. Write the structure of nigerose.

2. The structure of turanose is shown below.

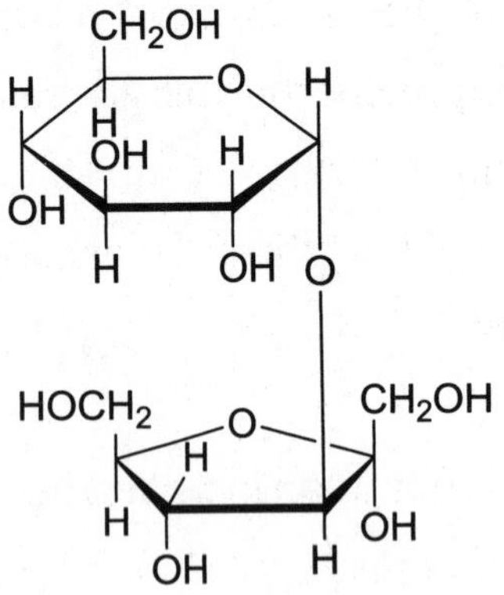

   a. Is turanose a disaccharide or polysaccharide?

   b. How many monosaccharides are produced when turanose is hydrolyzed? Are they the same or different sugars?

   c. What type of linkage connects the monosaccharides in turanose?

   d. Label the anomeric centers.

3. Why does cellulose have a layered, sheet-like appearance? Why does amylose have a helical shape?

11.4 Oligosaccharides as Cell Markers

Cells in our bodies contain chemical markers on their surface that allow them to host cells from foreign cells. Most cell markers are oligosaccharides, carbohydrates composed of a few monosaccharides that identify a cell. Covalently bonded to proteins and lipids in the cell membrane, oligosaccharide markers protrude out from the cell membrane into the extracellular fluid.

One example of cell markers is seen in the four basic blood types: A, B, AB, and O, which differ in the oligosaccharide marker on red blood cells. Type O blood cells have a trisaccharide (three monosaccharides) marker. Type A blood cells have a tetrasaccharide (four monosaccharides) marker containing the same three monosaccharide components as type O, but with an additional monosaccharide not present in any of the other blood types. Similarily, type B blood cells have a tetrasaccharide marker different from type A, but also containing the same three monosaccharide components as type O. Type AB blood cells contains a combination of both type A and type B markers.

People with type O blood are referred to as universal donors because they can donate blood to recipients with any of the other blood types. The chemical structure of the oligosaccharide marker on type O blood is common to all the other blood types, so it is recognized as self by people with any other blood type. However, people with type O blood will reject all other blood types except O because the fourth monosaccharide is not present in their own blood cells, so their immune system will recognize them as foreign. Conversely, individuals with type AB blood can receive type A, B, or O blood, but cannot donate their blood to anyone except others with type AB blood. People with type AB blood are referred to as universal recipients.

Worked Example #6

Can people with type A blood accept a blood donation from someone with type AB blood? Can people with type AB blood receive a blood donation from someone with type A blood? Explain.

Solution

People with type A blood cannot accept blood from someone with type AB blood. The bodies of people with type A blood would not recognize the type B oligosaccharide that is present in type AB blood. People with AB blood may receive a blood donation from someone with type A blood because the type A oligosaccharide is one of the oligosaccharides that makes up type AB blood. The body of the person with type AB blood would recognize the type A oligosaccharide.

Try It Yourself #6

How is type O blood different than type A blood? Can people with type O blood accept a blood donation from someone with type A blood? Explain.

Solution:

Practice Problems for Oligosaccharides as Cell Markers

1. How many monosaccharides are common to all four blood types?

2. Can someone with type B blood donate blood to someone with type O blood? Explain.

3. Can someone with type O blood donate blood to someone with type AB blood? Explain.

Chapter 11 Quiz

1. The Fischer projection for allose is shown below.

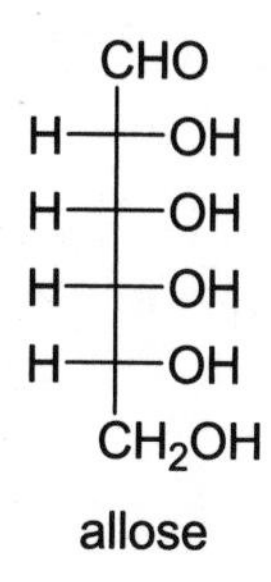

allose

   a. Is this sugar a D- or *l*-sugar? How can you tell?

   b. Draw the enantiomer of this sugar.

   c. What is the name of the enantiomer of this sugar?

2. The Fischer projections for three sugars are shown below:

L-gulose	D-mannose	D-gulose
CHO	CHO	CHO
HO—H	HO—H	H—OH
HO—H	HO—H	H—OH
H—OH	H—OH	HO—H
HO—H	H—OH	H—OH
CH_2OH	CH_2OH	CH_2OH

   a. Which sugars are enantiomers?

   b. Which sugars are diastereomers?

3. The Haworth projection for a sugar is shown below:

   a. Is the sugar a D- or L- sugar? How can you tell?

   b. Label the anomeric center.

   c. Is the α-anomer or β-anomer shown?

   d. Is the sugar shown a furanose or pyranose?

4. Trehalose, shown below, is found in fungi. It is formed between glucose monomers.

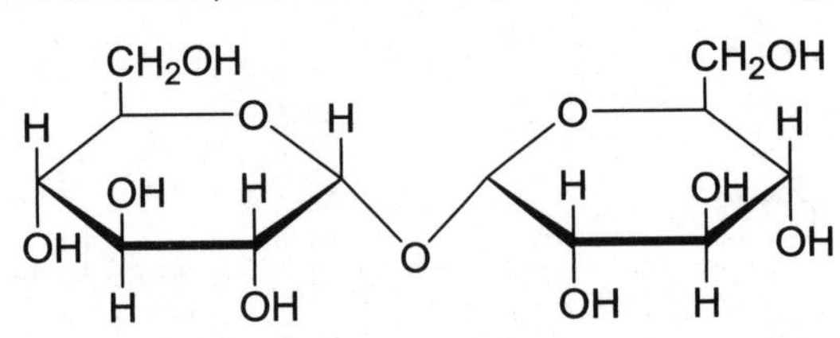

   a. Is trehalose a mono-, di-, or polysaccharide? Explain.

   b. Label all the anomeric centers in trehalose.

   c. Label and identify the glycosidic bond in trehalose.

5. Cellobiose, shown below, is formed between glucose monomers.

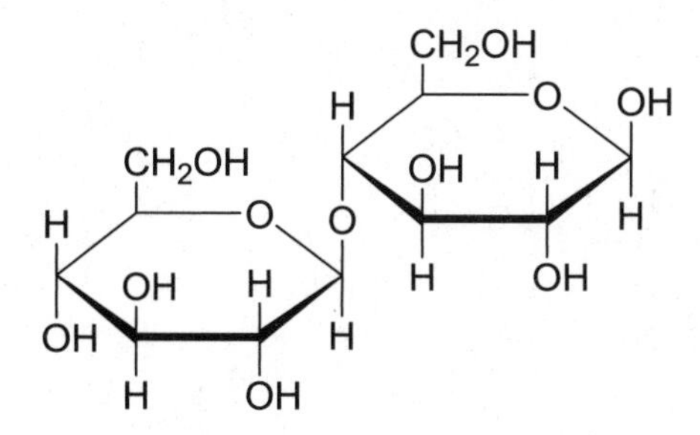

   a. Is cellobiose a mono-, di-, or polysaccharide? Explain.

   b. Label all the anomeric centers in cellobiose.

   c. Label and identify the glycosidic bond in cellobiose.

   d. What type of enzyme catalyzes the hydrolysis of cellobiose?

6. Amylose and cellulose are polymers that both have (1→4) glycosidic linkages between glucose monomers, yet amylose is helical in shape and cellulose has a flat, sheet-like appearance. Explain this difference in shape.

7. For each of the statements below, indicate the polysaccharides for which the statement is true. (More than one selection may be chosen.)

   amylose amylopectin glycogen cellulose

   a. ____ Is hydrolyzed by the enzyme *β-glycosidase*

   b. ____ Is hydrolyzed by an *α-glycosidase* enzyme

8. Gentianose, shown below, is found in the plant gentian.

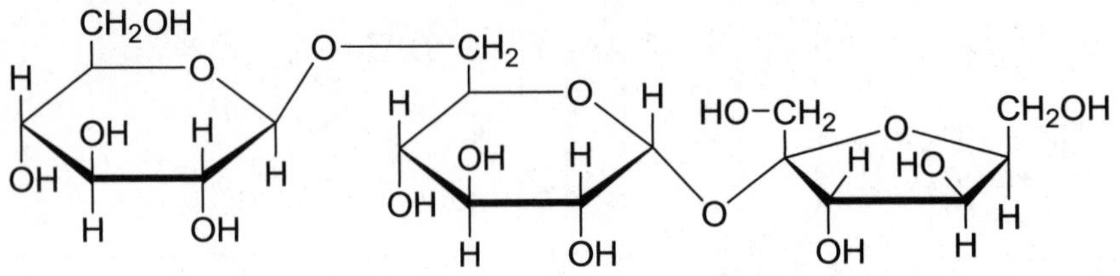

   a. Is gentianose a mono-, di-, or polysaccharide? Explain.

   b. Label all the anomeric centers in gentianose.

   c. Label and identify the glycosidic bonds in gentianose.

   d. How many monosaccharides are produced from the complete hydrolysis of gentianose into its monomer units?

   e. What type of enzyme catalyzes the hydrolysis of gentianose?

9. How do our cells distinguish host cells from foreign cells?

10. Can someone with type AB blood donate blood to someone with type A blood? Explain.

Chapter 11

Answers to Additional Exercises

21 Anaerobic bacteria present in the mouth metabolize glucose into lactic acid.

23 The metabolism of glucose into lactic acid by anaerobic bacteria in the mouth is the root cause of dental caries. Lactic acid dissolves tooth enamel.

25 Lactic acid is acidic and dissolves tooth enamel.

27 The four important biomolecules that provide the foundation for biochemistry are carbohydrates, lipids, proteins, and nucleic acids.

29 When we eat carbohydrates, the potential energy stored in the covalent bonds of the carbohydrates is converted into energy when they are metabolized.

31 Carbohydrates are metabolized into glucose in the mouth and in the small intestine.

33 Once glucose enters a cell, the cell uses cellular respiration to metabolize glucose into carbon dioxide, water, and energy.

35 Carbohydrates provide molecular building blocks for RNA, and DNA. They are important in cellular recognition.

37 When the structure of carbohydrates were first determined, scientists noted that their composition fit the molecular formula $(CH_2O)_n$. They hypothesized that carbohydrates might be hydrates of carbon.

39 Monosaccharides cannot be hydrolyzed into simpler carbohydrates. Disaccharides can be hydrolyzed into two monosaccharides. Polysaccharides are biological polymers of monosaccharides.

41 Monosaccharides are called simple sugars because they cannot be hydrolyzed into simpler carbohydrates.

43 Fructose is sweeter than sucrose.

45 Most monosaccharides are chiral. They contain one or more centers of chirality.

47 The common monosaccharides contain three to six carbon atoms.

49
- a. Arabinose contains five carbon atoms, while glucose contains six carbon atoms.
- b. These sugars contain an aldehyde as part of their structure as indicated by the CHO at the top of the Fischer projection.
- c. The OH group that is farthest away from the CHO group is pointing to the right.
- d. L-arabinose

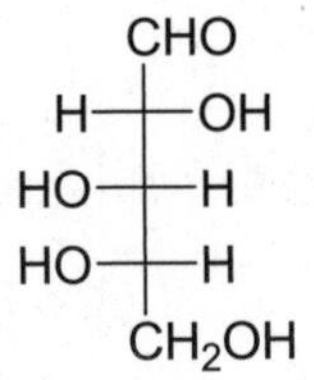

CHO
H—OH
HO—H
HO—H
CH_2OH

51 Diastereomers are nonsuperimposable, non-mirror image stereoisomers. Diastereomers have a different configuration at one or more centers of chirality, but not at every center of chirality. Enantiomers have a different configuration at every center of chirality.

53 In a Haworth structure, it is the carbon that is bonded to two oxygen atoms.

55 The α-anomer and the β-anomer are rapidly interconverting via the open chain form of the monosaccharide, creating an equilibrium mixture of all three forms of the monosaccharide.

57 a. furanose, five-membered ring, β-anomer, OH group is above the ring

anomeric center

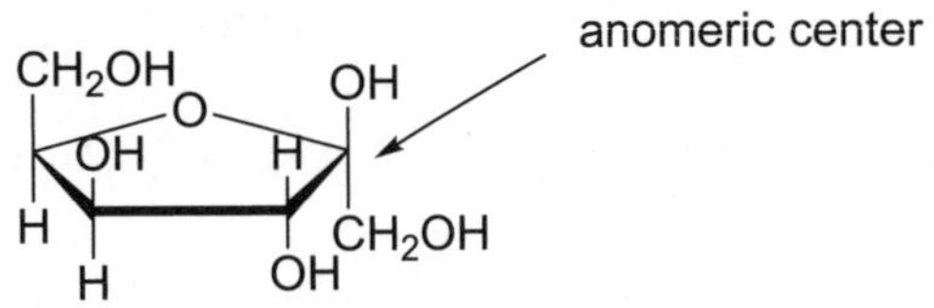

b. pyranose, six-membered ring, β-anomer, OH group is above the ring

anomeric center

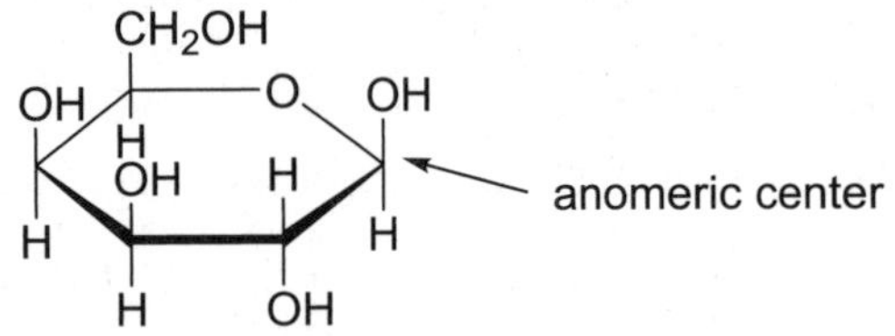

59 a.

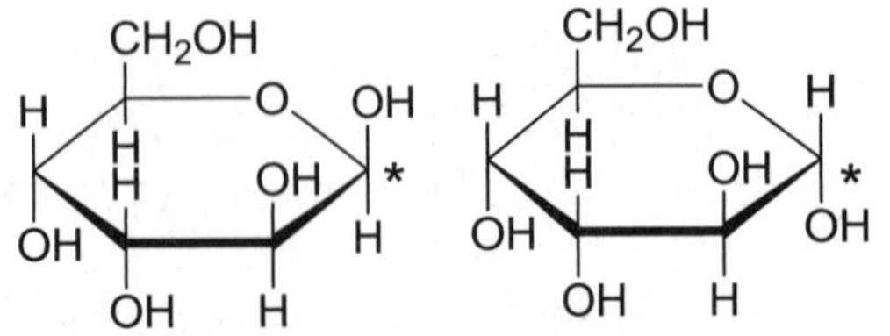

b. The sugar on the left is the β-anomer, the OH group on the anomeric center is above the ring. The sugar on the right is the α-α–anomer, the OH group on the anomeric center is below the ring.

c. These two structures are diastereomers.

d.

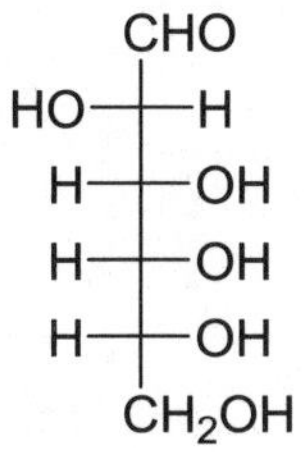

CHO
HO—H
H—OH
H—OH
H—OH
CH_2OH

e. One anomer ring opens up to the open chain form and then recloses to the other anomer ring form.

61 A disaccharide is a carbohydrate that when hydrolyzed yields two monosaccharides, either identical or different monosaccharides. In a disaccharide, the anomeric center at a glycosidic linkage is a stable functional group, so it does not open and close in aqueous solutions and cannot change configurations form the α-anomer to the β-anomer.

63 Maltose and cellobiose both consist of two glucose monomers connected by a 1→4 glycosidic bond. Maltose, however, has an α(1→4) bond, whereas cellobiose has a β(1→4) bond.

65 a. Maltulose is a disaccharide. There are two monomers that make up maltulose.

b. and c.

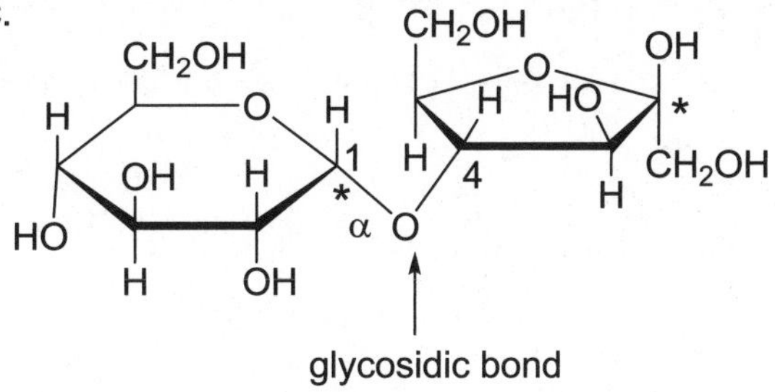

glycosidic bond

d. Maltulose has an α-glycosidic linkage. α(1→4)

e. The hydrolysis of maltulose produces two different monosaccharides.

67 Plants store energy in the form of starch.

69 Starch is hydrolyzed into glucose, which then provides energy for the body.

71 Cellulose plays an important role as dietary fiber. It adds bulk to the contents of the colon and facilitates the absorption of nutrients and water. It does not provide energy. Humans do not have enzymes that can hydrolyze β(1→ 4) linkages that are found in cellulose.

73 Glycogen, amylose, amylopectin all have α(1→ 4) glycosidic linkages.

75 Most cell markers are oligosaccharides that identify a cell like a fingerprint identifies a person. Oligosaccharide markers project out from the cell membrane into the

extracellular fluid. These cell markers allow the immune system to identify which cells are host cells and which ones are foreign.

77 Type A blood has an *N*-acetyl-D-galactosamine as a marker on red blood cells, whereas type B blood has a D-galactose marker.

79 When people with type O blood receive type A, B, or AB blood, their body rejects the donated blood because the fourth monosaccharide is unfamiliar. An immune response occurs, which can lead to life-threatening blood clotting. Type O blood has the trisaccharide component of all four blood types, so it is recognized as a part of the recipient's blood by all recipients. Therefore, people with type O blood can donate to everybody.

81 The girl's body did not recognize the fourth monosaccharide, *N*-acetyl-D-galactosamine, present in type A blood. An immune response occurred, which lead to life-threatening blood clotting.

83 Type I diabetes begins in childhood and is an autoimmune disease in which the body destroys its own β cells, the insulin-producing cells in the pancreas. There is not enough insulin present in the body. Type II diabetes generally appears in adulthood. Insulin is produced, but cells do not respond properly to the insulin present.

85 β cells produce insulin. They are found in the pancreas.

87 Insulin is a protein. If it were taken orally, it would be rendered useless by the hydrolysis of its peptide bonds in the stomach.

Chapter 12

Lipids: Structure and Function

Chapter Summary

In this chapter, you learned about lipids as a class of biological molecules that are insoluble in water but soluble in organic solvents. You learned that the function of a lipid depends on its structure. Lipids make up the cell membrane and serve as important hormones. The most abundant lipids are the triglycerides that serve as a supply of long-term energy for the cells. You learned how cholesterol and triglycerides are transported through the circulatory system.

12.1 Fatty Acids: A Component of Many Lipids

Lipids are one of the four major classes of biomolecules. Lipids, unlike the three other classes of biomolecules are defined by their physical properties. Lipids are biological compounds that do not dissolve in water but are soluble in organic solvents. The extensive hydrocarbons component of their structure gives them their hydrophobic properties. Lipids also differ from carbohydrates and the other major classes of biomolecules in that they are not polymers. Lipids serve a variety of functions, which depend on their structure. The different types of lipids include triglycerides; phospholipids and glycolipid; steroids; and eicosanoids.

Fatty acids are long straight-chain hydrocarbons, with a carboxylic acid at one end. Most naturally occurring fatty acids contain an even number of carbons. Fatty acids are further characterized by whether or not they contain carbon-carbon double bonds in their hydrocarbon chain. Saturated fatty acids contain no carbon-carbon double bonds, whereas unsaturated fatty acids contain one or more carbon-carbon double bonds. Fatty acids with one double bond are known as monounsaturated fats. Polyunsaturated fats contain more than one carbon-carbon double bond.

The overall shape of the a fatty acid depends on whether or not it is saturated or unsaturated, and in the case of an unsaturated fatty acid, whether it contains a *cis* or *trans* carbon-carbon double bond. Most naturally occurring unsaturated fatty acids contain *cis* double bonds. The geometry of a *cis* double bond creates a bend or a kink in the otherwise linear zigzag chain of the fatty acid. Unsaturated *trans* fatty acids adopt a linear zigag shape more like a saturated fatty acid than a *cis* unsaturated fatty acid.

The shape of a molecule greatly influences its physical properties, which is most evident in the melting points of fatty acids. Fatty acids with a greater number of carbon atoms will have higher melting points because they have more dispersion forces due to their greater surface area. The melting point of an unsaturated fatty acid is significantly lower than that of a saturated fatty acid, and it further decreases among a series of fatty acids with an increasing number of *cis* double bonds. Since every *cis* double bond creates an additional "bend" in the overall shape of the molecule, it reduces the number of points of contact between fatty acid molecules, thereby creating fewer dispersion forces, and hence a lower melting point. Consequently, most *cis* unsaturated fatty acids are liquids at room temperature, whereas *trans* unsaturated fatty acids are solids at room temperature.

Essential fatty acids are those fatty acids that cannot be synthesized in the body from other metabolic intermediates. They must be supplied through the diet. Linoleic and linolenic acid are both essential fatty acids.

Two naming systems are used to indicate the location of the double bonds in the hydrocarbon chain of a fatty acid: the omega system and the delta system. Nutritional scientists us the omega system, while chemists use the delta system. In the delta system, numbering begins with the carbonyl carbon (C1), and the location of first carbon of every double bond in the chain is indicated by a set of superscripts following the Greek capital letter delta (Δ). In the omega system, numbering begins at the end of the hydrocarbon chain, farthest from the carbonyl group and the location of only the first carbon atom containing the double bond is given. This locator number is placed after the Greek lowercase letter omega (ω).

Worked Example #1

The structure of myristoleic acid is shown on the next page.

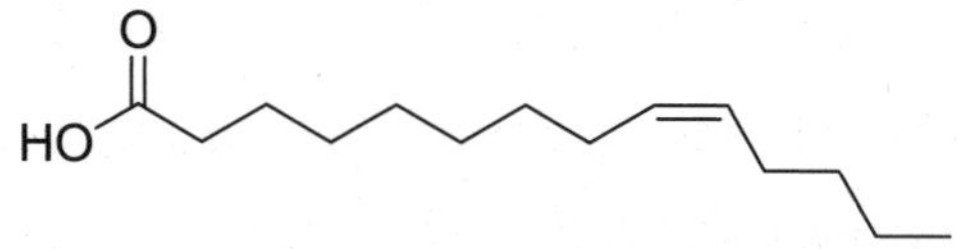

a. What type of fatty acid is myristoleic acid? Answer using the delta naming system and the omega naming system.
b. Myristic acid is a related fatty acid that also has 14 carbon atoms, but myristic acid is a saturated fatty acid. Which fatty acid, myristoleic acid or myristic acid, has the higher melting point? Explain.
c. Is myristoleic acid soluble in water?

a. *Myristoleic acid is a Δ^9 fatty acid. It is also an omega-5 (ω–5) fatty acid.*

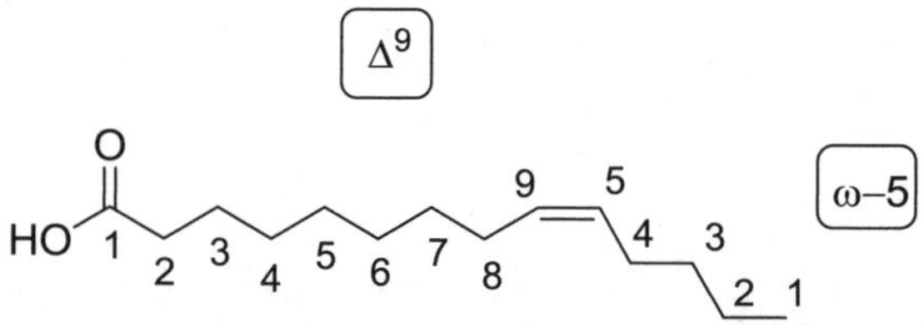

b. *Myristic acid has the higher boiling point. It is a saturated fatty acid; therefore, there are more dispersion forces present.*
c. *Myristoleic acid is not soluble in water. It is a nonpolar molecule.*

Try It Yourself #1

Docosahexaenoic acid, commonly known as DHA, is found in fish oil. Its structure is shown below.

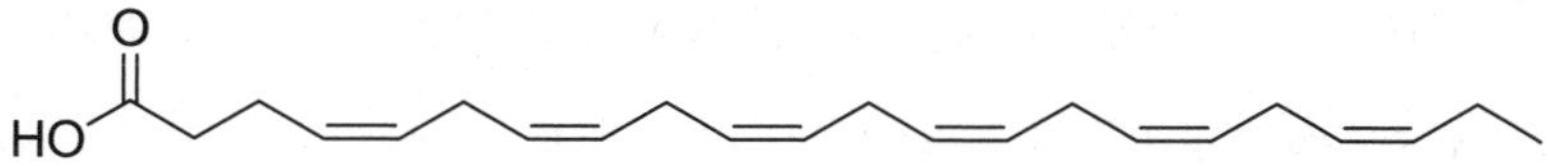

a. What type of fatty acid is DHA? Answer using the delta naming system and the omega naming system.
b. Is DHA a *trans* fatty acid?
c. Do you expect DHA to have a higher or lower melting point than a saturated fatty acid with 22 carbon atoms?

Solutions

a. *The delta naming system starts with the carbonyl carbon. The omega naming system starts with the end of the hydrocarbon chain farthest away from the carbonyl group.*

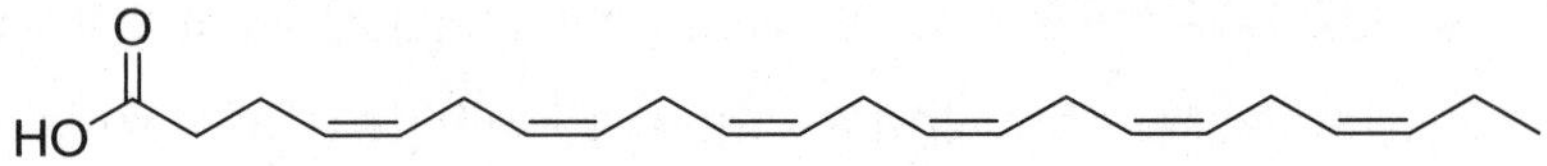

DHA is a _____ fatty acid. It is also an ______ fatty acid.

b. *The double bonds in DHA are _________ double bonds; therefore, DHA is a __________________ fatty acid.*

c. *DHA has more or fewer dispersion forces than a saturated fatty acid with 22 carbon atoms. The melting point of DHA should be ________ than the saturated fatty acid.*

12.2 Triglycerides: Energy Storage Lipids

The most abundant lipids in the body are triacylglycerols, also known as triglycerides, or as "fats." Triglycerides are obtained through diet from both animal and vegetable sources. They are stored in fat cells, known as adipocytes, and metabolized in muscle and liver cells. Triglycerides supply the body with energy over the long term, complementing glycogen, which is a source of energy over the short term.

Diets high in saturated fats and *trans* fats have been linked to cardiovascular disease, diabetes, and cancer. Vegetable oils generally have the lowest saturated fat content, while animal fats have the highest. The exceptions are palm oil and coconut oil with have the highest saturated fat content even though they are derived from plant sources.

In a Nutshell: The Chemical Structure of Triglycerides

Triglycerides are assembled from one glycerol molecule and three fatty acid molecules by esterification reactions at each of the three hydroxyl groups of glycerol and the carboxylic acid of three fatty acids. Although the ester functional group has a permanent dipole, the extensive hydrocarbon structure of a triglyceride makes it insoluble in water. The largely nonpolar structure of triglycerides makes them hydrophobic.

The physical state (solid or liquid) of a triglyceride at room temperature depends on whether it is derived from saturated or unsaturated fatty acids. Triglycerides derived from saturated fatty acids tend to be solids at room temperature and are commonly referred to as fats. These triglycerides tend to come from animal sources, though there are a few plant sources that contain saturated fatty acids. Triglycerides derived from unsaturated fatty acids, on the other hand, tend to be liquids at room temperature and are commonly referred to as oils. They are usually found in plants, though some come from fish.

Triglycerides have physical properties that resemble the fatty acids from which they are derived. Saturated fats tend to be solids at room temperature because they are more uniform in shape and therefore can pack more tightly. Tighter packing creates more intermolecular forces of attraction, resulting in a higher melting point and a waxy solid. In contrast, unsaturated fats tend to be liquids (oils) at room temperature because of their irregular shape, as a result of *cis* carbon-carbon double bonds, and therefore looser packing. Looser packing allows fewer intermolecular forces of attraction, thus resulting in a lower melting point and a liquid at room temperature.

Worked Example #2

Write the structure of the triglyceride composed of two molecules of oleic acid and one molecule of linoleic acid.

a. Would this triglyceride be considered a saturated or unsaturated fat?
b. Would you expect it to be a liquid or a solid at room temperature?

Solutions

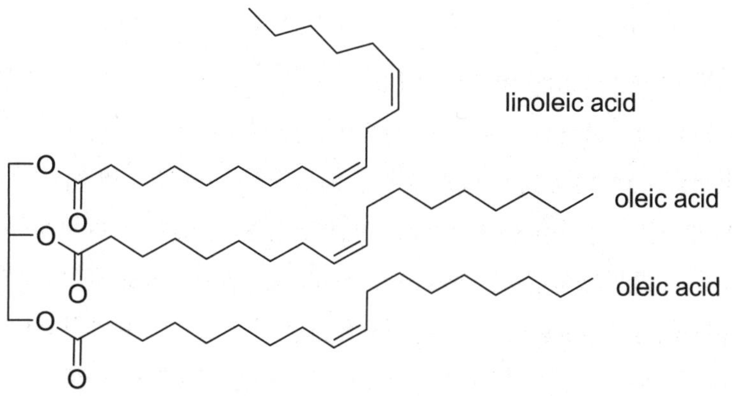

linoleic acid

oleic acid

oleic acid

a. *This triglyceride would be considered an unsaturated fat. The fatty acids are unsaturated fatty acids.*
b. *This triglyceride would be liquid at room temperature because there are* cis *double bonds present.*

Try It Yourself #2

Write the structure of a triglyceride composed of myristic acid, stearic acid, and palmitic acid.

a. Would this triglyceride be considered a saturated or an unsaturated fat?
b. Would this triglyceride be considered a fat or an oil?

Structure of triglyceride:

a. *Myristic acid, stearic acid, and palmitic acid are ____________________ fatty acids; therefore, this triglyceride is a ____________________ fat.*
b. *It would be considered a _______________.*

In a Nutshell: Digestion of Dietary Fats and Their Transport to Muscle and Fat Cells

Triglycerides stored in fat cells provide the body with energy over the long term. Fatty acid catabolism is the central energy producing pathway for many cells. The highly reduced (saturated or close to saturated) form of the hydrocarbon chain of the fatty acids makes it an excellent fuel. The digestion of dietary fats begins in the small intestine, where triglycerides from the diet enter as large globules of water insoluble fats. Before degradation reactions can begin, these globules of fat must be turned into a colloid of finely dispersed microscopic fat droplets suspended in water, a process known as emulsification. Emulsification brings water-insoluble triglycerides in contact with water-soluble lipase enzymes that catalyze the hydrolysis of triglycerides into glycerol and fatty acids. Emulsification is aided by bile acids, which are released into the small intestine from the gall bladder after a fatty meal. Triglycerides are too large to cross the cell membrane of the cells lining the small intestine known as the intestinal mucosa. Therefore, triglycerides must be hydrolyzed into fatty acids, glycerol, and partially hydrolyzed triglycerides, which can then enter intestinal mucosa.

Inside the intestinal mucosa, fatty acids and glycerol are esterified to reform triglycerides and packaged into chylomicrons, the largest of the lipoproteins for transport through the lymph system and bloodstream to their target cells (adipocytes, liver, or muscle cells).

Since triglycerides are not soluble in water, they must be transported through the blood and lymph systems by lipoproteins, spherical-shaped assemblies of lipids and proteins. Lipoproteins are hydrophilic on the outside and hydrophobic on the inside.

Upon arrival at the capillaries of their target tissues, triglycerides must be again hydrolyzed into fatty acids and glycerol, smaller molecules that can pass through the cell membrane of adipose and muscle cells. Once inside the adipocyte, fatty acids and glycerol are converted back into triglycerides and are stored until needed. In liver and muscle cells, fatty acids are degraded further.

Chylomicrons are spheres formed from a single layer of phospholipids. Since proteins are embedded throughout the lipid layer, chylomicrons are classified as a type of lipoprotein. The polar head of the lipid is positioned on the exterior of the chylomicron, where it interacts with the aqueous external environment. The nonpolar tails form a hydrophobic interior, which encapsulates the triglycerides.

Practice Problems for Triglycerides: Energy Storage Lipids

1. The structure of linolenic acid is shown below.

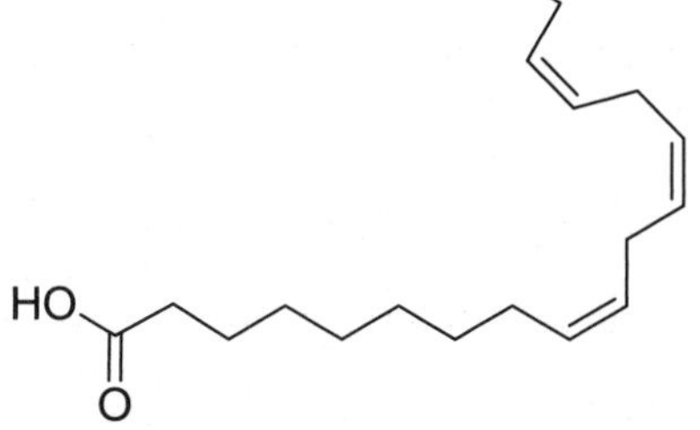

HO

O

a. What type of fatty acid is linolenic acid? Answer using the delta naming system and the omega naming system.

b. Is linolenic acid soluble in nonpolar solvents?

c. Do you expect stearic acid (a saturated fatty acid with 18 carbon atoms) to have a higher or lower melting point than linolenic acid?

d. Is linolenic acid an essential fatty acid?

2. Which is more likely to be a liquid at room temperature: oleic acid or palmitic acid?

3. Write the structure of a triglyceride composed of stearic acid, oleic acid, and palmitic acid.

a. Circle the ester functional groups in the structure.
b. Are the double bonds *cis* or *trans*?

12.3 Membrane Lipids: Phospholipids and Glycolipids

The cell membrane, part of all living cells, is a selectively permeable membrane. It separates the inside of a cell from the outside of the cell and controls the flow of ions and molecules in and out of the cell. While proteins are attached to and interspersed throughout the cell membrane and carbohydrates protrude from it, membrane lipids from the bilayer that is the cell membrane.

Our cell membranes are composed of two types of lipids: phospholipids and glycolipids. Cholesterol, another type of lipid, adds structural rigidity to the cell membrane. Membrane lipids and cholesterol have the unique characteristic of being amphipathic molecules; they contain both a polar (hydrophilic) and nonpolar (hydrophobic) region.

In a Nutshell: The Chemical Structure of Phospholipids and Glycolipids

Membrane lipids contain either a glycerol or a sphingosine backbone. Sphingosine is a straight chain of 18 carbon atoms with a *trans* carbon-carbon double bond at C(4)-C(5), two hydroxyl groups at C(1) and C(3), and an amino group at C(2). A large portion of the molecule is a hydrocarbon chain; thus, sphingosine itself supplies one of the two hydrocarbon "tails" in a sphingosine derived membrane lipid. The other "tail" is derived from a fatty acid. In contrast, the two hydrocarbon "tails" in a glycerol derived membrane lipid are both derived from fatty acids, much like triglycerides are derived from glycerol and three fatty acids.

Phospholipids are the most common type of membrane lipid, characterized by a polar phosphate group. The phosphate group is linked to the terminal hydroxyl group of sphingosine or glycerol as a phosphate ester. A phospholipid derived from glycerol is known as a glycerophospholipid, and a phospholipid derived from sphingosine is known as a sphingophospholipid. Sphingophospholipids, also known as sphingomyelins, insulate nerve fibers, creating the myelin sheath.

The two nonpolar tails in a glycerophospholipid are derived from two fatty acids that from ester linkages to the other hydroxyl groups in a glycerol molecule. In a sphingophospholipid one nonpolar tail is derived from a fatty acid that forms and amide linkage to the amine group in sphingosine. The phosphate group in all phospholipids forms two phosphate ester linkages: one to the terminal hydroxyl group of either glycerol or sphingosine and the second to a small amino alcohol, a molecule with an amine (—NH_2) and a hydroxyl group (—OH). The nitrogen atom in aminoalcohol has four bonds, and thus a positive charge, which together with the negative charge on the phosphate group creates the polar head of the phospholipid. The three amino alcohols from which most phospholipids are derived are ethanolamine, choline, and serine.

All glycolipids in human cells are derived from sphingosine, and thus, they are glycosphingolipids. The "glycol" in glycophospholipids refers to the carbohydrate component of these lipids, which furnishes the polar region of these amphipathic molecules. So instead of a phosphate group, glycolipids contain a carbohydrate joined at the anomeric center to the terminal hydroxyl group of sphingosine. The carbohydrate component of a glycosphingolipid can be a mono-, di-, or oligosaccharide. Like sphingolipids, one of the nonpolar tails in a glycolipid comes from the amide linkage formed between the amine of sphingosine and a fatty acid, while the other nonpolar tail is part of the long hydrocarbon chain on sphingosine itself.

Glycolipids containing a monosaccharide are known as cerebrosides. Cerebrosides derived from D-galactose are found primarily in cells of the central nervous system and the brain, while those containing D-glucose are found in most other tissues.

In a Nutshell: The Cell Membrane

Membrane lipids are amphipathic molecules: with two nonpolar hydrocarbon tails and a polar head. The polar head is the phosphate group with the positively charged nitrogen in a phospholipid or the carbohydrate in a glycolipid. The nonpolar tails are derived from fatty acids and in the case of a sphingolipid, sphingosine. Since membrane lipids have a complex chemical structure, they are often depicted as a sphere to show the polar head, attached to two wavy lines to show the nonpolar tails.

The cell membrane forms two layers of phospholipids or glycolipids assemble tail to tail in what is called a bilayer. The lipids that form the bilayer assemble so that their polar heads are in contact with the aqueous environment either on the inside or the outside of the cell, while the nonpolar tails are in the hydrophobic interior region of the bilayer.

The lipid bilayer is often described as a “fluid mosaic” because the phospholipid and glycolipid molecules are not covalently bonded to one another, but instead interact through weaker noncovalent forces of attraction. Embedded within the cell membrane are cholesterol molecules to provide added rigidity. The polar hydroxyl group of cholesterol lies among the polar heads and the hydrocarbon portion of the molecule is situated among the nonpolar tails. Various proteins are also found throughout the cell membrane. Some carbohydrate markers protrude from the surface of the cell.

Small nonpolar molecules, such as oxygen and carbon dioxide, pass freely through the cell membrane by simple diffusion, and water passes slowly by the process of osmosis. Ions and polar organic molecules cannot readily pass through the nonpolar interior of the cell membrane. Sometimes, special transport proteins are necessary to move ions from one side of the membrane to the other. Specific hormones regulate with ions move in and out of the cell along with other chemical messengers that interact with proteins on or within the cell membrane.

Cell membranes vary in their permeability depending on characteristics of the membrane lipids, in particular, the extent of unsaturation in the hydrocarbon tails. Cell membranes with more saturated phospholipids exhibit less permeability and less fluidity, while those with more unsaturated phospholipids have greater permeability. The saturated fats pack more closely than the unsaturated ones and affect the permeability of the membrane.

Worked Example #3

For the membrane lipid below, answer the following questions:

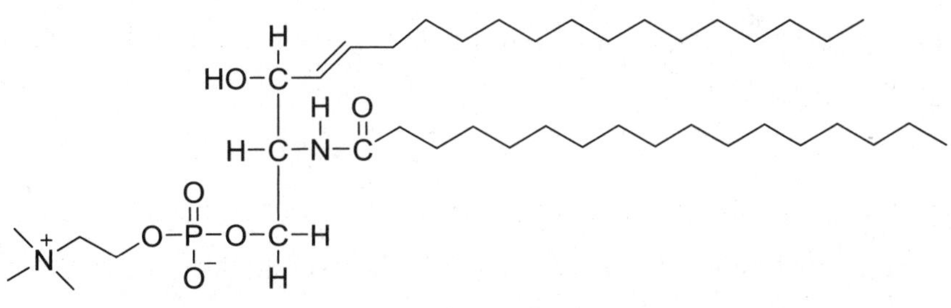

a. Is it a phospholipid or a glycolipid? How can you tell?
b. Is the backbone derived from sphingosine or glycerol? Circle the backbone.
c. What is the fatty acid from which the lipid is derived? Is it saturated or unsaturated? What functional group connects it to the backbone?

d. Is it a sphingomyelin or cerebroside?
e. If it is a phospholipid, what is the amino alcohol: serine, choline, or ethanolamine?
f. Circle the part of the molecule that defines the polar head.
g. Circle the part of the molecule that defines the two tails.

Solutions

a. *It is a phospholipid because it contains a phosphate group and not a carbohydrate.*
b. *It is a sphingosine.*

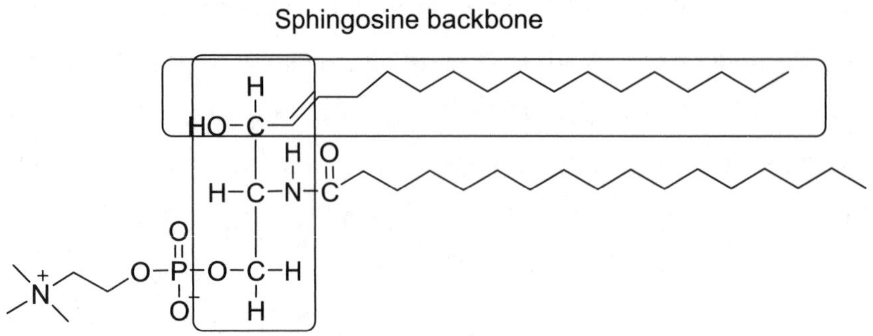

Sphingosine backbone

c. *The fatty acid is stearic acid—an 18-carbon saturated fatty acid. It is joined to the amine in sphingosine by an amide linkage.*
d. *It is a sphingomyelin. It is a phospholipid with a sphingosine backbone.*
e. *The amino alcohol is choline.*

f–g.

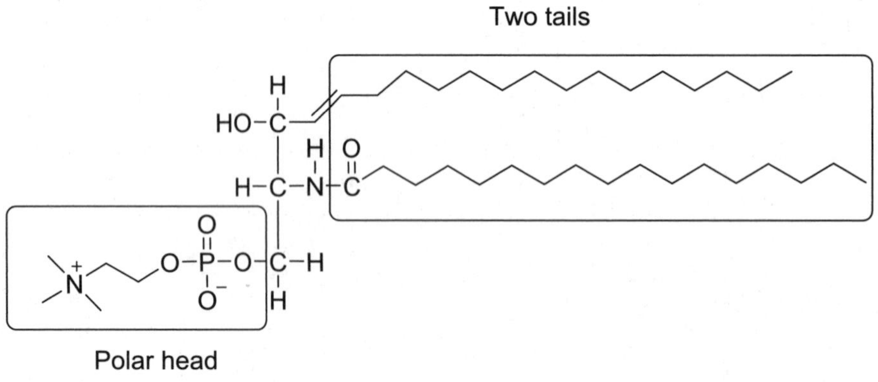

Two tails

Polar head

Try It Yourself #3

For the membrane lipid below, answer the following questions:

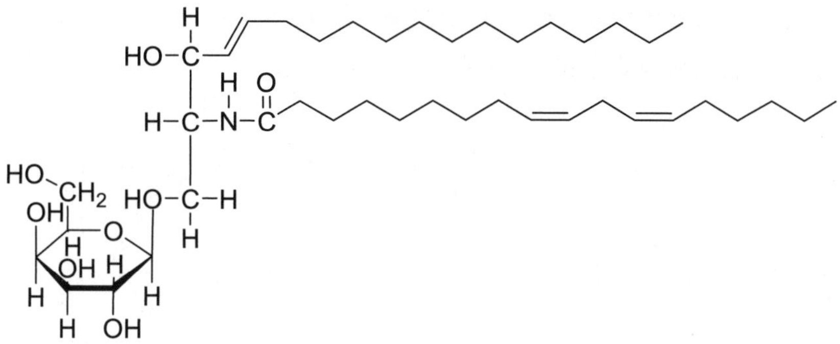

a. Is it a phospholipid or a glycolipid? How can you tell?

b. Is the backbone derived from sphingosine or glycerol? Circle the backbone.

c. What is the fatty acid from which the lipid is derived? Is it saturated or unsaturated? What functional group connects it to the backbone?

d. Is it a sphingomyelin or cerebroside? How can you tell?

e. If it is a glycolipid, is the glycosidic linkage α or β?

Solutions

a. *Is there a phosphate group or a carbohydrate attached to the backbone?*

   ____

b. *The backbone is derived from:* ______________________________.

c. *The name of the fatty acid is:* ______________________________. *Does the fatty acid contain any double bonds?* __________

   The fatty acid is (unsaturated or saturated) ______________________. *The fatty acid is connected to the back bone by a* ______________ *functional group.*

d. *If there is a carbohydrate attached to the backbone, is it a monosaccharide or an oligosaccharide?* ______________________________

   Therefore, the membrane lipid is a (sphingomyelin or cerebroside)

   ______________________________.

e. *The OH group on the anomeric center is (above or below)* ____________ *the ring. The glycosidic linkage is* _____.

Worked Example #4

Draw an illustration of 14 membrane lipids arranged to form a bilayer such as the cell membrane. What molecule provides rigidity to the cell membrane?

Solution

Cholesterol provides rigidity to the cell membrane.

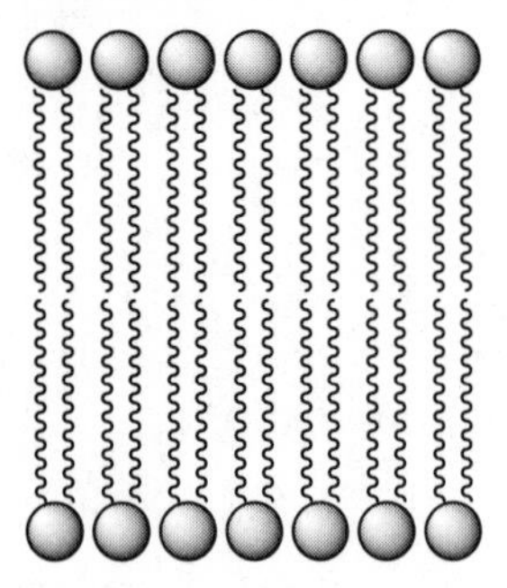

Try It Yourself #4

Draw an illustration of eight membrane lipids arranged to form a bilayer such as the cell membrane. Label the nonpolar part of the cell membrane. Why do polar organic molecules move slowly through the membrane, if at all?

Solution:

Practice Problems for Membrane Lipids: Phospholipids and Glycolipids

1. For the membrane lipid below, answer the following questions:

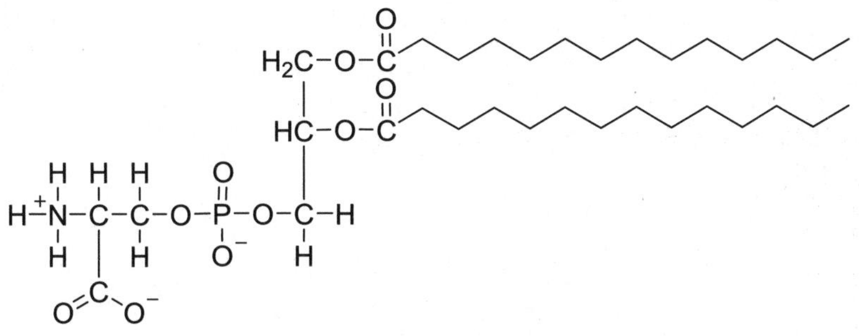

   a. Is it a phospholipid or a glycolipid? How can you tell?

   b. Is the backbone derived from sphingosine or glycerol? Circle the backbone.

   c. What are the fatty acids from which the lipid is derived? Are they saturated or unsaturated? What functional group connects them to the backbone?

   d. Is it a sphingomyelin, glycerophospholipid, or cerebroside?

   e. If it is a phospholipid, what is the amino alcohol: serine, choline, or ethanolamine?

   f. Circle the part of the molecule that defines the polar head.

   g. Circle the part of the molecule that defines the two tails.

2. For the membrane lipid below, answer the following questions:

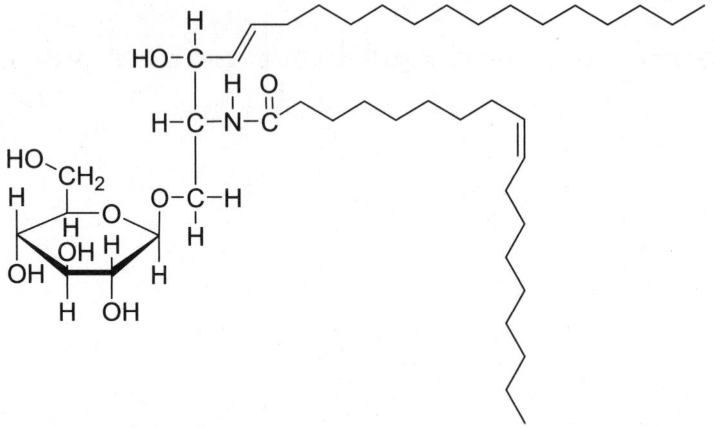

   a. Is it a phospholipid or a glycolipid? How can you tell?

   b. Is the backbone derived from sphingosine or glycerol? Circle the backbone.

   c. What is the fatty acid from which the lipid is derived? Is it saturated or unsaturated? What functional group connects them to the backbone?

   d. Is it a sphingomyelin, glycerophospholipid, or cerebroside?

   e. If it is a glycolipid, is the glycosidic linkage α or β?

   f. Circle the part of the molecule that defines the polar head.

g. Circle the part of the molecule that defines the two tails.

3. Draw an illustration of 10 membrane lipids arranged to form a bilayer such as the cell membrane. Label the polar part of the cell membrane.

12.4 Cholesterol, Steroids, and Lipoproteins

Steroids are molecules that contain the characteristic steroid ring system: three six-membered rings and one five-membered ring fused together. Steroids are classified as lipids because they are biomolecules insoluble in water and soluble in organic solvents. They are structurally different from triglycerides, phospholipids, and glycolipids because they are not derived from fatty acids. Their hydrophobic physical characteristics arise instead from the fact that the steroid ring system is a hydrocarbon.

Cholesterol is the most important steroid, which not only helps to maintain the integrity of the cell membrane, but serves as the biological precursor of all the other important steroids in the cell, including steroid hormones, vitamin D, and bile salts.

Bile salts, such as cholate, are amphipathic compounds that behave as emulsifying agents. The ionized carboxylic acid is the polar end, while the steroid ring system is the nonpolar end. Cholate and other bile salts break up globules of dietary fat in the first stage of digestion. Bile is produced in the liver and stored in the gall bladder.

In a Nutshell: Steroid Hormones

The important steroid hormones synthesized from cholesterol include: glucocorticoids, mineralocorticoids, progestins, androgens, and estrogens. The glucocorticoids are produced in the outer part of the adrenal gland and regulate glucose metabolism. They also have anti-inflammatory and immunosuppressant activity and can be used to treat autoimmune conditions.

Cortisol is an important glucocorticoid. Cortisol is released in response to stress. It also suppresses the immune system. Prednisone is another synthetic glucocorticoid that is used as an immunosuppressant.

The mineralocorticoids are produced in the adrenal gland and regulate electrolyte (Na^+, K^+, Cl^-) balance in tissues. Aldosterone, the primary mineralocorticoid, acts on the kidneys to increase the reabsorption of sodium ions and the release of potassium ions, as well as the retention of water.

Progestins, estrogens, and androgens represent the primary sex hormones. Progestins, such as progesterone, regulate the menstrual cycle and pregnancy; androgens (testosterone) and estrogens (estradiol) control the development of sexual characteristics in the developing fetus and secondary sexual characteristics during puberty.

In a Nutshell: Lipoproteins and Cholesterol

Since cholesterol is insoluble in aqueous solution, it cannot be transported to the cell through the circulatory system like blood sugar and other nutrients. Instead, cholesterol and triglycerides are transported through the blood and lymph system by liproproteins. Lipoproteins are assemblies composed of both phospholipids and proteins. They assemble as a spherically shaped monolayer of phospholipids with a polar exterior and a nonpolar interior. Triglycerides and cholesterol esters are found inside a lipoprotein in its nonpolar interior.

Lipoproteins are most readily differentiated by their density (m/V), which is dependent on the relative composition of proteins and lipids that make up the lipoprotein and its contents: The greater lipid content, the less dense the lipoprotein. The five basic types of lipoproteins include chylomicron, VLDL (very-low-density lipoprotein), IDL (intermediate-density lipoprotein), LDL (low-density lipoprotein), and HDL (high-density lipoprotein). Chylomicrons are the least dense of the lipoproteins (highest lipid content) because they contain mainly triglycerides. Low-density lipoproteins (LDLs) are also high in triglycerides, as well as cholesterol esters. High-density lipoproteins have the lowest lipid content.

Chylomicrons have the largest diameter of the lipoproteins and they function to transport triglycerides from the intestinal mucosa to muscle and fat cells via the circulatory and lymph systems. Once chylomicrons deliver their contents to cells, they return to the liver as chylomicron remnants (leftovers), where they are either degraded or packaged into very-low-density lipoproteins (VLDLs). VLDLs also travel through the bloodstream to deliver triglycerides to adipose and muscle cells. Eventually, as they unload cholesterol and triglycerides to cells, they become intermediate density lipoproteins (IDLs). High-density lipoproteins scavenge cholesterol form the blood and return it to the liver.

"Bad cholesterol" and "good cholesterol" refer to the lipoproteins that carry cholesterol through the bloodstream. High-density lipoproteins, HDLs, are considered good cholesterol because they remove cholesterol form the bloodstream. Bad cholesterol refers to low-density lipoproteins, LDLs, which are the primary carriers of excess triglycerides to adipose tissue.

Worked Example #5

What are the five classes of steroid hormones derived from cholesterol? What is the function of the mineralocorticoids?

Solution

The five classes of steroid hormones produced from cholesterol include glucocorticoids, mineralocorticoids, progestins, androgens, and estrogens. The mineralocorticoids regulate ion balance in tissues.

Try It Yourself #5

What are the functions of the glucocorticoids and estrogens?

Solution:

Worked Example #6

What is the shape of a lipoprotein? Where are triglycerides and cholesterol esters found on a lipoprotein?

Solution

A lipoprotein is a spherically shaped monolayer of phospholipids with a polar exterior and a nonpolar interior. Triglycerides are found inside a lipoprotein.

Try It Yourself #6

What are the functions of the low-density lipoproteins, LDLs, and high-density lipoproteins, HDLs?

Solution:

Practice Problems for Cholesterol, Steroids, and Lipoproteins

1. From what molecule are bile acids, vitamin D, and the steroid hormones derived?

2. What structural feature characterizes cholesterol and the steroid hormones?

3. Medrol, shown below, is commonly used to treat rheumatoid arthritis and severe allergic reactions, among other things. Is medrol a steroid?

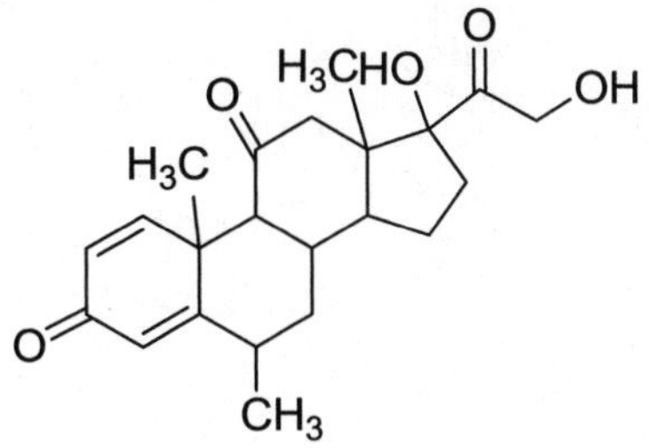

H_3CHO O OH O H_3C O CH_3

Chapter 12 Quiz

1. The structure of erucic acid, found in canola oil and mustard seed, is shown below.

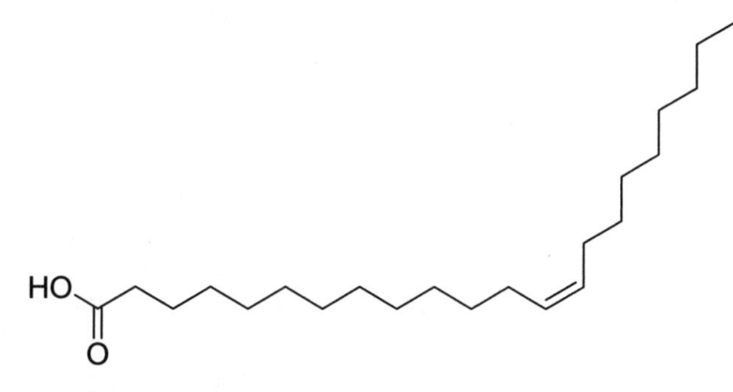

   a. What type of fatty acid is erucic acid? Answer using the delta naming system and the omega naming system.

   b. Is erucic acid a saturated or unsaturated fatty acid?

   c. What type of double bond is found in erucic acid?

   d. Is erucic acid soluble in water?

2. Explain why oleic acid has a lower melting point than stearic acid, even though both fatty acids contain 18 carbon atoms.

3. Write the structure of the triglyceride composed of one molecule of linolenic acid, one molecule of oleic acid, and one molecule of linoleic acid.

a. Do you expect this triglyceride to be a fat or an oil? Explain.

b. Is this triglyceride unsaturated or saturated?

4. For the membrane lipid below, answer the following questions:

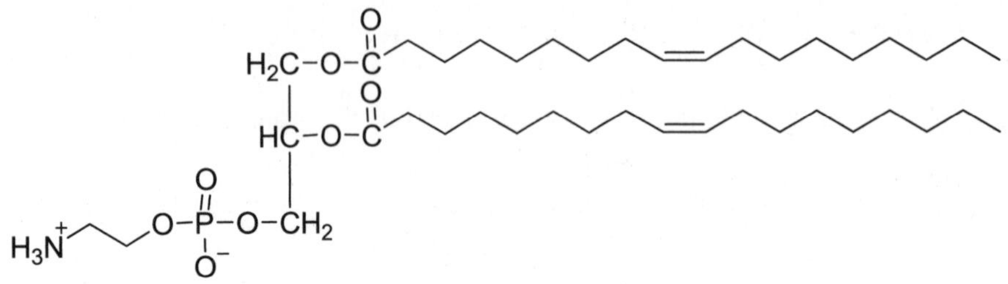

$H_2C-O-C(=O)$

$HC-O-C(=O)$

$H_3\overset{+}{N}$ $O-P(=O)(O^-)-O-CH_2$

a. Is it a phospholipid or a glycolipid? How can you tell?

b. Is the backbone derived from sphingosine or glycerol? Circle the backbone.

c. What is the fatty acid from which the lipid is derived? Is it a saturated or unsaturated fatty acid? What functional group connects it to the backbone?

d. Is it a sphingomyelin, glycerophospholipid, or cerebroside?

e. If it is a phospholipid, what is the amino alcohol: serine, choline, or ethanolamine?

f. Circle the part of the molecule that defines the polar head.

g. Circle the part of the molecule that defines the two tails.

5. For the membrane lipid below, answer the following questions:

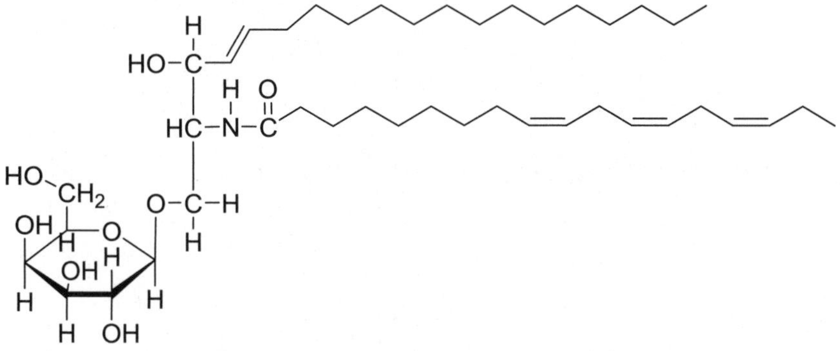

   a. Is it a phospholipid or a glycolipid? How can you tell?

   b. Is the backbone derived from sphingosine or glycerol? Circle the backbone.

   c. What is the fatty acid from which the lipid is derived? Is it a saturated or unsaturated fatty acid? What functional group connects it to the backbone?

   d. Is it a sphingomyelin, glycerophospholipid, or cerebroside?

   e. If it is a glycolipid, is the glycosidic linkage α or β?

   f. Circle the part of the molecule that defines the polar head.

   g. Circle the part of the molecule that defines the two tails.

6. Draw an illustration of 12 membrane lipids arranged to form a bilayer, such as the cell membrane. Label the polar and nonpolar parts of the bilayer.

7. Is the interior or exterior of a chylomicron polar? Explain.

8. Does a high-density lipoprotein contain more proteins than lipids or more lipids than proteins?

9. Explain how chylomicrons can transport nonpolar triglycerides through the aqueous bloodstream.

10. What do the structures of cholesterol and the steroid hormones have in common?

Chapter 12

Answers to Additional Exercises

35 Steroids belong to the lipid class of biomolecules.

37 The increase in muscle growth is the anabolic effect of steroids. It begins when the steroids bind to the androgen receptor and initiates a cascade of biochemical events that result in protein production that increases muscle growth.

39 Some side effects of anabolic steroid use are infertility, gynecomastia, testicular atrophy, and premature baldness. Prolonged steroid use can lead to liver damage, prostate cancer, and cardiovascular disease.

41 Anabolic steroids treat muscle wasting diseases such as AIDS; they aid in the recovery from burns and are an effective treatment for hypogonadism.

43 The extensive hydrocarbon component of lipids make them insoluble in water.

45 A fatty acid is a long straight-chain hydrocarbon with a carboxylic acid at one end. They all contain the carboxylic acid.

47 A monounsaturated fatty acid contains one carbon-carbon double bond. A polyunsaturated fatty acid contains more than one carbon-carbon double bond.

49 a. Unsaturated fatty acid. It has double bonds. b. Saturated fatty acids. It has no double bonds. c. Unsaturated fatty acid. It has double bonds.

51 Stearic acid has more intermolecular dispersion forces because it has more surface area to interact with other stearic acid molecules. Linoleic acid has two *cis* double bonds, which create bends in the overall shape of the molecule and reduces the number of contact points between fatty acid molecules, thereby causing fewer dispersion forces.

53 A *cis* double bond will cause a bend or kink in the structure, reducing the number of contact points and creating fewer dispersion forces, thus lowering the melting point.

55 If a *trans* double bond is present in the fatty acid, it is able to adopt a linear zigzag shape with more surface area and more dispersion forces and have a higher melting point. A *cis* double bond will cause a bend or kink in the structure, reducing the number of contact points and creating fewer dispersion forces, thus lowering the melting point.

57 A saturated fatty acid has stronger intermolecular forces of attraction and a higher melting point, therefore causing it to be a solid at room temperature.

59 Flaxseed, walnuts, salmon, and canola oil are good sources of fatty acids.

61 Oleic acid is a Δ^9 fatty acid and ω-9 fatty acid.

63 Linolenic acid is an ω-3 fatty acid.

65 a. EPA is a $\Delta^{5,8,11,14,17}$ fatty acid b. EPA is a ω-3 fatty acid. c. It should be a liquid at room temperature. It has five *cis* double bonds present. A *cis* double bond will cause a bend or kink in the structure, reducing the number of contact points and creating fewer dispersion forces, thus lowering the melting point.

67 Triglycerides are stored in fat cells, known as adipocytes. They are metabolized in muscle and liver cells.

69 Saturated fats and *trans* fats are unhealthy. They have been linked to cardiovascular disease, diabetes, and cancer.

71 A triglyceride is insoluble in water and therefore classified as a lipid. It is insoluble because the fatty acids give it an extensive hydrocarbon structure.

73 Triglycerides derived from saturated fatty acids are classified as fats. They are solids at room temperature because they are more uniform in shape and can pack more tightly, which creates more intermolecular forces of attraction which leads to a higher melting point. Fats are found in meat, whole milk, butter, cheese, coconut oil, and palm oil.

75 Coconut oil should be a solid at room temperature because it is mostly saturated fats, which pack together tightly.

77 Olive oil is considered a healthy fat because it consists mostly of unsaturated fats.

79 Safflower oil is lower in saturated fats.

81 Olive oil is a healthier food because it contains fewer saturated fats.

83 a. It is an unsaturated fat. There are carbon-carbon double bonds present.

b. It would be a liquid at room temperature.

c. It has three ester groups and four carbon-carbon double bonds.

d–e.

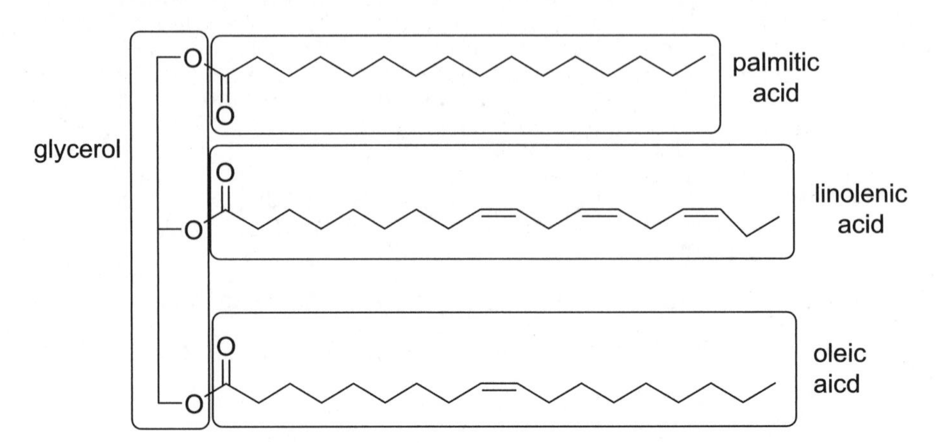

f. The carbon-carbon double bonds are *cis* double bonds. It is a healthy fat.

85

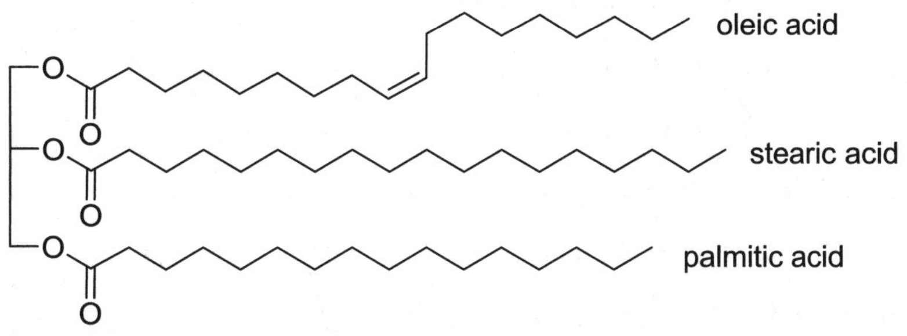

oleic acid

stearic acid

palmitic acid

87 Triglycerides provide long term energy storage.

89 Emulsification turns the globules of fat into a colloid of finely dispersed microscopic fat droplets suspended in water. This process brings the water insoluble triglycerides in contact with the water soluble lipase enzymes.

91 Triglycerides are too large to cross the cell membrane of the intestinal mucosa. The triglycerides are hydrolyzed into fatty acids, glycerol, and partially hydrolyzed triglycerides to cross the intestinal mucosa. Once inside the intestinal mucosa, the fatty acids and glycerol are esterified to reform the triglycerides and packaged into chylomicrons, which will carry them through the bloodstream.

93 Lipoproteins are hydrophobic on the interior.

95 Triglycerides are found on the inside of chylomicron. The interior of a chylomicron is hydrophobic. Triglycerides are also hydrophobic.

97 Phospholipids and glycolipids make up the cell membrane.

99 It amphipathic because it is both polar (the carboxylate end) and nonpolar (the hydrocarbon end).

101 Three fatty acids are needed to form a triglyceride. Two fatty acids are needed to form a glycerophospholipid. One fatty acid is needed to form a sphingolipid.

103 Sphingomyelins and sphingolipids are derived from sphingosine.

105 A glycolipid has a carbohydrate linked to the hydroxyl group farthest from the hydrocarbon chain, rather than an amino alcohol and a phosphate group. This difference is primarily in the polar head.

107
a. It is a phospholipid.
b. It is a glycerophospholipid. It has a glycerol backbone.

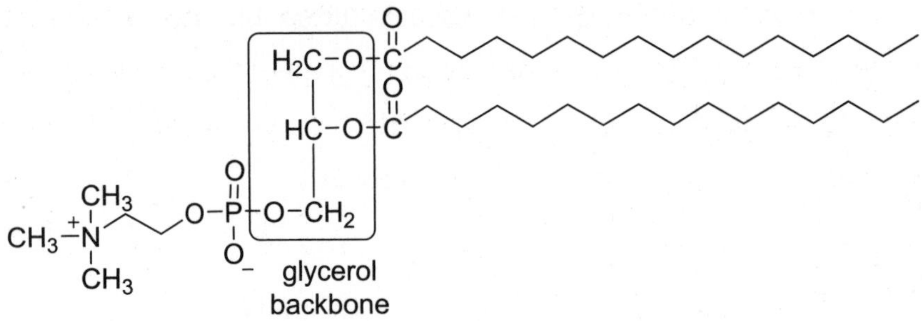

O
H_2C-O-C
O
$HC-O-C$
O
CH_3
$CH_3-\overset{+}{N}$ O–P–O–CH_2
CH_3 O^-
glycerol
backbone

c. The fatty acid derivatives are from palmitic acid. The fatty acid components are saturated. An ester group connects the fatty acids to the backbone.

d. The lipid contains a phosphate.

e. The choline group and the phosphate group are the polar head group.

f.

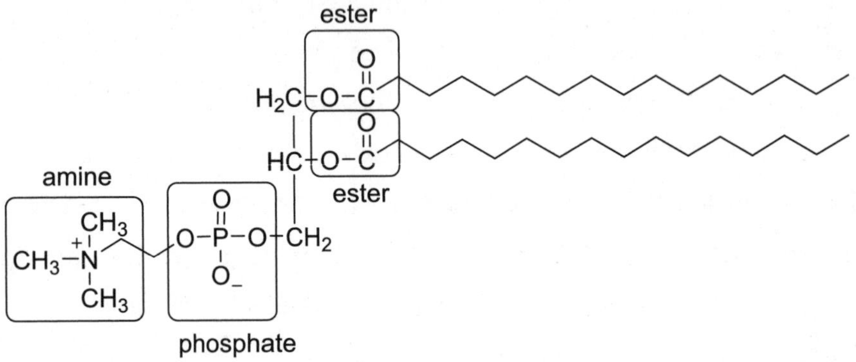

ester
O
H_2C-O-C
O
$HC-O-C$
amine
ester
O
CH_3
$CH_3-\overset{+}{N}$ O–P–O–CH_2
CH_3 O^-
phosphate

109 The phospholipid and glycolipid molecules interact through intermolecular forces of attraction which creates a flexible fluid like assembly. The membrane molecules are not covalently bonded to each other.

111 Small nonpolar molecules such as oxygen and carbon dioxide pass freely through the cell membrane.

113 The saturated phospholipids pack more closely than the unsaturated phospholipids; therefore, they are less fluid and more rigid.

115 Steroids are structurally different from triglycerides, phospholipids, and glycolipids because they are not derived from fatty acids.

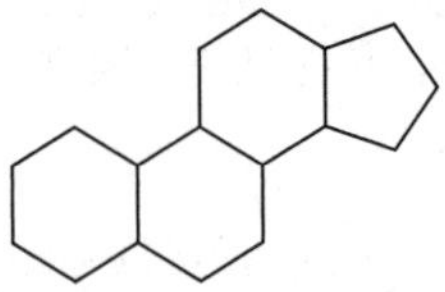

117 Cholesterol helps maintain the integrity of the cell membrane. It is also the biological precursor of all the other important steroids in the cell, including steroid hormones, vitamin D, and bile salts.

119 Cholate has a nonpolar end (the steroid ring system) and a polar end (the ionized carboxylic acid). The steroid ring system is part of the structure.

121 Prednisone is a glucocortoid. Glucocortoids are produced in the adrenal cortex. It has the steroid ring structure in the molecule.

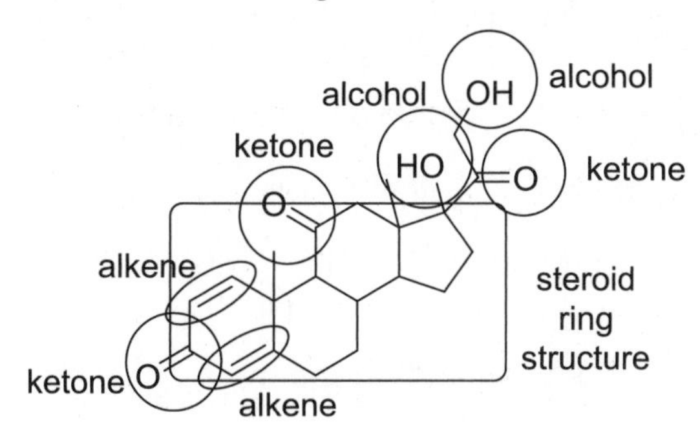

123 Cortisol is released in response to stress. When it is released, it causes an increase in blood glucose levels, and it facilitates fat and protein metabolism.

125

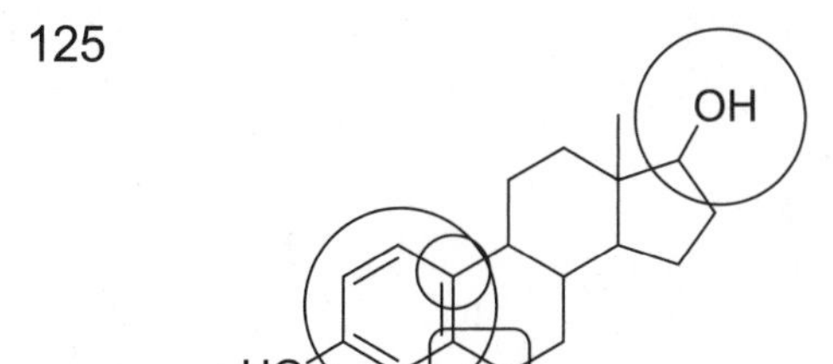

Estradiol

The six-membered ring attached to an OH group has been converted into an aromatic ring. The double bond in the adjacent six-membered ring has been removed. The hydrocarbon chain attached to the five-membered ring has been converted into an OH group. The methyl group between rings A and B has been removed.

127 Progesterone, testosterone, and estradiol are all produced from cholesterol. All three hormones have the steroid ring structure. Progesterone and testosterone both have a ketone and a double bond in ring A. Testosterone and estradiol both have an alcohol on ring D. Progesterone has a ketone attached to ring D, while the other two do not. Estradiol has a phenol group for ring A, while the other two do not.

129 Cell membranes have two layers, while lipoproteins have only one layer and are spherical in shape.

131 Chylomicrons have the largest diameter.

133 High-density lipoproteins are the densest.

135 HDLs are considered “good.” They remove cholesterol from the bloodstream and carry it back to the liver.

137 The three major classes of eicosanoids are prostaglandins, thromboxanes, and leukotrienes.

139 Arachidonic acid is an unsaturated fat. The double bonds are all *cis* double bonds.

141 Prostaglandin H_2 is formed by the action of *cyclooxygenase* on arachidonic acid.

143 *Phospholipase* A_2 catalyzes the release of arachidonic acid from glycerophospholipids. This enzyme catalyzes a hydrolysis reaction.

145 NSAIDs inhibit *cyclooxygenase*, the key enzyme involved in the formation of prostaglandin H_2 and the formation of prostaglandins and thromboxanes.

Chapter 13

Proteins: Structure and Function

Chapter Summary

In this chapter, you were introduced to one of the most important biomolecules, proteins. You have seen how proteins are built from the basic building blocks, the amino acids. You examined the structure of amino acids and learned how they form peptide chains. You studied how these peptide chains fold into more complex structures through different interactions based on the chemical structure of the amino acids in the peptide chain. You learned how enzymes work and how molecules can inhibit the function of the enzymes.

13.1 Amino Acids

Proteins have the most diverse functions of all biomolecules. They serve as enzymes, receptors, structural proteins, immunoglobins, transport proteins, dietary proteins, and protein motors. Proteins are biological polymers of amino acids.

All proteins are constructed from the same 20 amino acids. All amino acids contain an amine ($—NH_2$) and a carboxylic acid (—COOH) functional group as part of their structure. Both functional groups are covalently bonded to the α-carbon.

In a Nutshell: pH and Amino Acid Equilibria in Aqueous Solution

Amines are organic bases and carboxylic acids are organic acids; thus, in aqueous solution these groups can exist in either of two forms—neutral or ionized, depending on the pH. At physiological pH (pH = 7.3), the amine and the carboxylic acid functional groups of an amino acids are in their ionized forms: The amine is in its conjugate acid form ($—NH_3^+$) and the carboxylic acid is in its conjugate base form ($—COO^-$). Since both a positive and a negative charge are present within the same molecule, the net charge is neutral and this form of an amino acid is known as a zwitterion.

There are two other pH-dependent forms of an amino acid that are in equilibrium with the zwitterion form. At high pH, the conjugate acid of the amine loses a proton and becomes the neutral amine ($—NH_2$), while the carboxylic acid remains in its ionized from ($—COO^-$). At low pH, the carboxylate ion accepts a proton to become a neutral carboxylic acid, while the amine remains in its conjugate acid form (—NH_3^+).

In a Nutshell: Amino Acid Side Chains

Amino acids have an amine and a carboxylic acid functional group bonded to the α-carbon. The other two bonds to the α-carbon are a hydrogen atom and an R group. The R group, often referred to as the amino acid side chain, is a carbon chain of varying length and branching that in some amino acids also contains a functional group (except glycine, which has R = H). Since amino acids differ only in the structure of their side chain, the side chain is the distinguishing feature of an amino acid. The structure of the side chain determines whether an amino acid is classified as nonpolar or polar. There are 20 amino acids found in nature, each has a distinct side chain, name, and three letter abbreviation. The nonpolar amino acids have hydrocarbon side chains, as well as some sulfur- and nitrogen-containing functional groups that have relative nonpolar covalent bonds. The amino acids with nonpolar side chains are glysine, alanine, valine, leucine, isoleucine, proline, tryptophan, phenylalanine, and methionine.

The amino acids with polar side chains are further subdivided into three categories: acidic, basic, and neutral. Two amino acids, aspartic acid and glutamic acid, are polar acidic because their side chains contain a carboxylic acid functional group. At physiological pH, carboxylic acids exist in their conjugate base form. These amino acids have a net negative charge at physiological pH, even in their zwitterion form.

Three amino acids are polar basic: lysine, arginine, and histidine. At physiological pH, most basic side chains exist in their conjugate acid form, creating a net positive charge, even in their zwitterion form.

The other six amino acids are polar neutral because they have side chains that are polar but have no ionizable functional groups. They are neither acids nor base, but contain polar functional groups. The polar neutral amino acids are serine, threonine, tyrosine, cysteine, glutamine, and asparagine.

In a Nutshell: Essential Amino Acids

There are nine amino acids that cannot be synthesized in the body and must be obtained through diet. These amino acids, known as the essential amino acids, are as follows: arginine (required in children), histidine, isoleucine, leucine, lysine, methionine, phenylalanine, threonine, tryptophan, and valine. The essential amino acids can be found in both animal and vegetable sources. Animal sources contain all nine of the essential amino acids and are known as complete proteins. Single-vegetable sources of amino acids are usually missing one or more essential amino acid. A vegetarian diet should contain a complementary mixture of plant proteins.

Worked Example #1

Draw the structure of tyrosine at the following pH's:

a. pH = 1
b. physiological pH
c. pH = 10

Solutions

a. *At pH = 1, the amine is in its conjugate acid form and the carboxylic acid functional group is in its neutral form.*

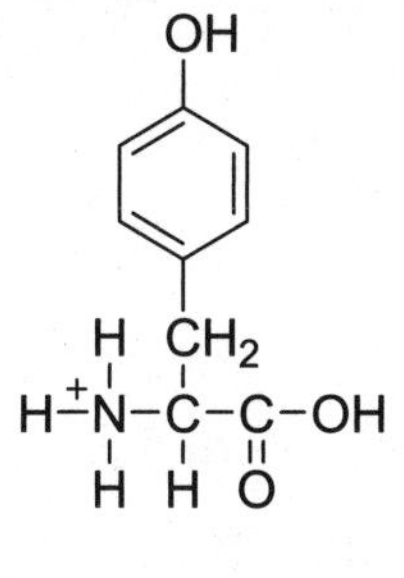

```
         OH
         |
      [benzene ring]
         |
   H    CH2
   |+   |
H–N–C–C–OH
   |    |   ||
   H    H   O
```

b. *At physiological pH, the amine is in its conjugate acid form and the carboxylic acid functional group is in its conjugate base form.*

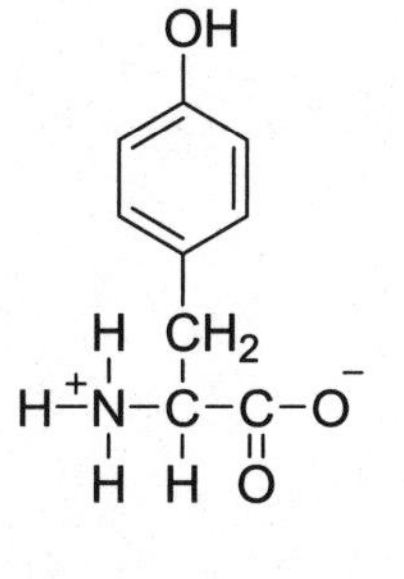

```
         OH
         |
      [benzene ring]
         |
   H    CH2
   |+   |
H–N–C–C–O⁻
   |    |   ||
   H    H   O
```

c. *At pH = 10, the amine is in its neutral form and the carboxylic acid functional group is in its conjugate base form.*

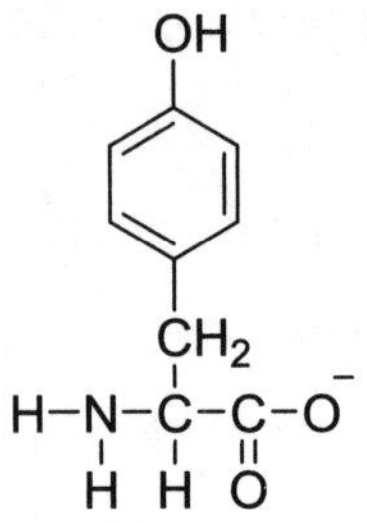

```
        OH
        |
     [benzene ring]
        |
       CH2
        |
  H-N - C - C - O⁻
    |   |   ||
    H   H   O
```

Try It Yourself #1

Draw the structure of threonine at the following pH's:

a. pH = 1
b. physiological pH
c. pH = 10

Solutions

a. *At pH 1, the amine functional group is in its (base or conjugate acid form)*

__________________________.

At pH 1, the carboxylic acid functional group is in its (acid or conjugate base form)

__________________________.

Structure of threonine at pH 1:

b. *At physiological pH, the amine functional group is in its (base or conjugate acid form)*

__________________________.

At physiological pH, the carboxylic acid functional group is in its (acid or conjugate base form) __________________________.

Structure of threonine at physiological pH:

c. *At pH = 10, the amine functional group is in its (base or conjugate acid form)*

_________________________.

At pH = 10, the carboxylic acid functional group is in its (acid or conjugate base form)

_________________________.

Structure of threonine at pH = 10:

Worked Example #2

What is the net charge on tyrosine at the pH's in Worked Example #1?

a. *At pH = 1, the amine is ionized with a +1 charge and the carboxylic acid is unionized with no charge. The amino acid has a net +1 charge.*
b. *At physiological pH, the amine is ionized with a +1 charge and the carboxylic acid is ionized with −1 charge. The amino acid has a net charge of zero.*
c. *At pH = 10, the amine is unionized with no charge and the carboxylic acid is ionized with −1 charge. The amino acid has a net −1 charge.*

Try It Yourself #2

What is the net charge on threonine at the pH's in Try It Yourself #1?

Solutions

a. *At pH = 1:*

   The amine has ____________ *charge.*

   The carboxylic acid has ________________________ *charge.*

   The net charge is ________________________________.

b. *At physiological pH:*

   The amine has ____________ *charge.*

   The carboxylic acid has ________________________ *charge.*

The net charge is ____________________________.

c. *At pH = 10:*

The amine has ____________ charge.

The carboxylic acid has _______________ charge.

The net charge is ____________________________.

Worked Example #3

One of the 20 natural amino acids is shown below.

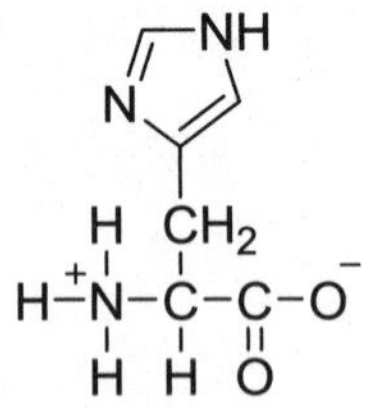

a. Circle and label the amine. In what form is this functional group at physiological pH?
b. Circle and label the carboxylic acid functional group that is present in all amino acids. In what form is this functional group at physiological pH?
c. Circle and label the side chain. What is the name of this amino acid?
d. Is the side chain polar or nonpolar? If it is polar, is it neutral, acidic, or basic?

Solutions

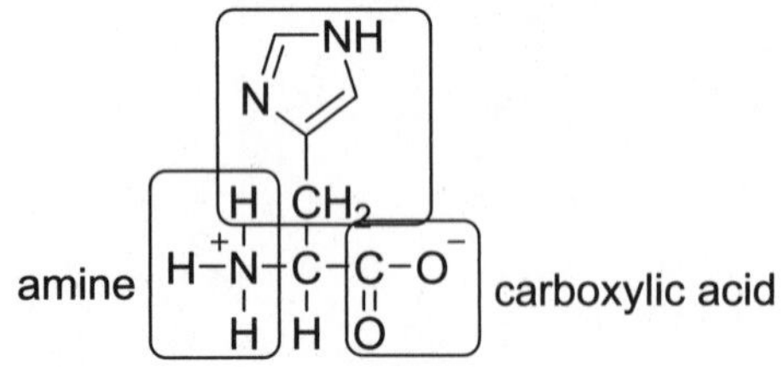

side chain

amine

carboxylic acid

a. *Normally, the amine is its conjugate acid form, —NH_3^+ at physiological pH.*
b. *Normally, the carboxylic acid is in its conjugate base form —CO_2^- at physiological pH.*
c. *The R group is shown above. This amino acid is histidine.*
d. *The side chain is polar and it is basic.*

Try It Yourself #3

One of the 20 natural amino acids is shown below.

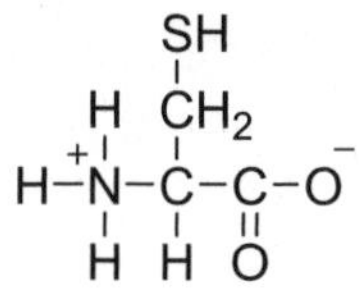

```
          SH
          |
      H  CH2
     +|   |       -
   H-N - C - C - O
      |   |   ||
      H   H   O
```

a. Circle and label the amine. Is it in its neutral or ionized form?

b. Circle and label the carboxylic acid functional group that is present in all amino acids. Is it in its neutral or ionized form?

c. Circle and label the side chain. What is the name of this amino acid?

d. Is the side chain polar or nonpolar? If it is polar, is it neutral, acidic, or basic?

Solutions

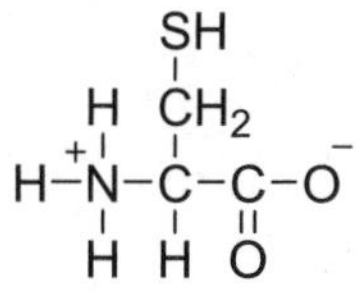

```
          SH
          |
      H  CH2
     +|   |       -
   H-N - C - C - O
      |   |   ||
      H   H   O
```

a. *The amine is in its ___________ form.*

b. *The carboxylic acid is in its ________________ form.*

c. *The name of the amino acid is: ___________________.*

d. *The side chain is:*

In a Nutshell: Chirality of the Amino Acids

Nineteen of the 20 natural amino acids are chiral because the α-carbon is a center of chirality. There are four different atoms or groups bonded to the α-carbon: the amine, the carboxylic acid, the R group, and a hydrogen atom. Glycine is the only natural amino acid that is achiral because it has two hydrogen atoms on the α-carbon. The 19 natural chiral amino acids are L-amino acids. In a Fischer projection, an L-amino acid has the amine positioned horizontally at the left and the H atom positioned horizontally at the right, while a D-amino acid has the reverse arrangement. By convention, the carboxylic acid is always drawn on the vertical bond at the top, and the side chain is drawn vertically at the bottom. The side chain is drawn in condensed notation, even if there is another center of chirality in the side chain.

Proteins are chiral because they are constructed from L-amino acids. Proteins provide a chiral environment, which accounts for many of the different physiological effects observed in enantiomeric drug molecules.

Worked Example #4

The Fischer projection of one of the natural amino acids is shown below.

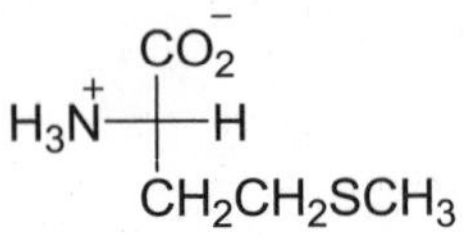

CO_2^-

$H_3\overset{+}{N}$—+—H

$CH_2CH_2SCH_3$

a. Which amino acid is shown?
b. Is this amino acid D- or L-? Explain.
c. Place an arrow pointing to the α-carbon.
d. Is this molecule chiral?
e. Write the Fischer projection of the enantiomer of this amino acid. How would it be named?

a. *The side chain of the amino acid is $CH_2CH_2SCH_3$. The amino acid is L-methionine.*
b. *The amino acid is the L-amino acid. The amine group is on the left of the α-carbon.*
c.

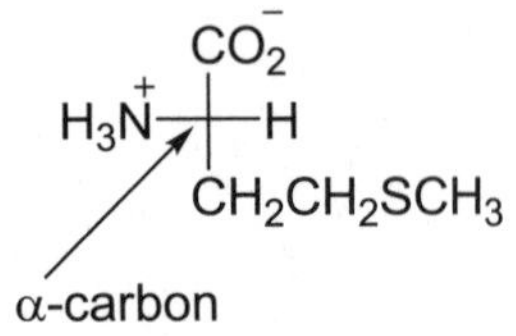

CO_2^-

$H_3\overset{+}{N}$—+—H

$CH_2CH_2SCH_3$

α-carbon

d. *The molecule is chiral because it is nonsuperimposable on its mirror image.*
e. *D-methionine*

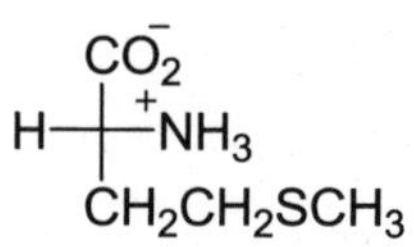

$$\begin{array}{c} CO_2^- \\ H-\!\!\!+\!\!\!-\overset{+}{N}H_3 \\ CH_2CH_2SCH_3 \end{array}$$

Try It Yourself #4

The Fischer projection of one of the natural amino acids is shown below.

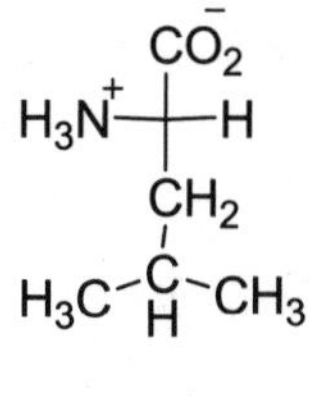

$$\begin{array}{c} CO_2^- \\ H_3\overset{+}{N}-\!\!\!+\!\!\!-H \\ CH_2 \\ | \\ H_3C-\underset{H}{C}-CH_3 \end{array}$$

a. Which amino acid is shown?

b. Is this amino acid D- or L-? Explain.

c. Place an arrow pointing to the α-carbon.

d. Is this molecule chiral?

e. Write the Fischer projection of the enantiomer of this amino acid. How would it be named?

Solutions

a. *The side chain for the amino acid is:*

The name of the amino acid is: ____________________.

b. *The amine group is on the (left or right) of the α-carbon.*

The amino acid is: ____________________.

c.

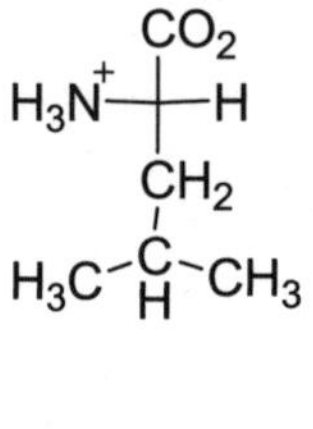

$$\begin{array}{c} CO_2^- \\ H_3\overset{+}{N}-\!\!\!+\!\!\!-H \\ CH_2 \\ | \\ H_3C-\underset{H}{C}-CH_3 \end{array}$$

d. *The mirror image of this amino acid is (superimposable or nonsuperimposable). The amino acid is:* __________________.

e. *The name of the enantiomer is:* __________________.

Structure of the enantiomer:

Practice Problems for Amino Acids

1. For each of the amino acids below, place an "X" in the boxes that apply.

Amino acid	Nonpolar	Polar	Acidic side chain	Basic side chain	Neutral side chain	Essential amino acid
Tryptophan						
Cysteine						
Aspartic acid						
Arginine						

2. Draw the structure of lysine at pH 10. What is the net charge on lysine at pH 10?

3. The Fischer projection for an amino acid is shown below:

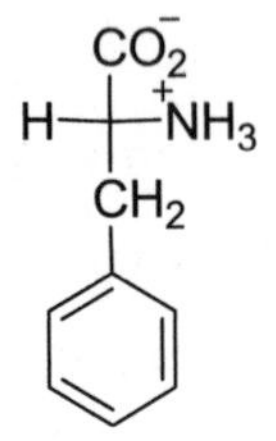

CO_2^-
H—+—$\overset{+}{N}H_3$
CH_2

a. Which amino acid is shown?

b. Is this amino acid D- or L-? Explain.

c. Place an arrow pointing to the α-carbon.

d. Write the Fischer projection of the enantiomer of this amino acid. How would it be named?

13.2 Peptides

Amino acids are the building blocks for all proteins. A dipeptide is formed when two amino acids undergo a reaction between the amine of one amino acid and the carboxylic acid of another amino acid to form a peptide bond joining the two amino acids. Polypeptide or peptides are polymers of amino acids, which are joined together by amide bonds.

In a Nutshell: Peptide Bonds

An amidation reaction forms an amide from a carboxylic acid and an amine. Amidation reactions form the carbon-nitrogen bond of an amide functional group. Similarly, when two amino acids undergo an amidation reaction, the carboxylic acid of one amino acid reacts with the amine of the other amino acids to form an amide functional group. The carbon-nitrogen bond formed between two amino acids is known as a peptide bond. Note that the reverse reaction forming two amino acids from a dipeptide and water is known as a hydrolysis reaction.

The reaction of a dipeptide with a third amino acid produces a tripeptide. Every amino acid is capable of forming two peptide bonds—one with its amine functional group and one with is carboxylic acid functional group. Thus, in a peptide, only the first and last amino acids in a peptide have a free amine and a free carboxylate group, respectively. The terminal end of a peptide containing the free amine group is known as the N-terminus; the terminal end containing the free carboxylate group is known as the C-terminus. The convention for writing the amino acid sequence for peptides is to list the amino acids in order from the N-terminus to the C-terminus, using the amino acid abbreviations.

Worked Example #5

A skeletal line structure for a tetrapeptide, tuftsin, which is used in immune system function, is shown below:

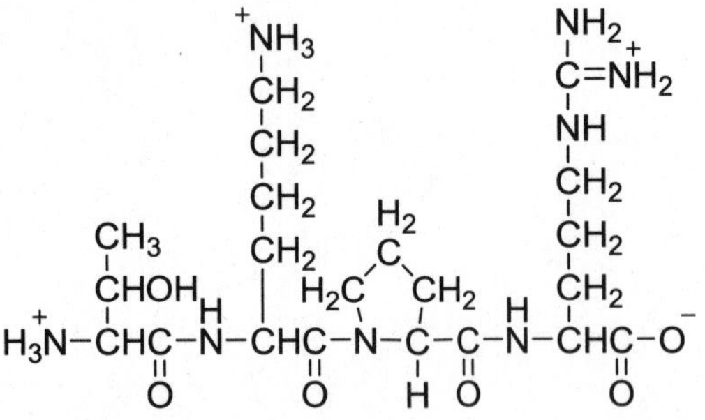

```
                +                         NH2 +
                 NH3                      |
                 |                        C=NH2
                 CH2                      |
                 |                        NH
                 CH2                      |
                 |                        CH2
                 CH2      H2              |
     CH3         |        C               CH2
     |           CH2    /   \             |
     CHOH   H    |    H2C    CH2    H     CH2
 +   |      |    |      \    |      |     |          -
 H3N-CHC----N----CHC-----N---C---C--N-----CHC--------O
        ||          ||       |   ||          ||
        O           O        H   O           O
```

a. Circle each amino acid in the peptide and identify it by its three-letter abbreviation.

b. Label the N-terminus and the C-terminus.

c. What is the amino acid sequence for this peptide?

d. Identify each peptide bond.

Solutions

a.

Thr *Lys* *Pro* *Arg*

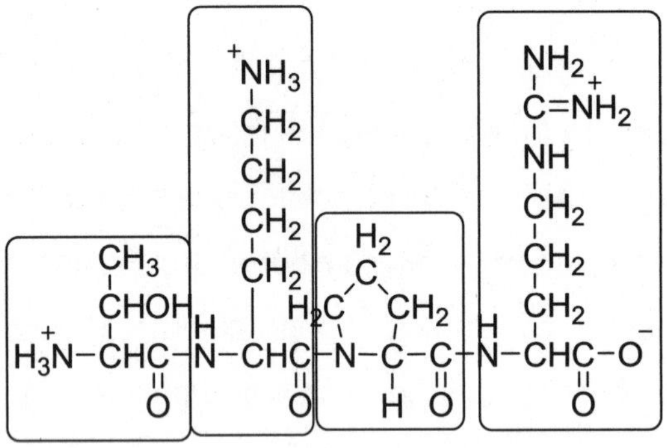

```
 Thr                 Lys           Pro          Arg
 [H3N-CHC(CHOH-CH3)] [N(H)-CHC(=O)-(CH2)4-NH3+] [N-C(H)-C(=O) ring] [N(H)-CHC(=O)-O- with (CH2)3-NH-C(=NH2+)-NH2]
```

b.

```
                +                         NH2 +
                 NH3                      |
                 |                        C=NH2
                 CH2                      |
                 |                        NH
                 CH2                      |
                 |                        CH2
                 CH2      H2              |
     CH3         |        C               CH2
     |           CH2    /   \             |
     CHOH   H    |    H2C    CH2    H     CH2
 +   |      |    |      \    |      |     |          -
 H3N-CHC----N----CHC-----N---C---C--N-----CHC--------O
        ||          ||       |   ||          ||
        O           O        H   O           O
```

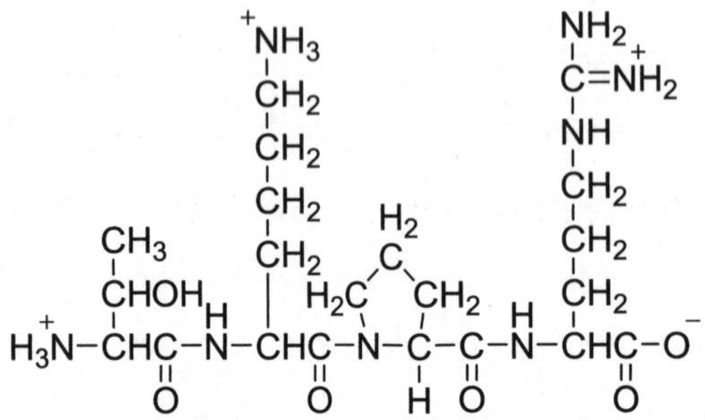

N-terminus *C-terminus*

c. *Thr-Lys-Pro-Arg*

d.

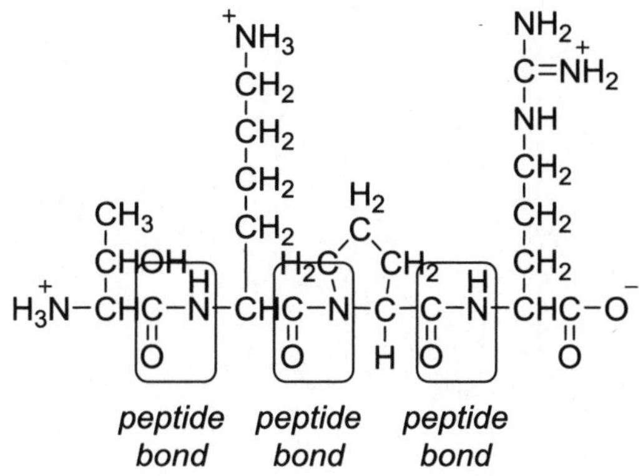

```
           +NH3              NH2
            |                 |  +
            CH2               C=NH2
            |                 |
            CH2               NH
            |                 |
            CH2     H2        CH2
   CH3      |       C         |
   |        CH2   /   \       CH2
   CHOH     |   H2C    CH2    |
 + |      H |     \     | H   CH2         -
H3N-CHC-N-CHC-N-C-C-N-CHC-O
       ||        ||    |  ||       ||
       O         O     H  O        O
```

peptide bond *peptide bond* *peptide bond*

Try It Yourself #5

The structure of a peptide is shown below:

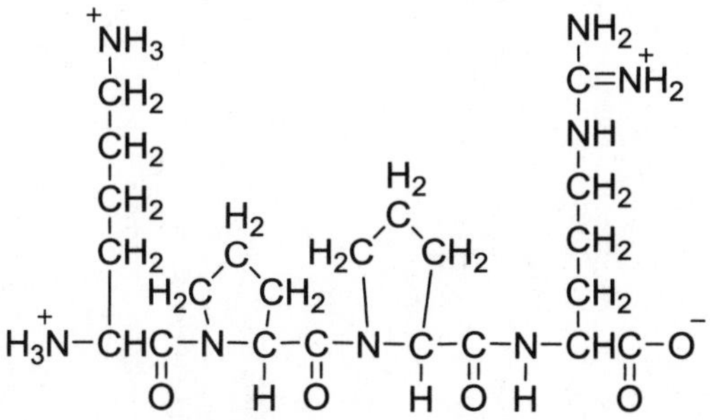

```
 +                            NH2
 NH3                           |  +
 |                             C=NH2
 CH2                           |
 |                             NH
 CH2                  H2       |
 |                    C        CH2
 CH2      H2        /   \      |
 |        C      H2C    CH2    CH2
 CH2    /   \     |     /      |
 |   H2C    CH2   |    /       CH2
+|      \    |    |   /        |         -
H3N-CHC-N-C-C-N-C-C-N-CHC-O
        ||      |  ||     |  ||  |        ||
        O       H  O      H  O   H        O
```

a. How many amino acids are in this peptide? Is this peptide a di-, tri-, tetra-, or pentapeptide?

b. Circle each amino acid in the peptide and identify it by its three-letter abbreviation.

c. What amino acid is the N-terminus?

d. What amino acid is the C-terminus?

e. How many peptide bonds are in this peptide?

f. Write the amino acid sequence for this peptide.

a. *There are ______ amino acids in this peptide. This peptide is a:*

________________________.

b.

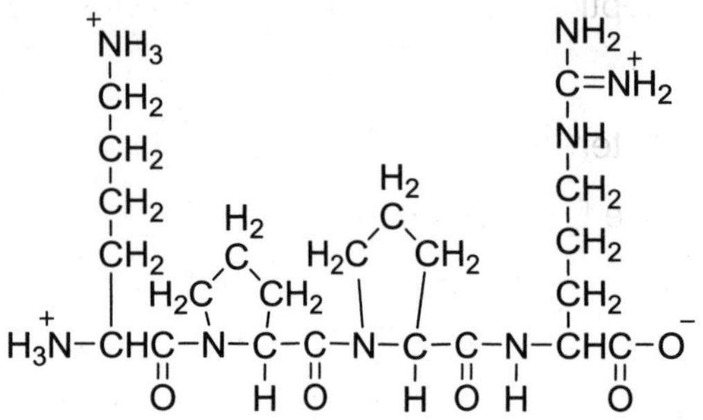

```
    +
    NH3                                  NH2
    |                                    |   +
    CH2                                  C=NH2
    |                                    |
    CH2                                  NH
    |                       H2           |
    CH2      H2             C            CH2
    |        C          H2C   CH2        |
    CH2    /   \         |     |         CH2
    |   H2C     CH2      |     |         |
+   |    \      |        |     |         CH2
H3N-CHC-N-------C-C------N-----C-C-N-----CHC-O-
      ||        | ||           | || |      ||
      O         H O            H O  H      O
```

c. *The N-terminus is:* ____________________.

d. *The C-terminus is:* ____________________.

e. *There are* __________ *peptide bonds in this peptide.*

f. *The amino acid sequence is:* ____________________________.

Practice Problems for Peptides

1. Write the structural formula for the tripeptide Gly-His-Lys.

2. Angiotensin II plays an important role in the biochemical system that controls blood volume and blood pressure in the body. Its structure is shown below:

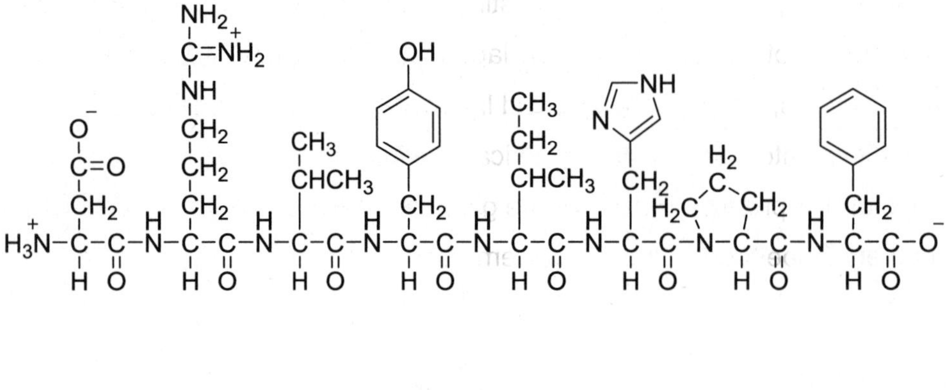

```
                NH2
                |  +
                C=NH2               OH
                |                                      -NH
                NH                                CH3  N
     O-         |                  [ring]         |     [ring]                     [ring]
     |          CH2       CH3       |             CH2     |             H2          |
     C=O        CH2       CHCH3     |             CHCH3  CH2            C           |
     |          |         |         CH2           |       |   H2C   CH2             CH2
     CH2     H  CH2    H  |      H  |        H    |    H  |    |     |     H        |
+    |       |  |      |  |      |  |        |    |    |  |    |     |     |        |      -
H3N-C-C-----N-C-C-----N-C-C-----N-C-C-------N-C-C-----N-C-C----N-----C-C---N--------C-C-O
    | ||      | ||      | ||      | ||        | ||      | ||         | ||           | ||
    H O       H O       H O       H O         H O       H O          H O            H O
```

a. Circle each amino acid in the peptide and identify it by its three-letter abbreviation.
b. Label the N-terminus and the C-terminus.
c. What is the amino acid sequence for this peptide?

3. How many amino acids are found in a pentapeptide? How many peptide bonds are in a pentapetide?

13.3 Types of Proteins and Protein Architecture

Proteins are composed of anywhere from 50 to thousands of amino acids. All proteins are built from the 20 L-amino acids. The information with the amino acid sequence for every protein the cell needs is encoded in our DNA.

In a Nutshell: Types of Proteins

There are three general classes of proteins based on their shape and solubility: fibrous proteins, globular proteins, and membrane proteins. Fibrous proteins tend to be strong and insoluble in water; therefore, they generally serve a structural role in nature. Three basic types of fibrous proteins include keratins, elastins, and collagens. Keratins provide structure to hair and fingernails. Elastins are fibrous proteins with elastic fibers that provide flexibility to skin, blood vessels, the heart, lungs, intestines, tendons, and ligaments. Collagens are the main structural proteins in the body. Collagens provide rigidity to the connective tissue found in skin, tendons, bones, cartilage, and ligaments. Globular proteins are soluble polypeptides folded into complex and spherical shapes. Hemoglobin and insulin are examples of globular proteins. Enzymes are globule proteins. Membrane proteins are found in part or entirely embedded with the cell membrane.

In a Nutshell: The Three-Dimensional Shape of Proteins

The three-dimensional shape of a protein is not linear; the polypeptide chain folds into a unique three-dimensional shape, known as the native conformation, which is necessary for the protein to be able to form its specific biological function. To accomplish their diverse array of functions, proteins come in a variety of different shapes and sizes. There are four hierarchical levels of architecture for protein folding: 1) primary structure, 2) secondary structure, 3) tertiary structure, and 4) quaternary structure.

In a Nutshell: The Primary Structure of Proteins

The primary structure of a protein is defined as its amino acid sequence or the sequence of amino acids for each of the polypeptides when a protein is composed of more than one polypeptide. Errors in the primary structure can have profound effects on its overall three-dimensional shape.

Worked Example #6

Thyroid stimulating hormone (TSH) is composed of two polypeptide chains. How many N-termini does it have? How many C-termini does it have?

Solution

There are two N-termini and two C-termini because there are two polypeptide chains.

Try It Yourself #6

What could happen to the shape of a protein if one of the amino acids in the protein sequence is changed?

Solution:

In a Nutshell: Secondary Structure

The secondary structure of a protein describes the regular folding patterns seen in localized regions of the polypeptide backbone. The polypeptide backbone of a protein is the repeating N—C_α—C=O atoms common to all polypeptides, basically all but the side chains. Since proteins are such large molecules, the polypeptide backbone is often depicted using ribbon diagrams, where the polypeptide backbone is represented as a flat, colored ribbon. Secondary protein structure arises from hydrogen bonding between amides in the polypeptide backbone. Hydrogen bonds are formed between the N—H of one amide and the carbonyl groups (C=O) of another amide along the polypeptide backbone.

The two most common types of secondary structure are the α-helix and the β-pleated sheet. An α-helix is a segment of polypeptide that coils in the shape of a helix. The helix is held in shape by hydrogen bonds between N—H and C=O groups on the polypeptide backbone. The carbonyl group of one amino acid forms a hydrogen bond with the hydrogen atom of an N—H bond on the amide of an amino acid three and a half amino acids farther down the polypeptide chain. The side chains extend outward from the α-helix and do not participate in the hydrogen bonding that creates and stabilizes the α-helix. In a ribbon drawing an α-helix appears as a coil.

A β-pleated sheet, also known as a β-sheet, is another type of secondary structure formed when two or more parallel or antiparallel sections of a polypeptide strand fold in a pattern that looks like a pleated skirt. Hydrogen bonding between N—H and C=O groups of aligned amides along two polypeptide backbones stabilizes the β-sheet. The side chains on the polypeptide in a β-pleated sheet project above and below. In a ribbon drawing the β-pleated sheet is represented by a straight, flat, wide arrow, with the arrow pointing in the direction of the C-terminus of the polypeptide chain. The segment of polypeptide forming a β-pleated sheet can have its arrows either pointing the same direction (parallel β-pleated sheet) or in the opposite direction (antiparallel β-pleated sheet).

Proteins vary in the extent and type of secondary structure they contain. One protein might have only α-helices throughout its structure, while another might contain some sections of α-helix and some sections of β-pleated sheet.

Worked Example #7

In a ribbon drawing, does a coil or a flat arrow represent a β-pleated sheet? What type of intermolecular force holds the β-pleated sheet together?

Solution

A wide, flat arrow is used to represent a β-pleated sheet in a ribbon drawing. Hydrogen bonding between the N—H and C═O groups holds the β-pleated sheet together.

Try It Yourself #7

In a ribbon drawing, does a wide, flat arrow or a coil represent an α-helix? Where are the side chains of an amino acid located in the α-helix?

An α-helix is represented by a: ______________________.

The side chains are located: ________________________.

In a Nutshell: Tertiary Structure

The tertiary structure of a protein describes the complex three-dimensional shape that results from the elaborate folding of a single polypeptide over and above its secondary structure. Tertiary structure arises from electrostatic interactions between the side chains of a single polypeptide as well as hydrophilic and hydrophobic interactions with the environment. For some proteins, tertiary structure also includes prosthetic groups: essential, non–peptide-containing organic molecules or metal ions that form ionic or covalent bonds to the polypeptide.

The fundamental driving force behind tertiary structure—protein folding—is lowering the potential energy of the protein and its surrounding environment. A folded protein and its surrounding environment are lower in potential energy than the unfolded protein and its surrounding environment. Proteins fold spontaneously into their native conformation under physiological conditions. The way a protein folds is unique and ideally suited to its biological function.

Folding occurs as a result of favorable electrostatic interactions between side chains located far apart in the primary structure as well as interactions of the side chains with the surrounding environment. Enzymes that exist in an aqueous medium will typically fold in a manner that places their hydrophilic polar side chains on the exterior of the protein, where they can interact with water molecules, while orienting their hydrophobic side chains on the interior of the protein, where they can avoid water can interact with other side chains through dispersion forces.

There are four electrostatic and covalent interactions within a single polypeptide chain that are contributes to the tertiary structure of a protein: 1) disulfide bridges (covalent bond between two cysteines), 2) salt bridges (ionic bonds, hydrophilic), 3) hydrogen bonding (electrostatic interaction, hydrophilic), and 4) dispersion forces (electrostatic interaction, hydrophobic).

Disulfide bridges, also known as disulfide bonds, are covalent bonds formed between the side chains on two cysteines located on either the same polypeptide or on different polypeptide chains of the same protein. Since a disulfide bridge is a covalent bond, it can only be formed or broken in a chemical reaction. The thiol (—S—H) functional groups on the side chains of two cysteines can undergo oxidation to form a disulfide bond (—S—S—), the strongest interaction between side chains.

Ionic bonds between the side chains of certain amino acids, also known as salt bridges, are formed between a positively charged polar, basic side chain and a negatively charged polar, acidic side chain. These interactions are stronger than hydrogen bonds.

Hydrogen bonding occurs between the side chains of the polypeptide containing O—H or N—H bonds and carbonyl groups. Although hydrogen bonds are weaker than disulfide and ionic bridges, there are usually so many of them in a protein that they contribute significantly to the structure of the protein. Tertiary structure is also influenced by hydrogen bonding interactions that can occur between polar side chains and the aqueous environment surrounding most proteins. This interaction is why proteins fold in a manner that places polar side chains on the surface of a protein, where they can form hydrogen bonds with water molecules, known as hydrophilic interactions.

Nonpolar side chains interact with other nonpolar side chains through dispersion forces. Nonpolar side chains are hydrophobic and avoid an aqueous medium. Thus, in an aqueous environment, proteins tend to fold so that sections of the protein with nonpolar side chains arrange themselves so that they are on the interior of the protein, away from the surrounding aqueous environment. In contrast, nonpolar side chains are found on the surface of membrane proteins because their external environment, the cell membrane, is nonpolar.

Worked Example #8

Label the following interactions as a disulfide bond, a salt bridge, hydrogen bonding, or dispersion forces.

a.

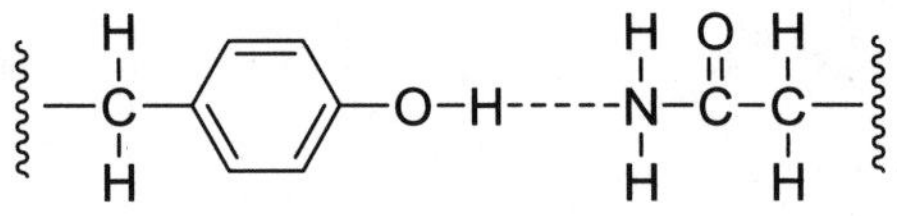

b.

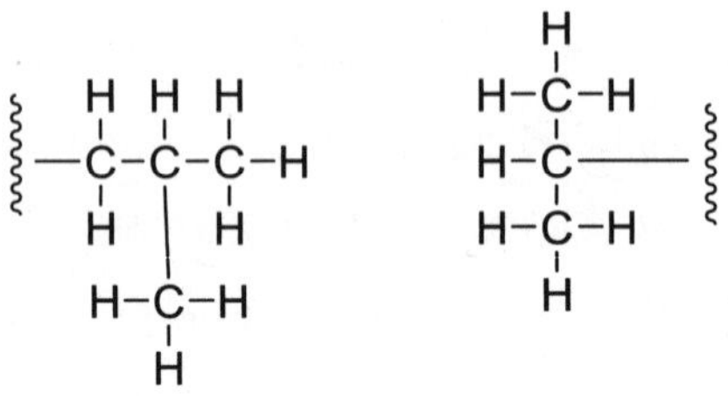

c.

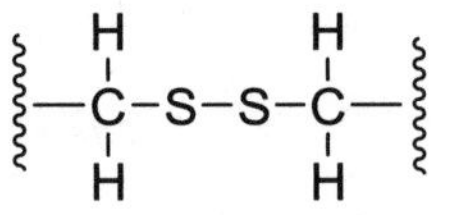

d.

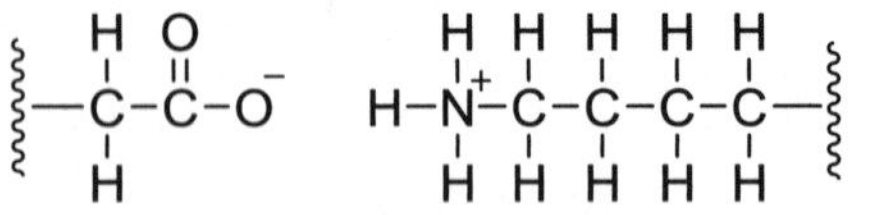

Solutions

a. *Hydrogen bonding. The bond is between an O—H group and an N—H group.*
b. *Dispersion forces. The interaction is between two nonpolar side chains.*
c. *Disulfide bond*
d. *Salt bridge. The interaction is between a negative charge, COO^-, and NH_3^+.*

Try It Yourself #8

Label the following interactions as a disulfide bond, a salt bridge, hydrogen bonding, or dispersion forces.

a.

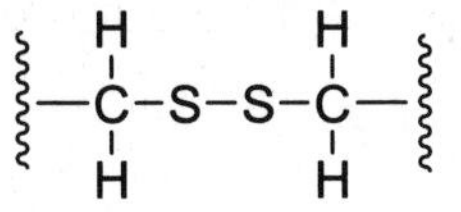

```
   H       H
   |       |
 —C–S–S–C—
   |       |
   H       H
```

b.

```
   H O             H     H H H H
   | ||           +|     | | | |
 —C–C–O⁻     H–N=C–N–C–C–C—
   |              |      | | |
   H             NH2     H H H
```

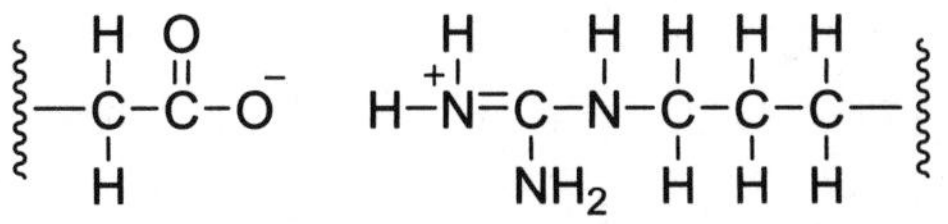

c.

```
   H H O H             H
   | | || |            |
 —C–C–C–N-----H–O–C—
   | |    |            |
   H H    H          H–C–H
                       |
                       H
```

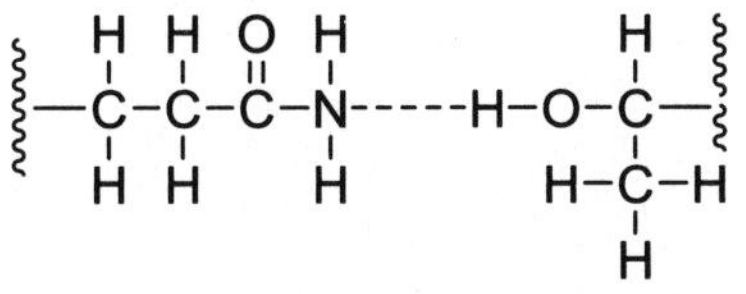

d.

```
                      H
                      |
     H H H          H–C–H
     | | |            |
 ——C–C–C–H        H–C——
   |   | |            |
   |   H H          H–C–H
   |                  |
 H–C–H                H
   |
   H
```

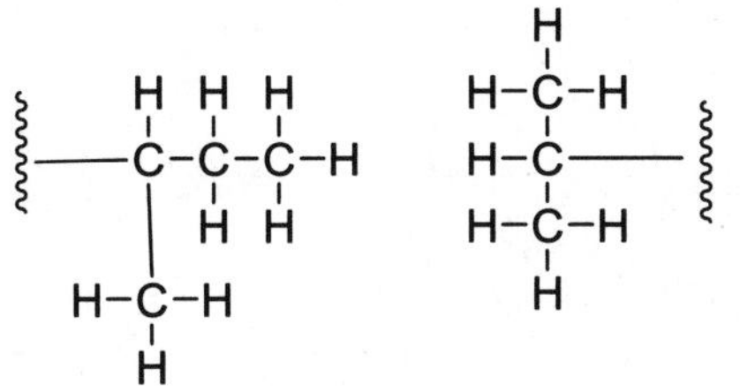

Solutions

a. *The interaction is between:* ____________________.

The interaction is: ____________________.

b. *The interaction is between:* ____________________.

The interaction is: ____________________.

c. *The interaction is between:* ____________________.

The interaction is: ____________________.

d. *The interaction is between:* ____________________.

The interaction is: ____________________.

In a Nutshell: Quaternary Structure

Many proteins are comprised of two or more polypeptide chains, giving rise to quaternary protein structure. The quaternary structure of a protein describes the interactions of two or more polypeptide chains that create the overall three-dimensional shape of the functional protein. The electrostatic and covalent interactions between polypeptide chains are the same kinds of interactions that stabilize the tertiary structure of a protein: disulfide bridges, salt bridges, hydrogen bonding, and dispersion forces, except they occur between polypeptide chains rather than within a single polypeptide chain.

Worked Example #9

For the following pairs of amino acids, indicate how they might interact to contribute to the tertiary or quaternary structure of a protein. Choose from the following interactions: disulfide bond, salt bridge, hydrogen bonding, or dispersion forces.

a. proline and alanine
b. threonine and glutamine
c. two cysteines
d. arginine and aspartic acid

a. *Dispersion forces. The two residues are nonpolar, so the interaction might be dispersion forces.*
b. *Hydrogen bonding. There is an O—H group in threonine and an N—H group in glutamine.*
c. *Disulfide bond. There is an S—H group on each cysteine.*
d. *Salt bridge. Arginine will have a positive charge on one of the nitrogen atoms in the side chain, and aspartic acid will have a negative charge on the carboxylate group in the side chain.*

Try It Yourself #9

For the following pairs of amino acids, indicate how they might interact to contribute to the tertiary or quaternary structure of a protein. Choose from the following interactions: disulfide bond, salt bridge, hydrogen bonding, or dispersion forces.

a. two cysteines
b. phenylalanine and valine

c. glutamic acid and histidine
d. asparagine and serine

a. The interaction is between: ________________.
The interaction is: ________________.
b. The interaction is between: ________________.
The interaction is: ________________.
c. The interaction is between: ________________.
The interaction is: ________________.
d. The interaction is between: ________________.
The interaction is: ________________.

Worked Example #10

A globular protein is found in an aqueous environment. What types of side chains would you expect to find on the interior of the protein? What types of side chains would you expect to find on the exterior of the protein?

Solution

The aqueous environment is polar; therefore, polar, hydrophilic side chains should be on the exterior of the protein. The hydrophobic, nonpolar side chains should be on the inside of the protein.

Try It Yourself #10

A membrane protein is found in the hydrophobic environment of a cell membrane. What types of side chains would you expect to find on the interior of the protein? What types of side chains would you expect to find on the exterior of the protein?

Solution

The cell membrane environment is: ________________.

The side chains on the exterior of a membrane protein are: ________________.
The side chains on the interior of a membrane protein are: ________________.

In a Nutshell: Denaturation of a Protein

Denaturation of a protein occurs when there is a disruption of the secondary, tertiary, or quaternary structure of a protein so that the protein can no longer perform its function. Denaturing a protein does not alter its primary structure, and in some instances can even be reversed. Proteins can be denatured in a number of ways including: addition of heat, pH changes, mechanical agitation, addition of detergents, and addition of certain metals.

A denatured protein loses its shape, and hence its biological activity as a result of the disruption in electrostatic interactions.

Practice Problems for Protein Architecture

1. In which levels of protein architecture are hydrogen bonding found?

2. What types of interactions are broken when a protein is denatured? What levels of protein architecture are affected when a protein is denatured?

3. A genetic mutation changed an aspartic acid residue that formed a salt bridge to a leucine residue in a protein. How might this change affect the secondary or tertiary structure of the protein?

13.4 Enzymes

In a Nutshell: How Do Enzymes Work?

Enzymes are essential for life because, in their absence, biochemical reactions are too slow. Nature's catalysts, enzymes, are specialized globular proteins, which increase the rate of biochemical reactions as much as 10^{16} times faster than in the absence of the enzyme. Generally, one enzyme catalyzes one reaction, acting specifically on one particular reactant, known as the substrate. Enzymes also ensure that the product of the enzyme catalyzed reaction is the correct stereoisomer when a chiral product is formed.

The enzyme itself is unchanged at the end of a biochemical reaction; thus, the same enzyme molecule may be used over and over again to catalyze the transformation of many substrate molecules into product molecules. In a chemical equation, the name of the enzyme is sometimes written above the reaction arrow.

The enzymes are generally named after the substrate or type of reaction they catalyze with a change in the ending to *-ase*. Furthermore, their names are usually italicized. Enzymes are classified by the types of reactions that they catalyze.

An enzyme-catalyzed reaction begins with the reversible begins with the reversible binding of the substrate (S) and the enzyme (E) to form an enzyme-substrate complex (ES), whereupon the reaction occurs and the enzyme releases the product, P, and free enzyme, E. The substrate binds to the enzyme, which is usually much larger in size than the substrate, at a pocket or groove somewhere on or within the protein, known as the active site or binding site. The three-dimensional shape of the active site is roughly complementary to the three-dimensional shape of the substrate, which is the reason we observe such selectivity of the enzyme for the substrate. Substrate selectivity is extremely important in biochemistry because there are so many other organic molecules present in solution at the same time. Within the active site of the enzyme, side chains bind the substrate through electrostatic interactions. An earlier model that described the binding of a substrate to an enzyme is the lock-and-key model. In this model, the enzyme and substrate have complementary shapes that allow them to fit together like a key fits into a lock. A more complementary refined model of how an enzyme binds to its substrate describes a conformational change in the enzyme following binding that improves the “fit” between the substrate and the enzyme, known as the induced-fit model. Thus, the enzyme appears to mold itself—induced fit—to the shape of the substrate.

While it is part of the enzyme-substrate complex (ES), the substrate and other reactants assume a position and orientation that places the reacting functional group(s) near one another, thereby facilitating the chemical reaction. It is the placement of the substrate in an optimal geometry for reaction that lowers the energy of activation, E_A, for the reaction, and thus, increases the rate of the reaction.

In a Nutshell: pH and Temperature Dependence of Enzymes

In order for an enzyme to perform its function, it must be in its native conformation, which is dependent on both temperature and pH. Changes in pH can alter the charges on side chains in acidic and basic amino acids of the enzyme, which in turn affects its shape. For most enzymes, the optimal pH is physiological pH (pH = 7.3). Acidosis, a serious medical condition caused by a drop in blood pH causes enzymes to denature.

Most enzymes also function best when around body temperature, 37 °C. Elevated body temperatures, as occurs with a high fever, can denature enzymes and decrease reaction rates. Lower body temperatures can also cause enzyme function to decrease and reaction rates to decrease.

Worked Example #11

How does an enzyme catalyze a reaction?

Solution

An enzyme catalyzes a reaction by holding the substrate in a position and orientation so that the reacting functional group(s) are near one another, thereby facilitating the chemical reaction. It is the placement of the substrate in an optimal geometry and proximity to the other reactant that lowers the energy of activation, E_A.

Try It Yourself #11

What interactions are responsible for binding an enzyme to a substrate?

The interactions are: ____.

In a Nutshell: Enzyme Inhibitors

Enzyme inhibitors are compounds that bind to an enzyme and prevent the enzyme from performing its function. Specific enzyme inhibitors target and inhibit a specific enzyme without affecting any other enzymes. There are two types of specific enzyme inhibitors: competitive inhibitors and noncompetitive inhibitors.

A competitive inhibitor is a substance that competes with the substrate for the active site of the enzyme because it too has a structure that is fits into the active site and can bind to the enzyme. By binding to the active side, a competitive inhibitor blocks the substrate from binding to the enzyme. As a result, the ES complex cannot form, and no reaction occurs. The effectiveness of a competitive inhibitor depends on the relative concentrations of substrate and inhibitor in solution. The higher the concentration of inhibitor, the less likely the substrate will find a free enzyme to bind to. Conversely, the higher the concentration of substrate, the more likely the enzyme-substrate complex will form, and therefore product is produced.

Noncompetitive inhibitors bind to the enzyme at a location other than the active site of the. Binding of a noncompetitive inhibitor affects the conformation of the enzyme, preventing it from properly binding the substrate (although the substrate can still bind) and thus preventing the formation of product. In contrast to competitive inhibitors, increasing the concentration of substrate will not restore enzyme activity when a noncompetitive inhibitor is bound to the enzyme. Enzyme activity can only be restored when the concentration of noncompetitive inhibitor decreases.

Worked Example #12

Celebrex® is an NSAID (nonsteroidal anti-inflammatory drug) that is a competitive inhibitor for the enzyme *cyclooxygenase 2,* COX-2. COX-2 is an enzyme that produces prostaglandins, which can cause pain and inflammation. Does Celebrex® bind to the active site of COX-2, or does it change the shape of COX-2?

Solution

Because Celebrex® is a competitive inhibitor for the COX-2 enzyme, it binds to the active site of the enzyme.

Try It Yourself #12

Cholesterol is a noncompetitive inhibitor of an enzyme required for its own synthesis, known as *HMG-CoA reductase.* Does cholesterol bind to the active site of *HMG-CoA reductase*, or does it change the shape of *HMG-CoA reductase*?

Solution:

Practice Problems for Enzymes

1. How is the rate of an enzyme-catalyzed reaction affected by the following changes?
   a. an increase in temperature

   b. an increase in pH

2. Is the active site of an enzyme a complementary shape to the substrate?

3. Do competitive inhibitors of an enzyme have the same shape as the substrate for that enzyme?

Chapter 13 Quiz

1. For each of the amino acids, place an "X" in the boxes that apply.

Amino acid	Nonpolar	Polar	Acidic side chain	Basic side chain	Neutral side chain	Essential amino acid
Glycine						
Serine						
Glutamic acid						
Histidine						

2. Which of the amino acids in Question 1 are chiral?

3. A peptide chain has the following sequence: Thr-Lys-Pro-Arg.
   a. What amino acid is the N-terminus?

   b. What amino acid is the C-terminus?

   c. How many peptide bonds are in this peptide?

   d. Which amino acids in this peptide are polar?

   e. How many amino acids in this peptide are chiral?

   f. Write the structure formula of this peptide.

4. Would you expect the amino acid aspartic acid to participate in hydrogen bonding, disulfide bridges, salt bridges, or dispersion forces in protein architecture? What level (primary, secondary, tertiary, or quaternary) of protein architecture would be involved?

5. Would you expect the amino acid serine to participate in hydrogen bonding, disulfide bridges, salt bridges, or dispersion forces in protein architecture? What level (primary, secondary, tertiary, or quaternary) of protein architecture would be involved?

6. What amino acid forms disulfide bridges? What functional group does this amino acid contain?

7. How does an enzyme catalyze a reaction?

8. Does a competitive inhibitor of an enzyme have the same shape as the substrate for the enzyme? How does a competitive inhibitor prevent an enzyme from performing its function?

9. Does a noncompetitive inhibitor of an enzyme have the same shape as the substrate for the enzyme? How does a noncompetitive inhibitor prevent an enzyme from performing its function?

10. Draw the structure of aspartic acid at pH 1. What is the net charge on aspartic acid at pH 1? Explain why low pH would denature a protein.

Chapter 13

Answers to Additional Exercises

45 Sickle-cell anemia is an inherited blood disorder. People who have the disease have crescent (sickle) shaped red blood cells. This shape affects the cells' ability to transport oxygen.

47 The symptoms of sickle cell anemia are severe pain, fatigue, and lethargy.

49 The β-chains of an individual with sickle-cell anemia have a valine as the sixth amino acid rather than a glutamic acid.

51 Treatment for sickle-cell anemia includes antibiotics taken for the first 5 years of life, periodic blood transfusions, and treatment with hydroxyurea.

53 a. Tyr

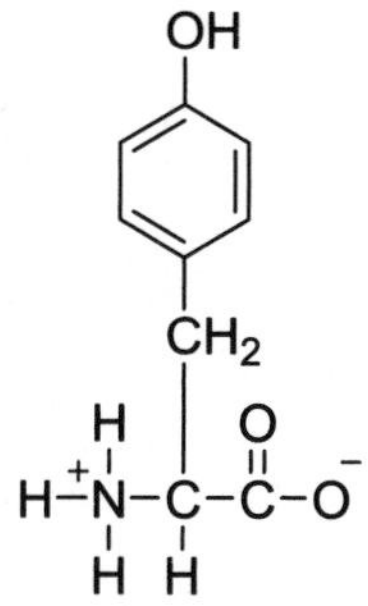

```
           OH
           |
        (benzene ring)
           |
          CH2
      H    |   O
     +|    |   ||   -
    H-N----C---C---O
      |    |
      H    H
```

b. Cys

```
          SH
          |
          CH2
      H   |   O
     +|   |   ||  -
    H-N---C---C---O
      |   |
      H   H
```

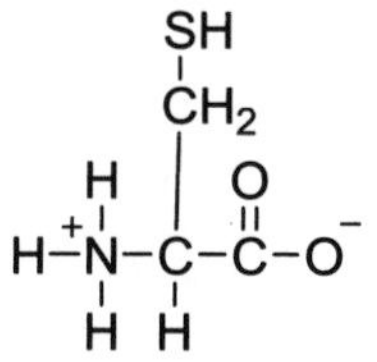

55. a. pH = 1

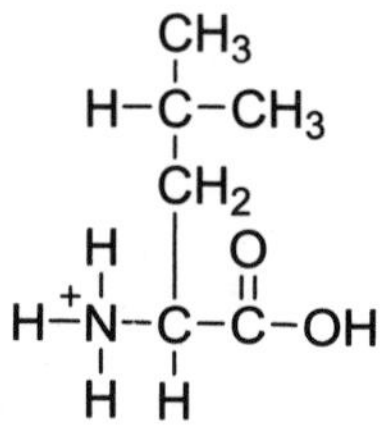

```
          CH3
          |
        H-C-CH3
          |
          CH2
      H   |   O
     +|   |   ||
    H-N---C---C-OH
      |   |
      H   H
```

b. physiological pH

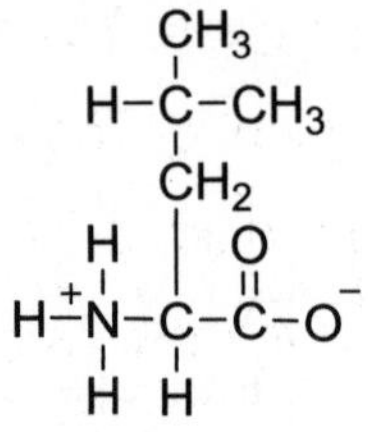

```
        CH3
        |
      H-C-CH3
        |
        CH2
   H    |  O
  +|    |  ||
 H-N----C--C-O⁻
   |    |
   H    H
```

c. pH = 10

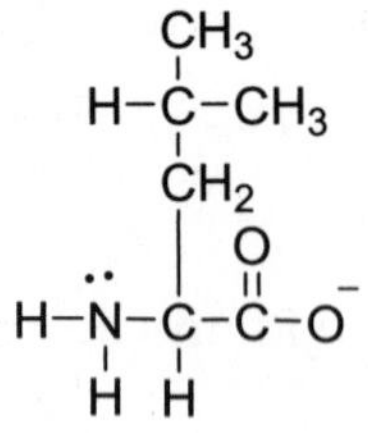

```
        CH3
        |
      H-C-CH3
        |
        CH2
        |  O
   ..   |  ||
 H-N----C--C-O⁻
   |    |
   H    H
```

57 a. pH = 1, charge +1

```
      (C6H5)
        |
        CH2
   H    |  O
  +|    |  ||
 H-N----C--C-OH
   |    |
   H    H
```

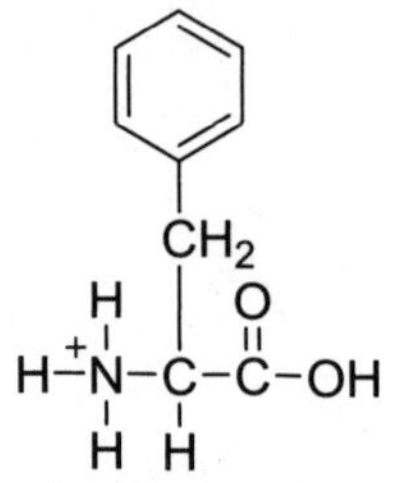

b. physiological pH, charge 0

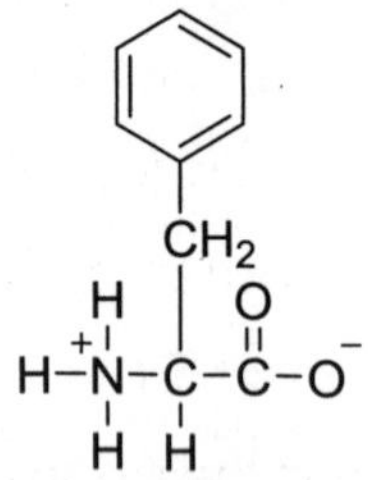

```
      (C6H5)
        |
        CH2
   H    |  O
  +|    |  ||
 H-N----C--C-O⁻
   |    |
   H    H
```

c. pH =10, charge −1

```
      (C6H5)
        |
        CH2
        |  O
   ..   |  ||
 H-N----C--C-O⁻
   |    |
   H    H
```

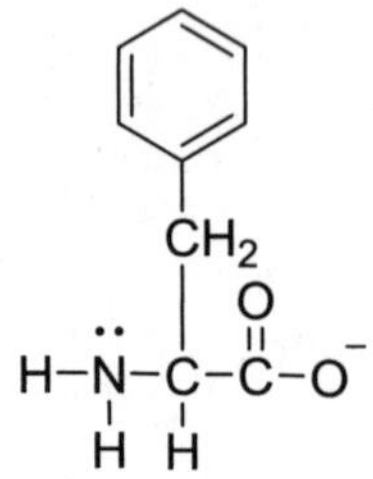

59 a. The zwitterion, the ionized amine form, water and hydronium ion are all present at equilibrium.

b. If more acid is added, the equilibrium will shift to the left.

61 a. The amino acids with nonpolar side chains are glycine, alanine, valine, leucine, isoleucine, proline, tryptophan, phenylalanine, and methionine. b. The amino acids with polar, basic side chains are lysine, arginine, and histidine. c. The amino acids with polar, acidic side chains are aspartic acid and glutamic acid. d. The amino acids with polar, neutral side chains are serine, threonine, tyrosine, cysteine, glutamine, and asparagine.

63 a. pH = 10

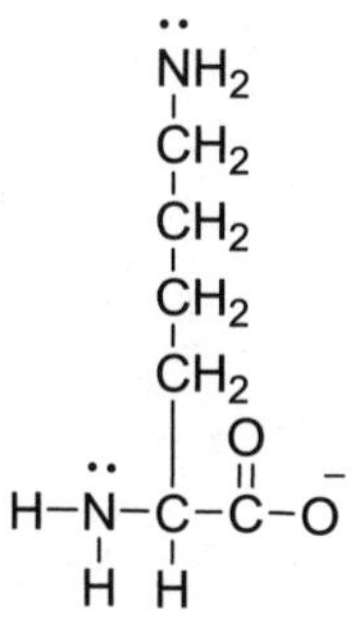

$$
\begin{array}{c}
\ddot{N}H_2 \\
| \\
CH_2 \\
| \\
CH_2 \\
| \\
CH_2 \\
| \\
CH_2 \\
| \\
H-\ddot{N}-C-\overset{\displaystyle O}{\overset{\|}{C}}-O^- \\
\quad\;\, | \quad\;\; | \\
\quad\;\, H \quad\; H
\end{array}
$$

b. physiological pH

$$
\begin{array}{c}
\overset{+}{N}H_3 \\
| \\
CH_2 \\
| \\
CH_2 \\
| \\
CH_2 \\
| \\
CH_2 \\
| \\
H-\overset{\displaystyle H}{\overset{|}{\overset{+}{N}}}-C-\overset{\displaystyle O}{\overset{\|}{C}}-O^- \\
\quad\;\, | \quad\;\; | \\
\quad\;\, H \quad\; H
\end{array}
$$

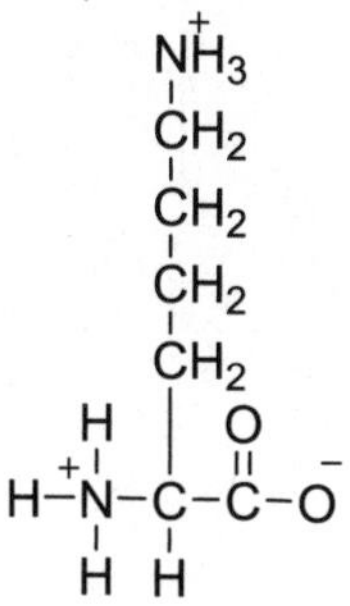

65 Essential amino acids are amino acids that are not synthesized by the body and must be supplied through the diet,

67 a. Aspartic acid and glutamic acid have carboxylic acids on their side chains. b. Glutamine and asparagine both have amides on their side chains.

69

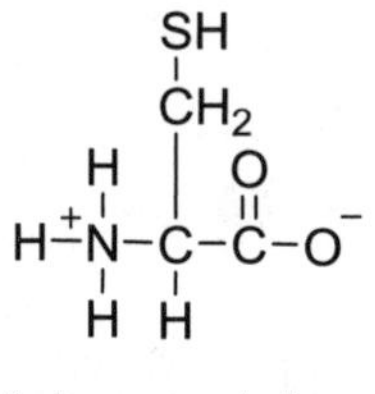

$$
\begin{array}{c}
SH \\
| \\
CH_2 \\
| \\
H-\overset{\displaystyle H}{\overset{|}{\overset{+}{N}}}-C-\overset{\displaystyle O}{\overset{\|}{C}}-O^- \\
\quad\;\, | \quad\;\; | \\
\quad\;\, H \quad\; H
\end{array}
$$

Cysteine contains a thiol functional group.

71

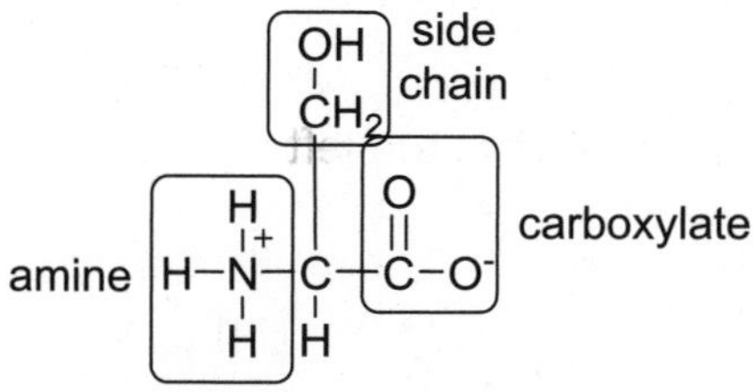

$$
\begin{array}{c}
\boxed{\begin{array}{c} OH \\ | \\ CH_2 \end{array}}\ \text{side chain} \\
| \\
\text{amine}\ \boxed{H-\overset{\displaystyle H}{\overset{|}{\overset{+}{N}}}-\underset{\displaystyle H}{\underset{|}{}}}\,C\,\boxed{-\overset{\displaystyle O}{\overset{\|}{C}}-O^-}\ \text{carboxylate} \\
\qquad\qquad\quad\ \ |\\
\qquad\qquad\quad\ \ H
\end{array}
$$

a. The amine is in its ionized form.

b. One of the oxygen atoms in the carboxylate group is missing a hydrogen atom and is negatively charged.

c. This amino acid is serine.

d. The side chain is polar. It contains an alcohol group.

e. At pH = 12 serine has the structure shown below. The net charge is −2.

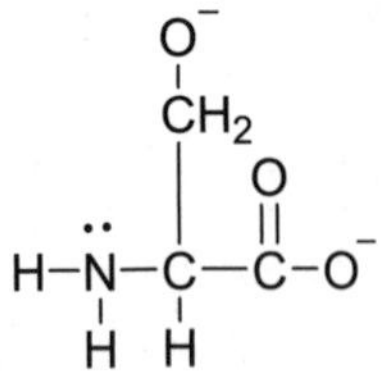

$$\mathrm{H{-}\ddot{N}H{-}CH(CH_2O^-){-}C(=O){-}O^-}$$

73

Amino acid	Nonpolar	Polar	Acidic	Basic	Neutral	Essential amino acid
Phe	x					x
Asn		x			x	
Met	x				x	x

75 Vegetarians need to eat a meal with a complementary mixture of plant proteins, such as grains with legumes.

77 a. Methionine

b. It is L-methionine. The —NH_3^+ group is pointing to the left in the Fischer projection.

c. Fischer projection: COO^- at top; $H_3\overset{+}{N}$ on left, H on right of the α carbon; below: CH_2–CH_2–S–CH_3

α carbon

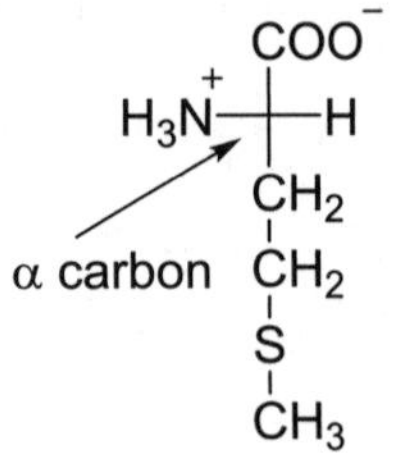

d. The molecule is chiral. There are four different groups attached to the α carbon.

e. D-methionine

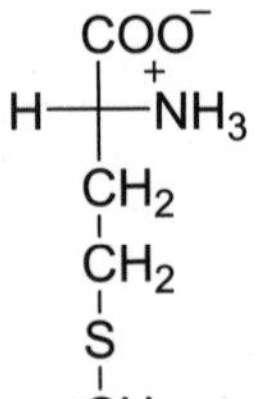

Fischer projection: COO^- at top; H on left, $\overset{+}{N}H_3$ on right; below: CH_2–CH_2–S–CH_3

79

Fischer projection: COO^- at top; $H_3\overset{+}{N}$ on left, H on right of the α carbon; below: CH bearing CH_3 and OH

α carbon

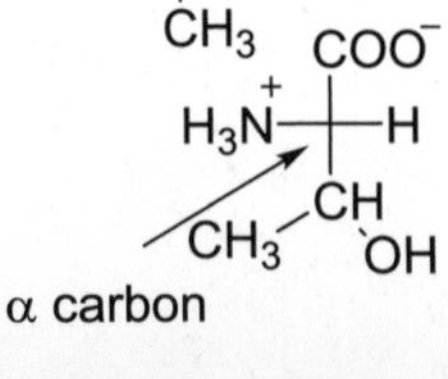

a. Yes this amino is found in nature.

b. See structure.

c. If the —NH_3^+ group is on the left it is an L- amino acid. If the —NH_3^+ group is on the right it is a D- amino acid.

d. D-threonine and L-threonine have the same four groups attached to the α-carbon, but the connectivity and spatial arrangement of the four groups is different.

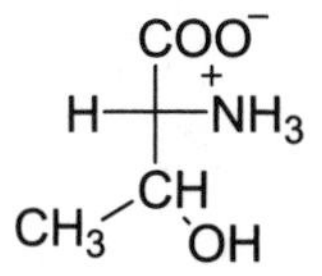

81 a. A hydrogen atom, a carboxylate group, a —NH_3^+ group and a $CH_2CH_2COO^-$ are attached to the α-carbon.

b. The carboxylate group and the —NH_3^+ group are present in all amino acids.

c. The $CH_2CH_2COO^-$ is unique to glutamic acid.

d. They are enantiomers. L-glutamic acid is found in nature.

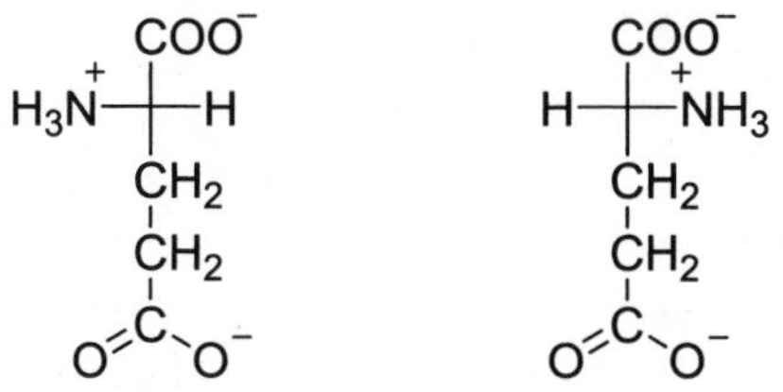

L-glutamic acid *D*-glutamic acid

83 A dipeptide contains two amino acids linked together by one amide (peptide) bond. A tripeptide contains three amino acids linked together with two amide (peptide) bonds. A polypeptide has more than 12 amino acids linked together by peptide bonds.

85 a.

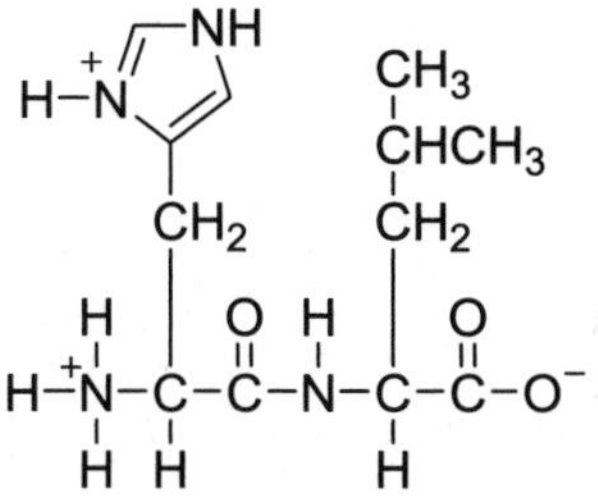

His-Leu

b.

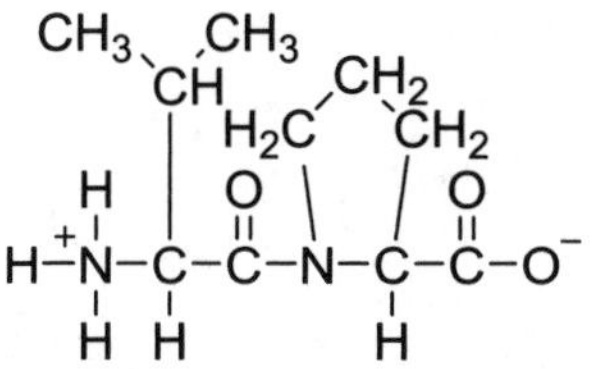

Val-Pro

c.

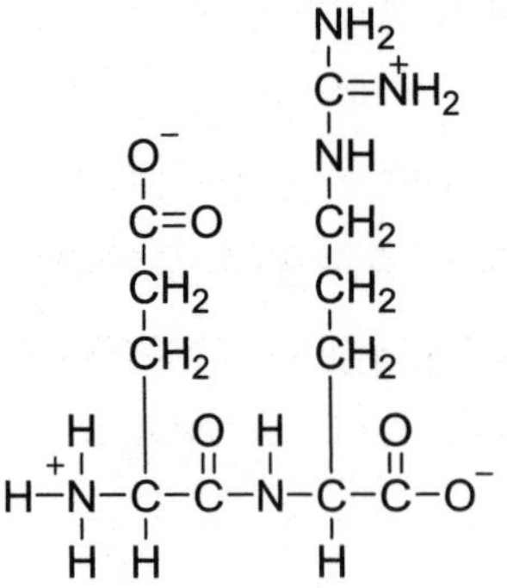

Glu-Arg

d.

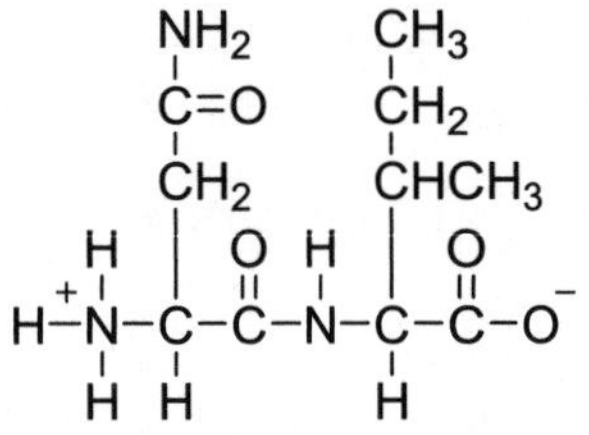

Asn-Ile

87

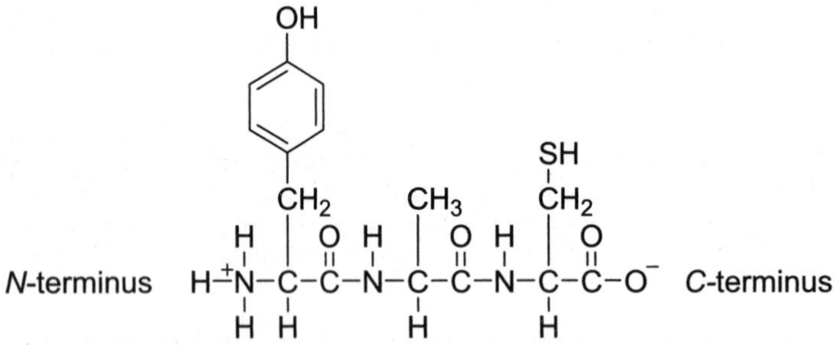

N-terminus *C*-terminus

89

a. Gln is the N-terminus.

b. Asp is the C-terminus.

c. This pentapetide contains four peptide bonds. This pentapeptide contains five amino acids.

d–e.

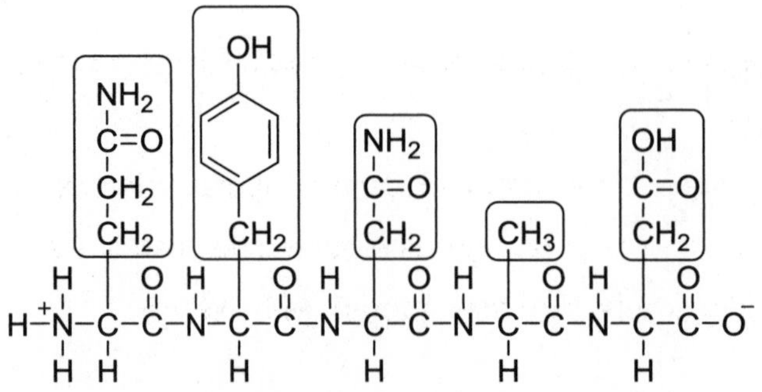

91 a. Bradykinin contains nine amino acids.

b. The amino acid sequence for bradykinin is Arg-Pro-Pro-Gly-Phe-Ser-Pro-Phe-Arg.

93 The native conformation of a protein is its unique three-dimensional shape, which is necessary for the protein to be able to perform its specific biological function. Intermolecular interactions, such as hydrogen bonding, electrostatic interactions, such as salt bridges, dispersion forces, and disulfide bridges, stabilize the protein in its native conformation.

95 In humans, hair and fingernails are rich in keratins. In animals, feathers, horns, claws, and hooves are rich in keratins.

97 Collagens are found in skin, tendons, bones, cartilage and ligaments. Collagens provide rigidity to the tissue.

99 Membrane proteins are found in part or entirely embedded in the cell membrane.

101 The primary structure of a protein is the sequence of amino acids that make up the protein.

103 The two most common forms of secondary structure are the α-helix and the β-pleated sheet. The secondary structure is formed by hydrogen bonding between different amino acids in the protein. The amide bonds in the polypeptide backbone are involved in the secondary structure. The side chains of a polypeptide are not involved in the secondary structure.

105 A prosthetic group is an essential non-peptide containing organic molecules or metal ions that form ionic or covalent bonds to the polypeptide. A prosthetic group is part of the tertiary structure.

107 Proteins fold because it lowers the potential energy of the protein and its surrounding environment. A folded protein and its surrounding environment are lower in potential energy than the unfolded protein and its surrounding environment.

109

```
       H  H          [O]        H  H       H  H
                                |  |       |  |
  2  H-C--C-S-H   -------->   H-C--C-S-S-C--C-H
       |  |                     |  |       |  |
       H  H                     H  H       H  H
```

111 Aspartic acid, glutamic acid, lysine, arginine, and histidine have side chains that can form salt bridges.

113 If the protein is an aqueous environment, the nonpolar side chains are most likely to be found on the interior of the protein. The polar side chains are most likely to be found on the exterior where “like may interact with like.”

115 b, d, and e. Isoleucine, valine, and proline are nonpolar and would be found in the interior of a protein in an aqueous environment.

117 Disulfide bridges, salt bridges, hydrogen bonding, and dispersion forces stabilize the quaternary structure of a protein.

119 a. Glutamic acid and lysine might form a salt bridge.

b. Tyrosine and lysine might hydrogen bond.

c. Isoleucine and alanine might interact through dispersion forces.

d. Two cysteines might form a disulfide bridge.

121 Heat, pH changes, mechanical agitation, detergents, and some metals may denature a protein. When a protein is denatured it loses its secondary, tertiary, and quaternary structure. Since the denatured protein has lost its shape, it has also lost its function. The primary structure of the protein is not affected by denaturation.

123

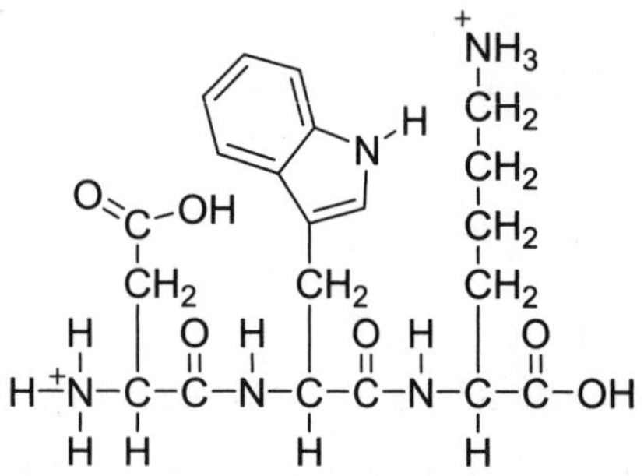

```
                                  +
                                  NH3
                                  |
                          N-H     CH2
                                  |
                                  CH2
      O=C-OH                      |
        |                         CH2
        CH2           CH2         CH2
   H    |    O   H    |    O   H  |    O
   |    |    ||  |    |    ||  |  |    ||
 H-+N - C -  C - N -  C -  C - N- C -  C - OH
   |    |              |              |
   H    H              H              H
```

125 Without enzymes, biological reactions are too slow for life.

127 energy of activation (E_A)

129 When a substrate binds to the active site, the enzyme changes its conformation to improve the “fit” between the substrate and the enzyme. This model is called the induced-fit model.

131 $E + S \rightleftharpoons ES \longrightarrow E + P$

E is the enzyme, S is the substrate, ES is the enzyme-substrate complex, and P is the product.

133 The substrate is also one particular reactant.

135 An enzyme is not chemically changed during an enzyme catalyzed reaction. The substrate is chemically changed to the product during an enzyme-catalyzed reaction.

137 Physiological pH (pH = 7.3) is the optimal pH, and body temperature (37 °C) is the optimal temperature.

139 Enzyme inhibitors are compounds that bind to an enzyme and prevent the enzyme from performing its function.

141 Noncompetitive inhibitors bind at a location on the enzyme other than the active site. The binding of a noncompetitive inhibitor causes the shape of the enzyme to change in such a way that the active site can no longer bind to the substrate.

143 Cholesterol inhibits is own synthesis when it binds to the enzyme *HMG-CoA reductase.* This type of inhibition is called feedback inhibition.

145 Hypertension is chronically elevated blood pressure.

147 Sodium ions cause water to be retained in the circulatory system through osmosis, which increases blood pressure.

149

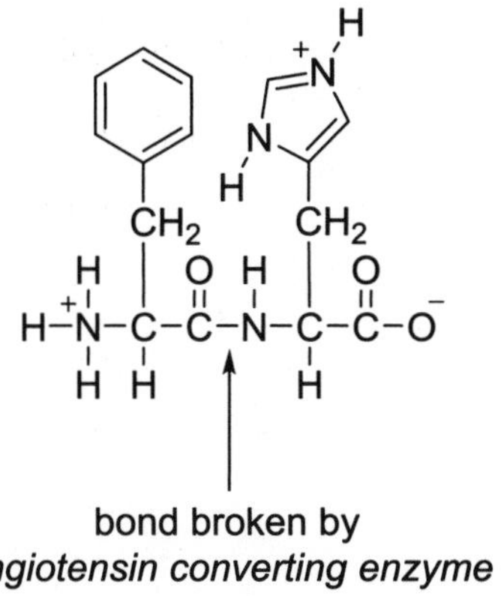

bond broken by
angiotensin converting enzyme

151 Captopril binds the zinc atom that is a cofactor at the active site of *ACE*. It is a competitive inhibitor since it prevents angiotensin I, the natural substrate, from binding to ACE.

Chapter 14

Nucleotides and Nucleic Acids

Chapter Summary

In this chapter, you learned how a cell uses the chemical information encoded within its DNA to build proteins, a process that requires RNA. You learned about the nucleotides that make up the nucleic acids, DNA and RNA. You observed how DNA is held together. You learned how DNA replicates itself so that the genetic information it contains can be passed on to daughter cells. You gained insight into how information flows from DNA to RNA to construct a protein from its individual amino acids. Understanding the structure and function of DNA helps people in the health field fight diseases.

14.1 The Chemical Structure of a Nucleotide and Polynucleotides

The ability to transmit genetic information is a fundamental condition for life. Nucleic acids are constructed from nucleotides. The biomolecules responsible for storing and transmitting genetic information are the nuclei acids deoxyribonucleic acid (DNA) and ribonucleic acid (RNA). The primary function of DNA is to serve as a blueprint for assembling all the proteins required by the individual throughout his or her life. Encoded within a segment of DNA, known as a gene, is the information for assembling the sequence of amino acids for a specific protein.

Nucleic acids are polymers (polynucleotides) constructed from nucleotides. There are only four different nucleotides used to build DNA or RNA. Chemically, a nucleotide contains three parts: 1) a monosaccharide (sugar), 2) a phosphate, and 3) a nitrogenous base.

In a Nutshell: The Monosaccharide

The monosaccharide component of DNA and RNA are slightly different: D-ribose is the sugar in RNA, while 2-deoxy-D-ribose is the sugar in DNA. Both monosaccharides are furanoses and contain five carbons, except that 2-deoxy-D-ribose lacks a hydroxyl group at the C(2) position. In the nucleotide, the carbon atoms of the monosaccharide are numbered starting at the anomeric center as 1' and proceeding clockwise around the ring to the 5' carbon, situated above the ring in the usual arrangement for a D-sugar.

At the anomeric center, the monosaccharide contains a β-*N*-glycosidic linkage to a nitrogenous base, which means the nitrogenous base is situated above the ring on the same side of the furanose ring as the 5' carbon, and a covalent bond joins the1' carbon of the sugar to a nitrogen atom in the nitrogenous base.

In a Nutshell: The Nitrogenous Base

A nucleotide contains a nitrogenous base derived from either pyrimidine or purine, planar, aromatic rings containing two or four nitrogen atoms in the rings, respectively. They are called nitrogenous bases because the rings contain nitrogen and bases because they are like amines that act as bases in aqueous solution. DNA nucleotides contain one of the four nitrogenous bases: adenine, A, guanine, G, cytosine, C, and thymine, T. RNA nucleotides contain one of the four nitrogenous bases: adenine, A, guanine, G, cytosine, C, and uracil, U. Both RNA and DNA contain A, G, and C, but DNA contains thymine, while RNA contains uracil. In a nucleotide, the nitrogenous base is covalently bonded to the anomeric center of the sugar in a β-glycosidic linkage to the nitrogen atom in the ring. A nucleotide without its phosphate group is known as a nucleoside and named according to the identity of its nitrogenous base.

In a Nutshell: The Phosphate Group

Nucleotides contain a phosphate group in the form of a monophosphate ester at the 5' hydroxyl group of the monosaccharide. A nucleotide has the same name as the corresponding nucleoside except that it has the added ending "5' phosphate." Since nucleotides are organic derivatives of phosphoric acid, the term acid appears in the names deoxyribonucleic acid (DNA) and ribonucleic acid (RNA).

Nucleotide triphosphates are common high energy forms of a nucleotide as a result of the high-energy phosphoanhydride bonds that join three phosphate groups. The potential energy in phosphoanhydride bonds is used to drive the reaction that joins a single nucleotide to a growing polynucleotide.

Worked Example #1

For the nucleotide shown below, circle and label the following:

a. The nitrogenous base. What is the name of the nitrogenous base?

b. The monosaccharide. What is the name of this monosaccharide?

c. The phosphate group

d. The 5’ carbon

e. The 3’ carbon

f. Would this nucleotide be found in DNA or RNA? Explain.

Solutions

a–e.

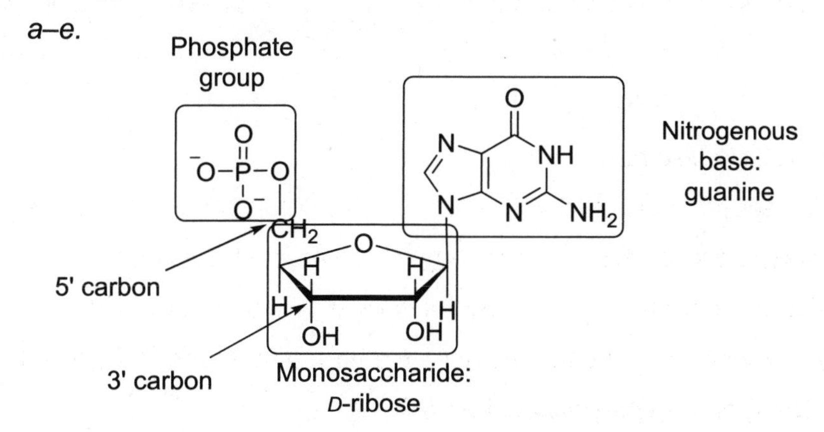

f. *This nucleotide would be found in RNA because the monosaccharide is D-ribose.*

Try It Yourself #1

For the nucleotide that follows, circle and label the following:

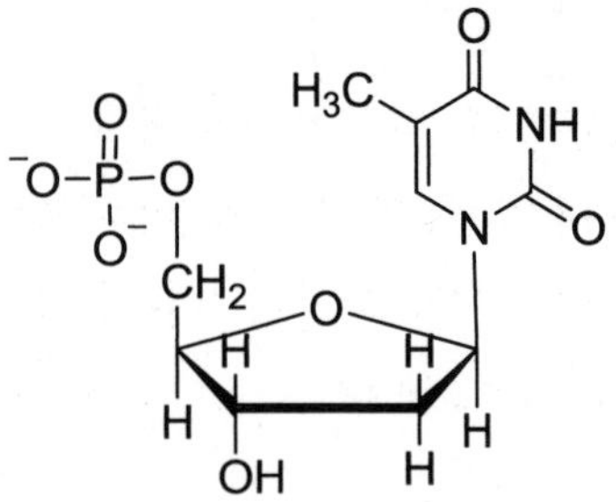

a. The nitrogenous base. What is the name of the nitrogenous base?

b. The monosaccharide. What is the name of this monosaccharide?

c. The phosphate group

d. The 2' carbon

e. The anomeric center

f. Would this nucleotide be found in DNA or RNA? Explain.

Solutions

a–e.

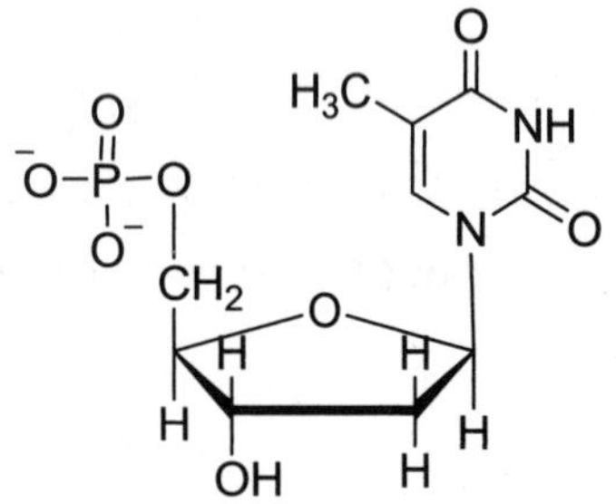

f. Does the 2' carbon atom have an OH group bonded to it? ________

The nucleotide is found in: ______________________________.

In a Nutshell: Constructing Polynucleotides from Nucleotides

RNA and DNA are linear polynucleotides formed when many nucleotides bond together in a head-to-tail fashion. A covalent bond joins the 3' hydroxyl group of one nucleotide to the 5' phosphate group of another nucleotide, forming a phosphate diester. Since the only difference between nucleotides in a polynucleotide is the nitrogenous base, a sequence of nucleotides is specified using the one letter abbreviation that corresponds to the nitrogenous base of each nucleotide. A lowercase “d” is inserted in front of the nucleotide sequence to indicate a DNA polynucleotide, otherwise it is assumed to be RNA.

Every nucleotide in a polynucleotide is joined to other nucleotides at the 5’ phosphate group and at the 3’ hydroxyl group with the exception of the terminal nucleotides, which have a free phosphate group on the 5’ carbon and a free 3’ hydroxyl group. The convention when writing a sequence of nucleotides is to list the nucleotides from the 5’ end to 3’ end (5’ → 3’).

The reaction that joins a nucleotide to a growing polynucleotide chain is a reaction much like an esterification reaction, but instead of an alcohol and a carboxylic acid reacting to form an ester, an alcohol and a phosphate react to form a phosphate ester. Since this type of reaction requires energy, nucleotide triphosphates are used when joining a nucleotide to the polynucleotide. In a cell, the reaction always takes place between the 5’ phosphate group of the incoming nucleotide triphosphate and 3’ hydroxyl group of the growing polynucleotide (5’ → 3’).

Worked Example #2

The two nucleotides that follow are T and G. Write the structure of the dinucleotide dTG and circle the new phosphate ester bond. Label the 5’ and the 3’ ends of the dinucleotide. Is the dinucleotide dTG the same as or different from dGT?

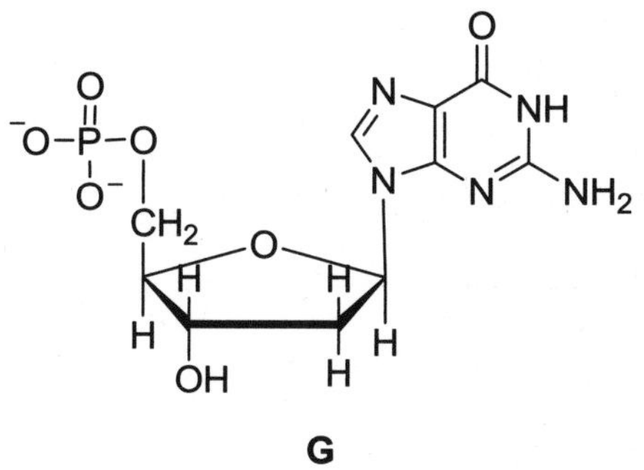

G 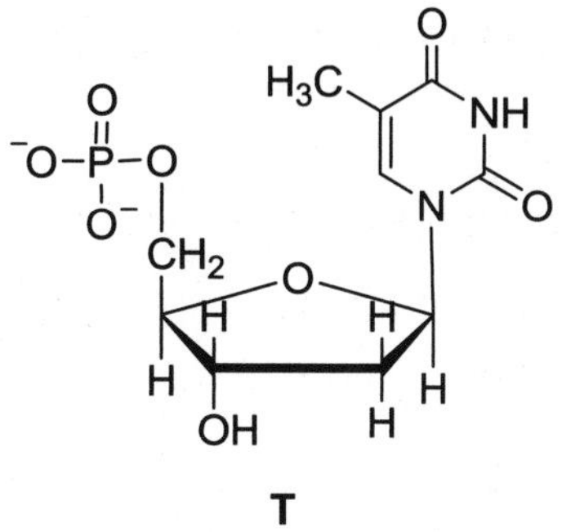 **T**

The structure of dTG is shown below:

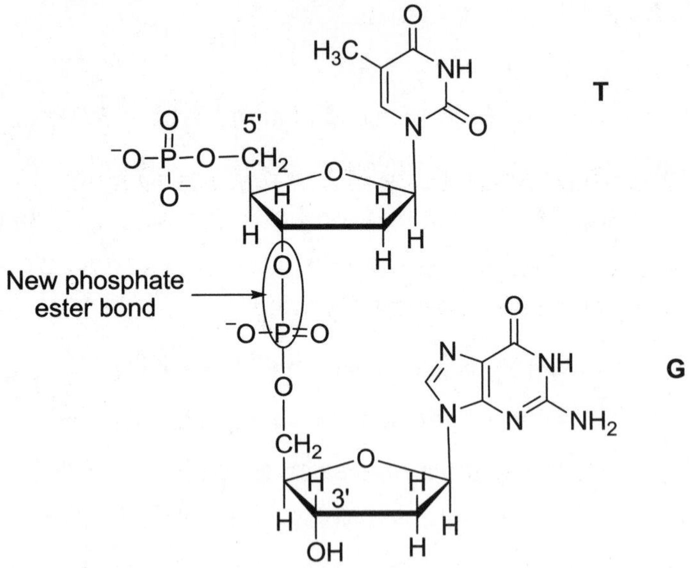

T

New phosphate ester bond

G

The molecule dTG is different from dGT.

Try It Yourself #2

The two nucleotides that follow are U and C. Write the structure of the dinucleotide CU and circle the new phosphate ester bond. Label the 5' and the 3' ends of the dinucleotide. Is the dinucleotide CU the same as or different from UC?

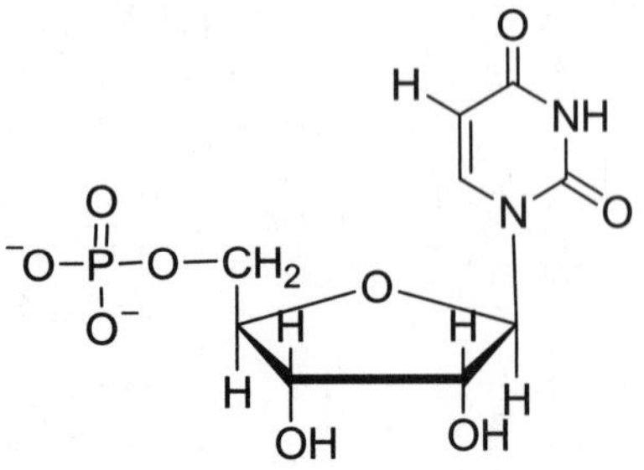

U

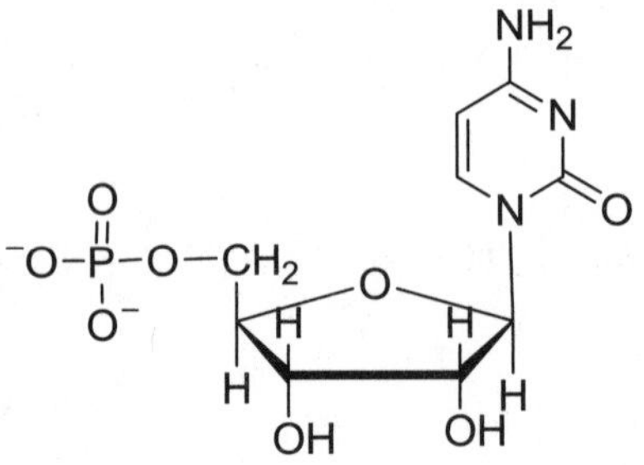

C

Structure of CU:

Is the nucleotide CU the same or different from UC? ______________

Practice Problems for the Chemical Structure of a Nucleotide and Polynucleotides

1. For the nucleotide shown below, circle and label the following:

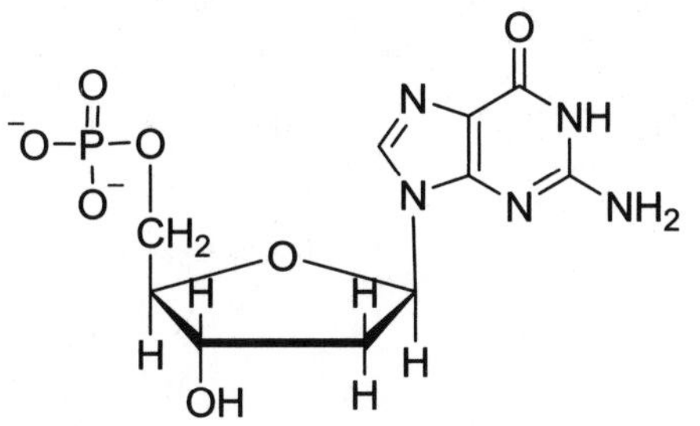

    a. The nitrogenous base. What is the name of the nitrogenous base?

    b. The monosaccharide. What is the name of this monosaccharide?

    c. The phosphate group

    d. The 5' carbon

    e. The anomeric carbon

    f. The 2' carbon

    g. The 3' carbon

    h. Would this nucleotide be found in DNA or RNA? Explain.

2. Which nitrogenous bases are derived from pyrimidine? Which ones are derived from purine?

3. Two nucleotides are shown below: T and A. Write the structure of the dinucleotide dAT and circle the new phosphate ester bond. Label the 5' and the 3' ends of the dinucleotide. Is the dinucleotide dAT the same as or different from dTA?

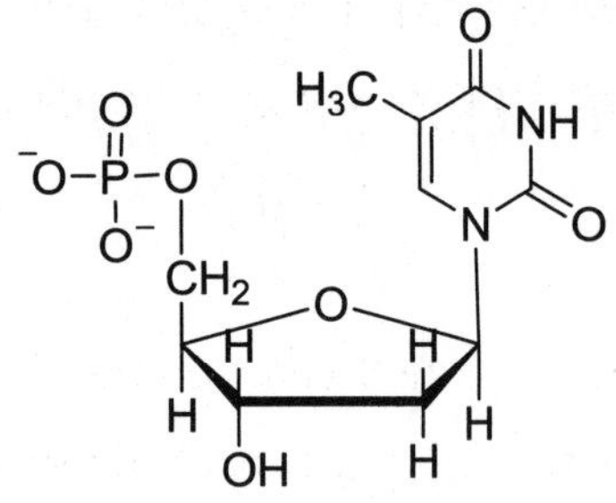

T

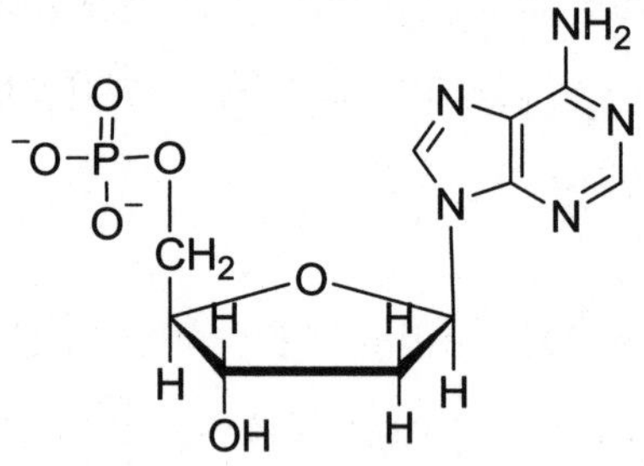

A

14.2 DNA Structure and Replication

In a Nutshell: Double Helix Structure of DNA

DNA is composed of two polynucleotides, each a linear sequence of millions of nucleotides, often referred to as two "strands." These two strands are twisted around each other in the form of a right-handed double helix. As a right-handed helix, DNA is chiral. The sugar-phosphate backbone forms the outer lengthwise portion of the DNA double helix. The two strands of a DNA double helix are arranged antiparallel, which means that their 5' and 3' ends are oriented in opposite directions. The hydrophilic monosaccharide and phosphate groups of the sugar-phosphate backbone are on the outside of the helix, where they can interact with the polar aqueous environment. The relatively nonpolar nitrogenous bases are projected toward the interior of the helix and perpendicular to the sugar-phosphate backbone.

A nitrogenous base forms two or three hydrogen bonds to a specific nitrogenous base on the opposite strand, creating a base pair. The base pairs in double-stranded DNA are complementary, which means that each nitrogenous base interacts with one specific base across from it on the other strand. The base pairs in DNA are A-T and G-C. There are two hydrogen bonds between adenine (A) and thymine (T) and three hydrogen bonds between guanine (G) and cytosine (C).

The two strands of DNA are held together in a double helix by the hydrogen bonds between the base pairs as well as base stacking, a stabilizing interaction that occurs between the delocalized electrons of the aromatic rings of the nitrogenous bases above and below. Although these electrostatic interactions are weaker than covalent bonds, their cumulative strength is significant. Yet because hydrogen bonding and base stacking are noncovalent forces of attraction, they are more easily disrupted.

Worked Example #3

The sequences of nucleotides located on a segment of one DNA strand are shown below. Indicate the complementary sequence of base pairs that would appear on the other DNA strand. How many hydrogen bonds hold this section of DNA together?

5' dTAGC 3'

The complementary base pairing would be:

5' dTAGC 3'

3' dATCG 5'

Each A-T pair has two hydrogen bonds and each G-C pair has three hydrogen bonds: (2 A-T pairs × 2 hydrogen bonds) + (2 G-C pairs × 3 hydrogen bonds) = 10 hydrogen bonds.

Try It Yourself #3

The sequences of nucleotides located on a segment of one DNA strand are shown below. Indicate the complementary sequence of base pairs that would appear on the other DNA strand. How many hydrogen bonds hold this section of DNA together?

5' dGGACT 3'

DNA strand: *5' dGGACT 3'*

Complementary strand: ____________

Number of A-T base pairs: ____________

Number of G-C base pairs: ____________

Total number of hydrogen bonds: __________

Worked Example #4

The sequences of nucleotides located on a segment of one DNA strand are shown below. Indicate the complementary sequence of base pairs that would appear on the other DNA strand.

a. 5' dTAGGCCAT 3'
b. 5' dGGCCAATT 3'

The complementary base pairings would be:

a. *5' dTAGGCCAT 3'*
   3' dATCCGGTA 5'

b. *5' dGGCCAATT 3'*
   3' dCCGGTTAA 5'

Try It Yourself #4

The sequences of nucleotides located on a segment of one DNA strand are shown below. Indicate the complementary sequence of base pairs that would appear on the other DNA strand.

a. 5' dGACGTGGCA 3'
b. 5' dAATCCGAGG 3'

a. *DNA strand:* *5' dGACGTGGCA 3'*
   Complementary strand: ________________

b. DNA strand: 5' dAATCCGAGG 3'

Complementary strand: ___________________

In a Nutshell: Genes and the Human Genome

DNA, together with specialized proteins, coils into highly compact structures known as chromosomes. A DNA double helix is coiled around a core of proteins, known as histones, creating a compact structure known as nucleosomes. Since histones have a net positive charge, they facilitate the packing of negatively charged DNA into these compact structures. Nucleosomes coil further to create even longer fibrous structures, which create the familiar X-shape of a chromosome. A DNA double helix is about 2 nm wide, while a chromosome is about 1400 nm wide and visible under a light microscope.

The complete sequence of nucleotides in your DNA, distributed over 46 chromosomes, is known as your genome. The human genome contains about 3 million base pairs. The human genome is 99.9 percent the same in all people. Only about 2 percent of the human genome contains sequences of proteins that code for proteins. A gene is a segment of DNA that contains the instructions for making a protein. The average human gene contains about 3000 base pairs, and we have approximately 30,000 genes.

The entire nucleotide sequence and arrangement of all the genes in the human genome has been determined and the genes associated with traits and disorders were identified. Genetic modification, GM, refers to DNA technologies that alter the genetic makeup of an organism, such as animals, plants, or bacteria. Combining genes from different organisms is known as recombinant DNA technology, creating genetically modified, genetically engineered, or transgenic products.

In a Nutshell: DNA Replication

The ability to reproduce requires that an organism or a cell be able to replicate itself, a process that begins with DNA replication. DNA replication requires several enzymes and begins with the unwinding of a section of the DNA double helix to expose the two strands of DNA, referred to as parent strands. Free nucleotides in solution assemble along the exposed parent strands forming base pairs. The enzyme, *DNA polymerase*, catalyzes the formation of phosphate ester linkages between the 5' phosphate of a nucleotide and the 3' hydroxyl of the growing nucleotide chain. Both parent strands serve as the template for a new daughter strand, complementary to the parent strand. At the end of the replication process two new double stranded DNA molecules, identical to the original DNA are produced. Each new double helix contains one strand from the original double helix and one new daughter strand.

Both strands of DNA are copied simultaneously and always in the 5' → 3' direction. Since the two strands are antiparallel, this means that the two daughter strands are synthesized in the opposite direction. *DNA polymerase* can only synthesize a nucleic acid in one direction 5' → 3'. Replication begins where the two strands of the double helix have been separated known as the replication fork. Since *DNA polymerase* can only work in the 5' → 3' direction, only one strand, known as the leading strand, can be synthesized continuously from the 5' to 3' end. The other strand, known as the lagging strand, which is also synthesized from the 5' to the 3' end, must be synthesized discontinuously, forming polynucleotide segments that are then later joined.

As the leading strand approaches the replication fork, the enzyme *helicase* further unwinds the two DNA parent strands, so the replication fork moves in the same direction as the synthesis of the leading strand.

Synthesis of the lagging strand is discontinuous because it is synthesized in the opposite direction, opposite the movement of the replication fork. As a result, new sections of parent strand are constantly being exposed that have not been replicated. *DNA polymerase* begins to assemble another lagging strand fragment from the site of the new replication fork to the start of the previous fragment. The lagging strand is built in fragments of 100–200 base pairs. The enzyme *DNA ligase* then catalyzes the reaction that joins adjacent fragments to form the lagging strand.

DNA polymerase is not only fast, but accurate. *DNA polymerase* also proofreads the daughter strands for mistakes. When errors are detected, it signals other enzymes to replace and repair incorrectly placed nucleotides. Natural mistakes in replication do occur, and these mistakes account for some of the mutations that exist in our DNA.

Worked Example #5

Consider a portion of double-stranded DNA with the following complementary sequence of base pairs:

5' dATATGCGGCCATA 3'

3' dTATACGCCGGTAT 5'

Write the sequence of nucleotides found in each new replicated DNA segment and label the original parent strands and the two new daughter strands.

Parent strands:	*5' dATATGCGGCCATA 3'*	*3' dTATACGCCGGTAT 5'*
Daughter strands:	*3' dTATACGCCGGTAT 5'*	*5' dATATGCGGCCATA 3'*

Try It Yourself #5

Consider a portion of double stranded DNA with the following complementary sequence of base pairs:

5' dGAGCCTTCCAACG 3'

3' dCTCGGAAGGTTGC 5'

Write the sequence of nucleotides found in each new replicated DNA segment and label the original parent strands and the two new daughter strands.

Parent strands: ____

Daughter strands: ____

Practice Problems for DNA Structure and Replication

1. The sequences of nucleotides located on a segment of one DNA strand are shown below. Indicate the complementary sequence of base pairs that would appear on the other DNA strand. How many hydrogen bonds hold this section of DNA together?

   5' dGCCAATGACT 3'

2. Explain how DNA coils to fit inside a cell.

3. Consider a portion of double-stranded DNA with the following complementary sequence of base pairs:

   5' dGCCTCTAGAATGAG 3'
   3' dCGGAGATCTTACTC 5'

   Write the sequence of nucleotides found in each new replicated DNA segment and label the original parent strands and the two new daughter strands.

14.3 The Role of DNA and RNA in Protein Synthesis

The primary function of DNA is as a repository of genetic information. The basic flow of information is from DNA to messenger RNA to protein. Protein synthesis requires both DNA and RNA. RNA is found in three forms, each with a different role in protein synthesis: 1) ribosomal RNA (rRNA), 2) messenger RNA (mRNA), and 3) transfer RNA (tRNA).

Amino acids are assembled into proteins at ribosomes, structures located throughout the cytoplasm of the cell. Ribosomes are composed of both ribosomal RNA (rRNA) and proteins. Ribosomal RNA is composed of polynucleotides containing over 1000 nucleotides. Although ribosomes are located outside the nucleus, DNA never leaves the nucleus. Instead, messenger RNA (mRNA) carries the information encoded in a segment of DNA from the nucleus to the ribosomes, where protein synthesis takes place.

The process of protein synthesis involves two steps, transcription and translation. Transcription takes place in the nucleus of the cell, where a segment of one strand of DNA (a gene) is copied as a complementary single-stranded messenger RNA. Translation takes place at the ribosome, where mRNA serves as a template for assembling tRNAs containing the requisite amino acids, which are joined by peptide bonds to form a polypeptide

In a Nutshell: Transcription: DNA to mRNA

When a particular protein is needed by the cell, the appropriate gene is expressed and the protein is synthesized. Gene expression is a complex process that is carefully regulated by the cell. Gene expression begins with transcription. During transcription, a nucleotide sequence—the gene on a section of one strand of DNA—known as the template strand or (+) strand, is copied as a complementary single-stranded mRNA transcript. The DNA strand that is not copied is known as the coding strand or the (−) strand because it has the same nucleotide sequence as the mRNA transcript.

As with DNA replication, the mRNA transcript is synthesized in the 5' → 3' direction by reading the DNA template strand in the 3' → 5' direction. The nucleotide sequence of the mRNA transcript is complementary to the nucleotide sequence on the DNA template strand. Note that an A on DNA is copied as U on mRNA because RNA contains uracil instead of thymine.

In our cells, *RNA polymerase II* catalyzes the reaction between the 5' phosphate group on an incoming nucleotide and the hydroxyl group at the 3' end of the growing mRNA chain. This reaction is similar to the reaction catalyzed by *DNA polymerase*, except the nucleotides contain D-ribose as a sugar and the nitrogenous base uracil rather than thymine.

Worked Example #6

For the following sequences on DNA, write the sequence of nucleotides from the 5' to the 3' end in the mRNA transcript after transcription.

a. 3' dGAGTACCTACCC 5'
b. 3' dATGGCTCAACGG 5'

a. *DNA strand:* *3' dGAGTACCTACCC 5'*
   mRNA strand: *5' CUCAUGGAUGGG 3'*
b. *DNA strand:* *3' dATGGCTCAACGG 5'*
   mRNA strand: *5' UACCGAGUUGCC 3'*

Try It Yourself #6

For the following sequences on DNA, write the sequence of nucleotides from the 5' to the 3' end in the mRNA transcript after transcription.

a. 3' dTAAGGACTTTGG 5'
b. 3' dTTTTCATTGGTT 5'

a. *DNA strand:* ____________________
   mRNA strand: ____________________
b. *DNA strand:* ____________________
   mRNA strand: ____________________

In a Nutshell: Translation

Translating the nucleotide sequence on mRNA into the sequence of amino acids of a polypeptide to form a protein is a process known as translation. The term translation is used because a nucleotide sequence serves as the blueprint for an amino acid sequence. Every three non-overlapping nucleotides on an mRNA molecule, known as a codon, uniquely specifies one amino acid. The 64 codons and each of the amino acids that they specify are known as the genetic code. All living organisms have the same genetic code. The codon AUG specifies methionine and also the start of an amino acid sequence. Three of the 64 codons are stop codons (UAA, UAG, UGA), which signal the end of the polypeptide. The remaining 61 codons code for the 20 natural amino acids. Since there are more codons than there are amino acids, most amino acids are specified by more than one codon. The redundancy of the genetic code is important because it minimizes the effect of transcription errors. Note that while there is redundancy in the genetic code, a codon always specifies only one amino acid.

Translation occurs at ribosomes. Ribosomes are enormous structures that consist of two subunits, both composed of rRNA and various proteins, one larger than the other. During protein synthesis, the two subunits lock together with the messenger RNA locked in the space between them. The larger subunit contains the active site, where peptide bonds are made. During translation, the ribosome moves along the mRNA in the 5' → 3' direction, reading one codon at a time.

Assembling amino acids in sequence, according to the codons specified by mRNA, is the job of transfer RNA (tRNA). Transfer RNAs (tRNAs) are smaller nucleic acids composed of 75–85 nucleotides, with a characteristic cloverleaf shape, containing sections of both double-stranded and single-stranded loops. tRNA contains two important regions: the anticodon loop, a sequence of three nucleotides, that varies for the different tRNAs and the 3' end, which forms a covalent bond (ester) to a specific amino acid associated with the anticodon of that tRNA.

During translation, three base pairs are formed between an mRNA codon and the complementary tRNA anticodon. Note that the mRNA codons are always written in the 5' → 3' direction and tRNA anticodons are written in the 3' → 5' direction.

Although there are 61 mRNA codons that specify an amino acid, most cells contain fewer than 45 tRNAs. Fewer tRNAs are needed because tRNAs can base pair to more than one mRNA codon. This explanation is known as the wobble hypothesis. In the wobble hypothesis, nonstandard base pairings in the third position of an mRNA codon are allowed if it codes for the same amino acid.

Base pairing between the anticodon on a tRNA and a codon on the mRNA brings an amino acid to the ribosome, in proximity to the growing polypeptide chain. Next, a reaction (acyl transfer) occurs between the ester of the growing polypeptide chain on the previous tRNA and the free amine of the amino acid of the most recently recruited tRNA to form a peptide bond (an amide). Thus, the growing polypeptide chain moves to the most recently recruited tRNA and the previous tRNA molecule, with its amino acid no longer attached and now containing a free 3' hydroxyl group, and diffuses away from the ribosome to find another amino acid. Note the location of the ribosome shifted by one codon in the 5' → 3' direction. The next codon on mRNA is read, another tRNA carrying the appropriate amino acid is recruited, and the process repeats itself until the entire polypeptide has been assembled and a stop codon has been reached.

The polypeptide must still undergo additional modifications to become a functional protein: folding, forming disulfide bonds, and electrostatic interactions with other polypeptides to form quaternary structure.

Worked Example #7

Using the genetic code shown in Table 14-2, determine the amino acid specified by the following codons on mRNA:

a. UGG
b. ACG
c. GUG

a. The codon UGG codes for tryptophan.
b. The codon ACG codes for threonine.
c. The codon GUG codes for valine.

Try It Yourself #7

Using the genetic code shown in Table 14-2, determine the amino acid specified by the following codons on mRNA:

a. AUC
b. CAA
c. CGC

a. The codon AUC codes for: __________.
b. The codon CAA codes for: ______________.
c. The codon CGC codes for: _______________.

Worked Example #8

For the following mRNA sequences, indicate the anticodons on the three tRNA molecules recruited. What is the amino acid sequence formed? Consult Table 14-2.

a. 5' UGUUACGAA 3'
b. 5' UUCCAAGCG 3'

a.	*mRNA strand:*	*5' UGUUACGAA 3'*
	Anticodons on tRNA:	*3' ACA 5' 3' AUG 5' 3' CUU 5'*
	Amino acids:	*Cys Tyr Glu*
	The tripeptide will be Cys-Tyr-Glu.	
b.	*mRNA strand:*	*5' UUCCAAGCG 3'*
	Anticodons on tRNA:	*3' AAG 5' 3' GUU 5' 3' CGC 5'*
	Amino acids indicated by codons on mRNA:	*Phe Gln Ala*
	The tripeptide will be Phe-Gln-Ala.	

Try It Yourself #8

For the following mRNA sequences, indicate the anticodons on the three tRNA molecules recruited. What is the amino acid sequence formed?

a. 5' GAUAUUAUG 3'
b. 5' CAUCGACAA 3'

a. *mRNA strand:* ______________________________

Anticodons on tRNA molecules: ________ ________ ________

Amino acids indicated by codons on mRNA: _____ _____ _____

Tripeptide: ______________________________

b. *mRNA strand:* ______________________________

Anticodons on tRNA molecules: ________ ________ ________

Amino acids indicated by codons on mRNA: _____ _____ _____

Tripeptide: ______________________________

In a Nutshell: Genetic Mutations

A genetic mutation is any permanent chemical change to one or more nucleotides in a gene (DNA) that affects the primary structure of a protein. Two basic types of mutations include: 1) substitutions, where one nucleotide in a gene has been substituted by another, and 2) frameshift, where one nucleotide is added to or deleted from the gene. Both types of mutations have the potential to change the amino acid sequence as a result of transcription and translation. The effects of DNA mutations are minimized by the redundancy of the genetic code, as more than one codon exists for every amino acid.

Worked Example #9

Indicate whether the following normal mRNA sequence would produce the same or a different peptide sequence if the mutation shown occurred (consult Table 14-2). If so, indicated the new amino acid sequence:

DNA 3' dAATGTA 5'

mRNA 5' UUACAU 3'

a. The third nucleotide on DNA is substituted with C.
b. The fifth nucleotide on DNA is substituted with A
c. The fourth nucleotide is deleted.

Solutions

DNA: 3' dAATGTA 5'

mRNA: 5' UUACAU 3'

Amino acid sequence: Leu-His

a. *DNA mutation: 3' dAA**C**GTA 5'*

   mRNA: 5' UUGCAU 3'

   Dipeptide: Leu-His

   There is no change in the sequence.

b. *DNA mutation: 3' dAATG**A**A 5'*

   mRNA: 5' UUACUU 3'

   Dipeptide: Leu-Leu

   The amino acid sequnce is different.

c. *DNA mutation: 3' dAATTA 5'*

   mRNA: 5' UUGAU 3'

Dipeptide: *Leu-?*

The sequence will be different because the first nucleotide for the codon for the second amino acid is not C. The codon for histidine begins with C.

Try It Yourself #9

Indicate whether the following normal mRNA sequence would produce the same or a different peptide sequence if the mutation shown occurred (consult Table 14-2). If so, indicated the new amino acid sequence:

DNA 3' dTGATCA 5'

mRNA 5' ACUAGU 3'

a. The third nucleotide on DNA is substituted with C.
b. The sixth nucleotide on DNA is substituted with G.
c. The second nucleotide is deleted.

Solutions

DNA *3' dTGATCA 5'*

mRNA *5' ACUAGU 3'*

Amino acid sequence: ____________

a. *DNA mutation:* ________________

mRNA: ________________________

Amino acid sequence: ___________________

Effect on sequence: ___________________

b. *DNA mutation:* ________________

mRNA: ________________________

Amino acid sequence: ___________________

Effect on sequence: ___________________

c. *DNA mutation:* ________________

*mRNA:*________________________

Amino acid sequence: ___________________

Effect on sequence: ___________________

Practice Problems for the Role of DNA and RNA in Protein Synthesis

1. For the following sequences on DNA, write the sequence of nucleotides from the 5' to the 3' end in the mRNA transcript after transcription.

   3' dCTTATTATAACG 5'

2. For the following mRNA sequences, indicate the anticodons on the three tRNA molecules recruited. What is the amino acid sequence formed? Consult Table 14-2.

   a. 5' CGCCAAGGC 3'

   b. 5' AUAACAUCC 3'

3. Indicate whether the following normal mRNA sequence would produce the same or a different peptide sequence if the mutation shown occurred (consult Table 14-2). If so, indicated the new amino acid sequence:

   DNA 3' dGGGCGA 5'

   mRNA 5' CCCGCU 3'

a. The third nucleotide on DNA is substituted with C.

b. The sixth nucleotide on DNA is substituted with G.

c. A nucleotide, T, is added between the first and second nucleotides on DNA

Chapter 14 Quiz

1. The three basic parts of a nucleotide are given below.

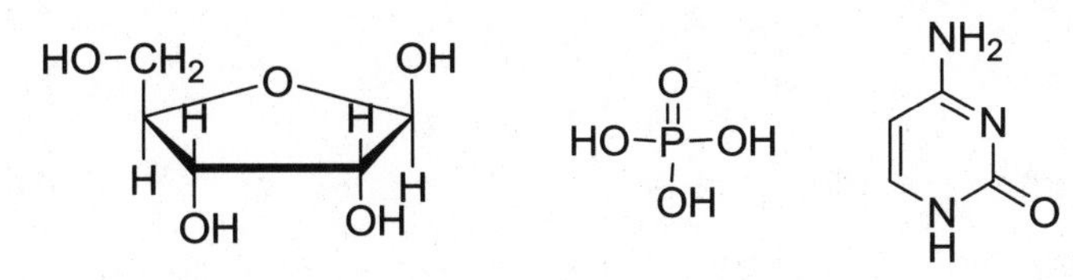

a. What are each of these molecules called?

b. Construct a nucleotide from these three parts.

c. Label the 3' and the 5' carbon atoms in your nucleotide.

d. Would this nucleotide be found in DNA or RNA? How can you tell?

2. How many hydrogen bonds are found in the complementary base pair G-C in DNA?

3. How many hydrogen bonds are found in the complementary base pair A-T in DNA?

4. Below are the sequences of nucleotides located on a segment of one DNA strand. Indicate the complementary sequence of base pairs that would appear on the other DNA strand.

   a. 5' dATCGGCAATT 3'

   b. 5' dGCATGCCATAG 3'

5. What is a gene?

6. Consider a portion of double stranded DNA with the following complementary sequence of base pairs:

   5' dTGCCATCATG 3'

   3' dACGGTAGTAC 5'

   Write the sequence of nucleotides found in each new replicated DNA segment and label the original parent strands and the two new daughter strands.

7. What role does *DNA polymerase* play in DNA replication?

8. For the following sequences on DNA, write the sequence of nucleotides from the 5' to the 3' end in the mRNA transcript after transcription.

   a. 3' dGGGACAATGTAA 5'

   b. 3' dGTACGGCCCAAT 5'

9. For the following nucleotide sequences on mRNA, indicate the anticodons on the three tRNA molecules recruited. What is the amino acid sequence formed?

   a. 5' CGCAUGACU 3'

   b. 5' GUCGAUAAG 3'

10. Indicate whether the following normal mRNA sequence would produce the same or a different amino acid sequence if the mutation shown occurred:

    DNA 3' dGCCCGA 5'

    mRNA 5' CGGGCU 3'

    a. The first nucleotide on DNA is substituted with T.

    b. The fifth nucleotide on DNA is substituted with A.

    c. The third nucleotide is deleted.

Chapter 14

Answers to Additional Exercises

29 If the genes contain mutations, they can predispose a person to disease.

31 Acquired mutations are alterations to the genes that take place during a person's lifetime, while inherited mutations are the genes with which a person is born.

33 Carcinogens, radiation, and viral infections can trigger mutations.

35 DNA and RNA are responsible for storing and transmitting genetic information.

37 A nucleotide consists of a monosaccharide, a phosphate, and a nitrogenous base.

39 The anomeric carbon in the monosaccharide has a covalent bond to the nitrogenous base. The nitrogen atom in the nitrogenous base is bonded to the monosaccharide. The nitrogenous base is situated above the ring (β), and a covalent bond joins the 1' carbon of the sugar to a nitrogen atom on the nitrogenous (*N*) base.

41 Adenine (A), guanine (G), cytosine (C), and uracil (U) are found in RNA.

43 a–f.

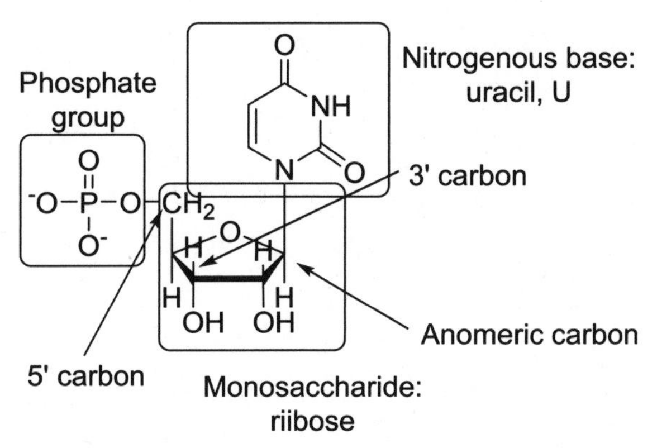

g. The nucleotide would be found in RNA. The second carbon atom in the monosaccharide has an OH group.

h. uridine 5' monophosphate

45

Phosphate group

Nitrogenous base: adenine, A

Monosaccharide: deoxyribose

47

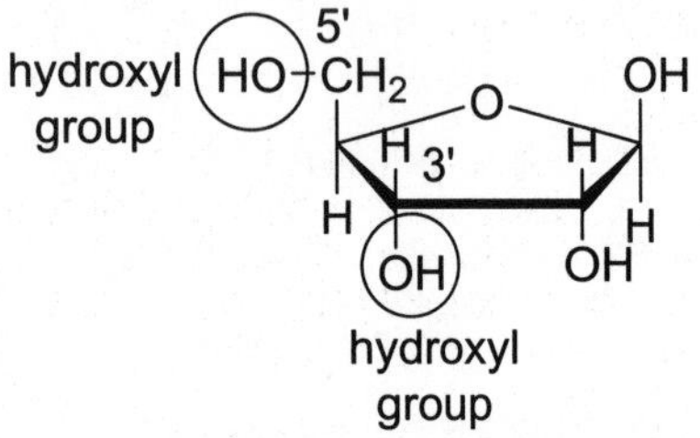

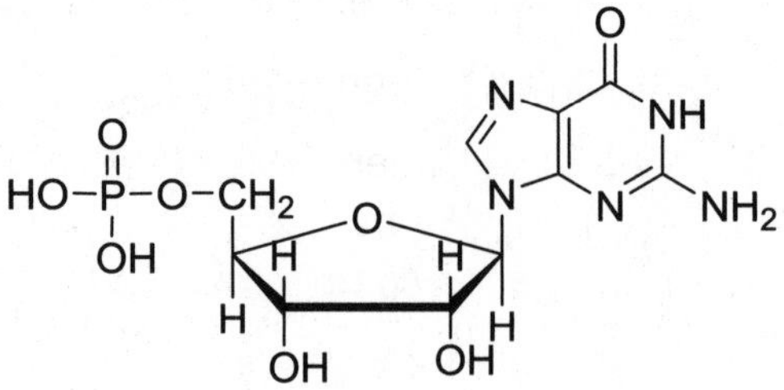

Guanosine 5’ monophosphate. This nucleotide would be found in RNA. The second carbon atom in the monosaccharide has an OH group.

49 The 3’ hydroxyl group of one nucleotide and the 5’ phosphate group of the other nucleotide react to form a phosphate diester.

51

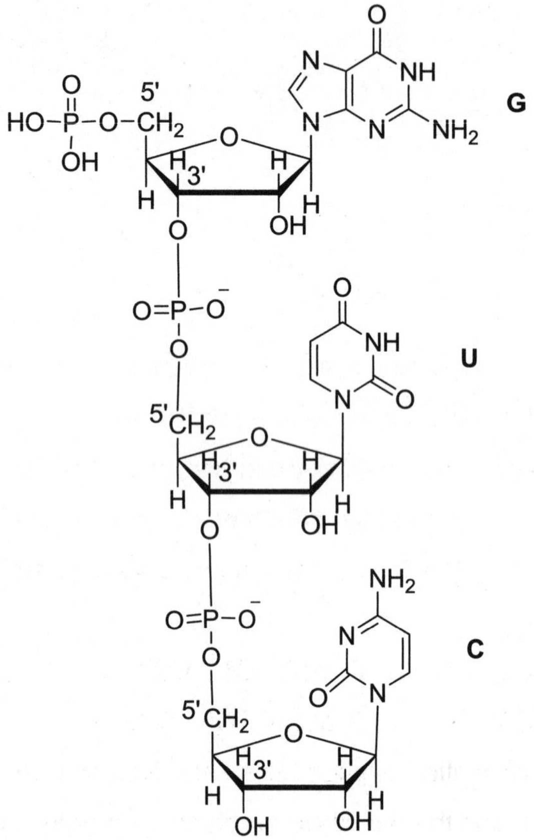

53 A nucleotide contains a nitrogenous base, a monosaccharide, and a phosphate group. A nucleoside has a nitrogenous base and a monosaccharide, but no phosphate group.

55 DNA is negatively charged, while the histones are positively charged.

57 Most human cells contain 46 chromosomes.

59 Leukemia and epilepsy are associated with a protein that is coded for by a gene located on chromosome 6.

61 The three-dimensional shape of DNA is a double helix.

63 DNA is a right-handed double helix. It is chiral.

65 The monosaccharide and phosphate groups are located on outside of the three-dimensional DNA structure. These groups are hydrophilic and can interact with the polar aqueous environment.

67 Hydrogen bonding and base stacking hold the two strands of DNA in the double helix.

69 Base stacking is a stabilizing interaction.

71 An A-T base pair is linked by two hydrogen bonds.

73 The complementary sequence is dATAGCG. Each A-T pair has two hydrogen bonds and each C-G pair has three hydrogen bonds: (3 A-T pairs × 2 hydrogen bonds) + (3 C-G pairs × 3 hydrogen bonds) = 15 hydrogen bonds.

75
a. dGATCCG
b. dTGACTT
c. dAACCTT
d. dCCATGA

77 A single cell grows and divides to become a human being with one trillion cells. Before the first cell can divide, DNA must be able to replicate itself.

79 *DNA polymerase* catalyzes the formation of a new daughter strand of DNA from a parent strand. The enzyme also proofreads the daughter strand for mistakes.

81 The DNA parent strands are unwound ahead of the leading strand by the enzyme *helicase.*

83

Parent:	5' dCATTAAGCCG 3'	5' dCGGCTTAATG 3'
Daughter:	3' dGTAATTCGGC 5'	3' dGCCGAATTAC 5'

85 *DNA polymerase* catalyzes the formation of phosphate ester linkages between the 5' phosphate group of a nucleotide and the 3' hydroxyl group on the growing nucleotide chain.

87
a. 5' dATTCCGTA 3': Parent strand
3' d**C**AAGG**T**AT 5': Daughter strand (Errors in daughter strand are bolded.)
3' dTAAGGCAT 5': Corrected daughter strand
b. 5' dGGGCCCTTTAA 3': Parent strand
3' dCCC**A**GGAA**G**TT 5': Daughter strand (Errors in daughter strand are bolded.)

3' dCCCGGGAAATT 5:' Corrected daughter strand

89 *DNA polymerase* only works in the 5' → 3' direction.

91 DNA polymerase can only work in one direction, so only one strand, known as the leading strand can be synthesized continuously from the 5' to 3' end. The lagging strand, which is also synthesized in the 5' to 3' direction, must be synthesized discontinuously.

93 The leading strand moves in the same direction as the replication fork. The enzyme *helicase* unwinds the two DNA parent strands ahead of the leading strand, so the replication fork moves in the same direction as the synthesis of the leading strand.

95 The central dogma of molecular biology is that the basic flow of information is from DNA to messenger RNA to protein.

97 The three major forms of RNA are ribosomal RNA (rRNA), messenger RNA (mRNA), and transfer RNA (tRNA).

99 Translation occurs at the ribosome. mRNA and tRNA are involved.

101 RNA contains the monosaccharide D-ribose. RNA contains the nitrogenous base U, uracil.

103 Ribosomes are composed of rRNA and various proteins.

105 Translation and transcription are involved in synthesizing a protein from a gene.

107 *RNA polymerase II* catalyzes the reaction between the 5' phosphate group on an incoming nucleotide and the hydroxyl group at the 3' end of the growing mRNA chain.

109
a. 3' dCCGGAATATA5': DNA strand
   5' GGCCUUAUAU 3': mRNA strand
b. 3' dAAGGCCAATT 5': DNA strand
   5' UUCCGGUUAA 3': mRNA strand
c. 3' dGTACACGTCG 5': DNA strand
   5' CAUGUGCAGC 3': mRNA strand

111 The genetic code is the 64 codons and each of the amino acids they specify.

113 Translation begins when the mRNA codon forms base pairs with the complimentary anticodon of a tRNA molecule carrying the correct amino acid at its other end. Another matching tRNA molecule arrives and a peptide bond is formed between the amino acids on adjacent tRNA molecules.

115 A genetic mutation is any permanent chemical change to one or more nucleotides in a gene (DNA) that affects the primary structure of a protein.

117 The anticodon AAA will correspond to the codon UUU; therefore, the amino acid will be phenylalanine.

119 No more amino acids will be added to the growing peptide chain when a stop codon is reached.

121 CUU, CUC, CUA, and CUG all code for leucine. The effects of a genetic mutation are minimized by the presence of more than one codon for leucine.

123 Yes, an amino acid can be specified by more than one codon.

125

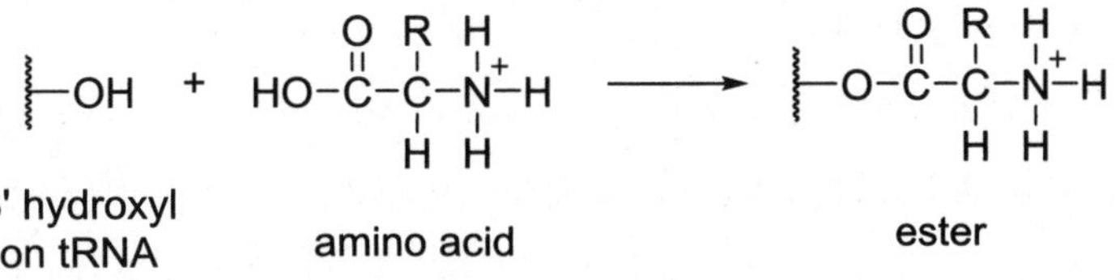

$$\text{⌇}\!-\!OH + HO\!-\!\overset{O}{\overset{\|}{C}}\!-\!\underset{H}{\overset{R}{C}}\!-\!\underset{H}{\overset{H}{\overset{|}{N^{+}}}}\!-\!H \longrightarrow \text{⌇}\!-\!O\!-\!\overset{O}{\overset{\|}{C}}\!-\!\underset{H}{\overset{R}{C}}\!-\!\underset{H}{\overset{H}{\overset{|}{N^{+}}}}\!-\!H$$

3' hydroxyl on tRNA — amino acid — ester

The carboxylic acid of the amino acid is involved. An ester is formed.

127 a. mRNA: 5' AAUAGUGUG 3'

Amino acids: Asn Ser Val

The amino acid sequence would be Asn-Ser-Val.

b. mRNA:5' CCCUUUGGG 3'

Amino acids: Pro Phe Gly

The amino acid sequence would be Pro-Phe-Gly.

c. mRNA: 5' CACCGGUGG 3'

Amino acids: His Arg Trp

The amino acid sequence would be His-Arg-Trp.

d. mRNA: 5' UCCUUAGCA 3'

Amino acids: Ser Leu Ala

The amino acid sequence would be Ser-Leu-Ala.

129 DNA template strand 3' dGGTGCT 5'

mRNA: 5' CCACGA 3'

Amino acid sequence: Pro-Arg

a. DNA mutation: 3' dGGCGCT 5'

mRNA: 5' CCGCGA 3'

Amino acid sequence: Pro-Arg

The same amino acid sequence is produced.

b. DNA mutation: 3' dGGTGCA 5'

mRNA: 5' CCACGU 3'

Amino acid sequence: Pro-Arg

The same amino acid sequence is produced.

c. DNA mutation: 3' dGTGCA 5'

mRNA: 5' CACGU 3'

Amino acid sequence: His-

The amino acid sequence is different.

d. DNA mutation: 3' dGGTCT 5'

mRNA: 5' CCAGA 3'

Amino acid sequence: Pro-

The amino acid sequence will be different. If the next nucleotide in the mRNA sequence is U or C, the amino acid sequence will be Pro-Asp. If the next nucleotide in the mRNA sequence is A or G, the amino acid sequence will be Pro-Glu.

131 No, you could not predict the unique sequence of nucleotides on mRNA that coded that protein. The amino acids have more than one sequence of mRNA that codes for them; therefore, you cannot tell which nucleotide was used in mRNA.

133 A virus is a nucleic acid—RNA or DNA—encapsulated in a protein coat.

135 The HIV virus uses *reverse transcriptase* and *protease* in its life cycle.

137 HIV infects T-lymphocytes.

139 Reverse transcription allows a single strand of DNA to be transcribed in the opposite direction of normal transcription; therefore, the single strand of DNA can serve as the template for the second strand. *Reverse transcriptase* is the enzyme that catalyzes this process.

141 HIV is a rapidly mutating viral DNA.

143 Gene expression in T cells occurs when the T cell is needed to fight off an infection.

145 Some of the viral mRNA produced is hydrolyzed by the viral enzyme HIV *protease* to produce genomic RNA and some viral mRNA is transcribed into proteins, such as the key viral enzymes.

147 Since the HIV virus rapidly mutates, it eventually becomes resistant to any one type of drug therapy. Therefore, a combination of drug therapies is needed.

Chapter 15

Energy and Metabolism

Chapter Summary

In this chapter, you learned about the catabolic pathways that break down carbohydrates and triglycerides, the compounds in food that supply the body with most of its energy. You examined the biochemical pathway, glycolysis that generates energy from carbohydrates. You learned how ATP carries potential energy for these biochemical pathways. You learned how the citric acid cycle produces NADH and $FADH_2$. You observed how the electrons carried by these coenzymes drive the formation of ATP from ADP through a series of oxidation-reduction reactions known as oxidative phosphorylation. You learned that energy in the forms of ATP, NADH, and $FADH_2$ is produced during the catabolism of fatty acids in a biochemical process called β-oxidation. The energy produced in these catabolic pathways, enables the body to perform the anabolic pathways that allow mechanical work as well as the anabolic process that build proteins, DNA, and cell membranes.

15.1 Energy and Metabolism: An Overview

Extracting energy from our food and storing it in the form of ATP is the primary function of cellular respiration, the biochemical pathways that convert carbohydrates, fats, and sometimes dietary proteins into CO_2, water, and energy. These catabolic biochemical pathways supply the energy for the energy-demanding anabolic biochemical pathways, such as building proteins, synthesizing DNA, and operating pumps that transport ions and molecules across membranes. Catabolic pathways are generally divided into three stages.

In stage 1, each class of the biomolecules is hydrolyzed into their respective building blocks: Polysaccharides are hydrolyzed into glucose; proteins are hydrolyzed into amino acids; and triglycerides are hydrolyzed into fatty acids and glycerol.

Stage 2 occurs inside the cell, where the various hydrolysis products are further degraded by separate biochemical pathways into acetyl CoA, the central molecule of metabolism. Glucose is converted to pyruvate by glycolysis. In the presence of oxygen, pyruvate is further oxidized to acetyl CoA. Fatty acids are degraded into acetyl CoA, through β-oxidation.

Amino acids can be degraded into either pyruvate, acetyl CoA, or intermediates of the citric acid cycle depending on the number of carbon atoms in the amino acid. However, amino acids are not a primary source of energy.

In stage 3, acetyl CoA enters the citric acid cycle, a biochemical pathway that oxidizes acetyl CoA to carbon dioxide, while reducing coenzymes, NAD^+ and FAD to NADH and $FADH_2$. The electrons carried by NADH and $FADH_2$, which are produced in stage 2 and the citric acid cycle, are transferred to the electron transport chain, a series of oxidation-reduction reactions concluding with the reduction of oxygen to water. The energy transferred in the electron transport chain drives the phosphorylation of ADP to ATP, a process known as oxidative phosphorylation.

In a Nutshell: Stage 1: Hydrolysis of Carbohydrates, Fats, and Proteins

In the first stage of catabolism, also known as digestion, the large biomolecules that make up food react with water in hydrolysis reactions that break them down into their smaller components.

Digestion of carbohydrates begins in the mouth. The enzyme *amylase*, found in saliva, catalyzes the hydrolysis of starch into: glucose, a monosaccharide; maltose, a disaccharide composed of two glucose monomers; and dextrins, oligosaccharides composed of 3–12 glucose monomers. In the small intestine, maltose and dextrins are completely hydrolyzed into glucose, while other important disaccharides, lactose and sucrose, are hydrolyzed into their respective monosaccharides. Hydrolysis of complex carbohydrates (di-, oligo-, and polysaccharides) occurs at the glycosidic linkages. Complete hydrolysis of starch yields many glucose molecules.

Monosaccharides diffuse through the small intestine and enter the bloodstream. Glucose, soluble in water, is distributed to cells via the bloodstream, hence the reference blood sugar. The monosaccharides, fructose and galactose, are carried to the liver, where they are converted into glucose and then reenter the blood.

Triglycerides, stored as adipocytes, serve as the body's long term supply of energy. To extract the energy stored in triglycerides, they must be first hydrolyzed into fatty acids, a process that occurs in skeletal muscle cells and liver cells. Hydrolysis of a triglyceride occurs at the ester bonds in the triglyceride to produce one molecule of glycerol and three fatty acids.

Hydrolysis of proteins occurs in the stomach. Hydrolysis of proteins occurs at the peptide bonds (amides) producing amino acids, a process that is facilitated by the low pH in the stomach. Generally, amino acids are not metabolized to supply energy, except in circumstances, such as starvation.

In a Nutshell: ATP and Energy Exchange

In a biological cell, the energy released in catabolic reactions is not transferred as heat, but rather as chemical potential energy in the form of high energy bonds, the most important of which are phosphoanhydride bonds of nucleotide triphosphates, such as adenosine triphosphate (ATP).

Hydrolysis of ATP to form adenosine diphosphate (ADP) and inorganic phosphate (P_i) breaks one of these high energy phosphoanhydride bonds, releasing30.5 kJ/mol of energy. By convention, energy released is shown as a negative value—a downhill reaction—and energy absorbed as a positive value—an uphill reaction. Hydrolysis of ATP can also occur at both phosphoanhydride bonds to from adenosine monophosphate (AMP) and two molecules of inorganic phosphate and releasing 61 kJ/mol of energy. The reverse of these reactions requires an input of energy of the same amount.

Biochemical reactions that require an input of energy do not happen on their own. In biochemistry, reactions that require an input of energy are coupled to reactions that release energy, such as ATP hydrolysis. The energy released in a coupled reaction is the sum of the energy transferred in each individual reaction. If the sum of the energy transferred is a negative value, the reaction will go forward. The convention used in biochemistry for showing that two or more reactions are coupled is to write a curved arrow(s) intersecting a straight arrow. The straight arrow represents the transformation of the substrate into product, while the curved arrow represents the coupled reactant and product.

In a Nutshell: Coenzymes: NADH, FADH$_2$, and Acetyl CoA

In addition to ATP, other high energy molecules in the cell include the reduced coenzymes NADH and FADH$_2$, which are formed from NAD$^+$ and FAD, respectively. These reduced coenzymes represent potential energy because they deliver electrons to the electron transport chain that ultimately provide the energy to phosphorylate ADP to ATP.

Acetyl coenzyme A, abbreviated acetyl CoA, is one of the most important intermediates of metabolism because it is central to so many biochemical pathways. Coenzyme A is a biological acyl carrier molecule, involved in transferring acyl groups in various biochemical pathways. Acetyl CoA has a chemical structure derived from the nucleoside adenosine, the vitamin pantothenic acid (vitamin B$_5$), the thiol amine $HSCH_2CH_2NH_2$, and acetic acid (CH_3CO_2H). A high energy thioester bond joins the two-carbon acetyl group to the thiol coenzyme A.

Worked Example #1

When glycogen is broken down in muscle, the following reactions occur, resulting in the transformation of glucose-1-phosphate into fructose-6-phosphate.

Reaction	
Glucose-1-phosphate → Glucose-6-phosphate	−7.28 kJ/mol
Glucose-6-phosphate → Fructose-6-phosphate	+1.67 kJ/mol
Glucose-1-phosphate → Fructose-6-phosphate	

a. How much energy is released or absorbed by the coupled reactions? Show your calculation.

b. Will the coupled reactions provide enough energy to convert glucose-1-phosphate to fructose-6-phosphate?

Solutions

a. *Reaction*

Glucose-1-phosphate → Glucose-6-phosphate	*−7.28 kJ/mol*
Glucose-6-phosphate → Fructose-6-phosphate	*+1.67 kJ/mol*
Glucose-1-phosphate → Fructose-6-phosphate	*−5.61 kJ/mol*

The coupled reactions will release −5.61 kJ/mol.

b. *Yes, coupling these reactions will provide the energy required to convert glucose-1-phosphate to fructose-6-phosphate.*

Try It Yourself #1

In glycolysis, 1,3-bisphosphoglycerate is converted to 3-phosphoglycerate. This reaction is coupled to the formation of ATP from ADP. The reactions are shown below.

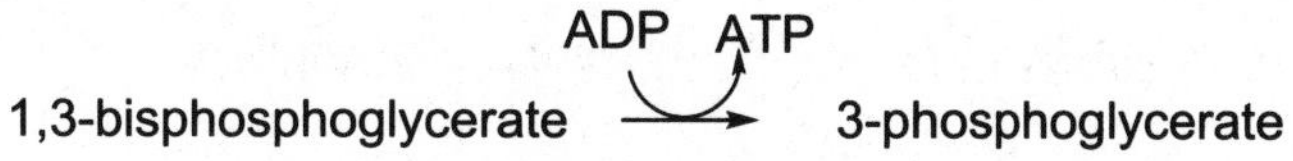

$$\text{1,3-bisphosphoglycerate} \xrightarrow{\text{ADP} \curvearrowright \text{ATP}} \text{3-phosphoglycerate}$$

Do the coupled reactions release or absorb energy? Note: The formation of ATP requires 30.5 kJ/mol, and the formation of 3-phosphoglycerate from 1,3 bisphosphoglycerate releases −49.8 kJ/mol. Show your calculation.

Calculation:

The amount of energy from the coupled reactions: ____________________

The value of the amount of energy is (positive or negative): ____________.

Energy is (released or absorbed) ____________________ *from the coupled reactions.*

Practice Problems for Energy and Metabolism: an Overview

1. What are carbohydrates, triglycerides, and amino acids hydrolyzed into in the first stage of catabolism?

2. Where is potential energy stored in ATP? Why do NADH and $FADH_2$ represent potential energy for biochemical pathways?

3. In the construction of cell membranes, the conversion of glycerol to glycerol-3-phosphate is coupled to the hydrolysis of ATP to ADP.

ATP ADP

glycerol → glycerol-3-phosphate

The formation of glycerol-3-phosphate from glycerol requires 9.2 kJ/mol and the hydrolysis of ATP to form ADP releases 30.5 kJ/mol. Does the coupled reaction absorb or release energy? Show your calculation.

15.2 Carbohydrate Catabolism: Glycolysis and Pyruvate Metabolism

In a Nutshell: Glycolysis

Glycolysis is the primary energy producing metabolic pathways for some cells. Glycolysis is an anaerobic pathway, which means that it does not require oxygen. Many microorganisms that live in an anaerobic environment use glycolysis to produce energy.

Glycolysis occurs in the cytosol, the fluid part of the cell outside the nucleus. Glycolysis is a ten-step biochemical pathway that converts glucose, a molecule containing six carbon atoms, into two molecules of pyruvate, a molecule containing three carbon atoms. For every glucose molecule that enters glycolysis, two ADP are phosphorylated to two ATP, two NAD^+ are reduced to two NADH + $2H^+$, and two molecules of pyruvate are produced.

Every intermediate in glycolysis contains one or more phosphate groups in its structure. The −2 charge on these monophosphate ester groups prevents the intermediates of glycolysis from diffusing out of the cell since polyatomic ions cannot diffuse through the cell membrane. In step 4, the six-carbon molecule fructose-1,6-biphosphate is split into two three-carbon structural isomers: glyceraldehyde-3-phosphate and dihydroxyacetone-3-phosphate. In step 5, dihydroxyacetone-3-phosphate is converted into glyceraldehyde-3-phosphate in an isomerization reaction. Thus, in the first half of glycolysis, one molecule of glucose is split into two molecules of glyceraldehyde-3-phosphate. In the second half of glycolysis, the two molecules of glyceraldehyde-3-phosphate are converted into two molecules of pyruvate. Every step in the second half of glycolysis is doubled.

The first half of glycolysis requires an input of energy: Two ATP molecules are required for every one glucose molecule that undergoes glycolysis (steps 1 and 3). In the second half of glycolysis, four ADP are phosphorylated to four ATP, two for each glyceraldehyde-3-phosphate that is converted into pyruvate (steps 7 and 10). By subtracting the two ATP used in the first half of glycolysis, the net output of ATP can be calculated as $4 - 2 = 2$ ATP per glucose. In addition, two molecules of NAD^+ are reduced to NADH during glycolysis (step 6).

Worked Example #2

Which steps in glycolysis are isomerization reactions?

Solution

Steps 2 and 5 are isomerization reactions. In step 2, a pyranose ring is isomerized to a furanose ring. In step 5, dihydroxyacetone-3-phosphate is isomerized into glyceraldehyde-3-phosphate.

Try It Yourself #2

Which steps of glycolysis require energy? Which steps of glycolysis produce energy?

Solution:

In a Nutshell: The Metabolism of Pyruvate

In human cells, pyruvate produced from glycolysis can follow one of two catabolic pathways, depending on whether or not oxygen is present. Under aerobic conditions, pyruvate is oxidized into acetyl CoA and carbon dioxide and NAD^+ is reduced to NADH. Under anaerobic conditions, pyruvate is reduced to lactic acid and NADH is oxidized to NAD^+.

In the presence of oxygen, pyruvate is converted into acetyl CoA and CO_2. This reaction requires coenzyme A, as an acyl carrier molecule and NAD^+ as an oxidizing agent. Thus, the three-carbon pyruvate molecule is converted into carbon dioxide and a two-carbon acetyl group, carried by coenzyme A (acetyl CoA). During the oxidation of pyruvate, electrons are transferred to NAD^+, reducing it to NADH and H^+. Oxygen is necessary for the regeneration of NAD^+ from NADH, an oxidation.

In a Nutshell: Reduction of Pyruvate to Lactic Acid

When oxygen levels in the cell are low, as they are during strenuous exercise, anaerobic conditions exist. Under anaerobic conditions, pyruvate is reduced to lactate and NADH + H^+ is oxidized to NAD^+, in a process known as lactic acid fermentation. Reduction of pyruvate to lactate ensures that NAD^+ is regenerated so that glycolysis can continue; step 6 of glycolysis requires NAD^+.

Worked Example #3

Does the formation of acetyl CoA from pyruvate or the formation of lactic acid from pyruvate produce NAD^+? Is pyruvate oxidized or reduced in the reaction that produces NAD^+?

Solution

The formation of lactic acid from pyruvate produces NAD^+. Pyruvate is reduced in that reaction.

Try It Yourself #3

What structural changes occur when pyruvate is converted to acetyl CoA under aerobic conditions?

Solution:

Practice Problems for Carbohydrate Catabolism: Glycolysis and Pyruvate Metabolism

1. What molecules are the end products of glycolysis? What molecules store the energy produced in glycolysis?

2. Is lactic acid fermentation an aerobic or an anaerobic biochemical pathway? How does lactic acid fermentation help the glycolysis pathway?

3. What happens to pyruvate under aerobic conditions?

15.3 The Citric Acid Cycle

In a Nutshell: The Citric Acid Cycle

The last stage of catabolism for glucose, fatty acids, and some amino acid begins with acetyl CoA entering the citric acid cycle. The citric acid cycle is an eight-step biochemical pathway that takes place in mitochondria, the energy producing organelles of the cell. The citric acid cycle is an aerobic pathway that produces much more energy than glycolysis, in the form of coenzymes NADH and $FADH_2$, as well as the nucleotide guanidine triphosphate, GTP.

For every acetyl CoA molecule that enters the citric acid cycle, one GDP is phosphorylated to GTP, three NAD^+ are reduced to NADH, one FAD is reduced to $FADH_2$, and two molecules of CO_2 and one molecule of coenzyme A are produced. These enzymes deposit their electrons into the electron transport chain where they provide the energy to phosphorylate ADP to ATP during oxidative phosphorylation. NADH provides the energy to phosphorylate approximately 2.5 ADP to 2.5 ATP, and $FADH_2$ provides the energy to phosphorylate approximately 1.5 ADP to 1.5 ATP.

The citric acid cycle is written as a circular pathway because, in the first step, acetyl CoA reacts with an oxaloacetate molecule that is the product of the last step. The two carbon atoms from oxaloacetate are the carbon atoms in the two molecules of carbon dioxide that are expelled in the first half of the citric acid cycle. The expulsion of carbon dioxide causes the six-carbon-containing citric acid molecule to be shortened to five and then four carbons. In the second half of the citric acid cycle, this four-carbon intermediate is converted to oxaloacetate, also a four-carbon molecule.

Step 1 of the citric acid cycle is a condensation reaction that forms a carbon-carbon bond between the acetyl group and oxaloacetate, to produce citrate, a molecule with six carbon atoms, while releasing coenzyme A, the acyl carrier molecule. The biochemical pathway is named after citric acid, the product of this first step. The six-carbon citric acid molecule is shortened to five-carbon-molecule α-ketoglutarate, and then to succinyl CoA, a four-carbon acyl group carried by coenzyme A. Steps 3 and 4 are the reactions that expel carbon dioxide (CO_2). Over the remaining four steps, succinyl CoA is converted into oxalacetate.

The energy-producing steps in the citric acid cycle are the four oxidation-reduction steps, 3, 4, 6, and 8, and the phosphorylation of GDP to GTP in step 5. Steps 3, 4, and 8 each reduce a molecule of NAD^+ to NADH. Step 6 reduces FAD to $FADH_2$. Every acetyl CoA molecule that goes through the citric acid cycle provides the energy to phosphorylate 10 ADP to 10 ATP, either by direct phosphorylation or during oxidative phosphorylation. In summary, one acetyl CoA molecule that enters the citric acid cycle provides the energy to phosphorylate 10 ADP to produce 10 ATP.

Worked Example #4

The unionized form of oxaloacetate is shown below:

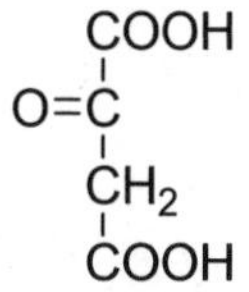

```
    COOH
     |
 O=C
     |
    CH2
     |
    COOH
```

a. What structure would this compound have at physiological pH?
b. During the citric acid cycle what happens to two of the carbon atoms of oxaloacetate?

a. *At physiological pH, the carboxylic acid functional groups would be in their ionized form.*

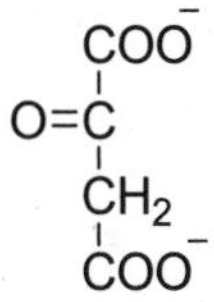

```
    COO-
     |
 O=C
     |
    CH2
     |
    COO-
```

b. *During the citric acid cycle, two of the four carbon atoms of oxaloacetate are released as two molecules of carbon dioxide.*

Try It Yourself #4

The structure of α-ketoglutarate, produced in the third step of the citric acid cycle, in its unionized form is shown below.

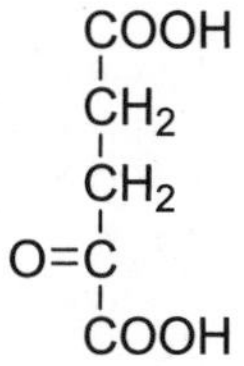

```
    COOH
     |
    CH2
     |
    CH2
     |
 O=C
     |
    COOH
```

a. What structure would this compound have at physiological pH?

b. How many carbon atoms are in α-ketoglutarate?

a. *The structure of α-ketoglutarate at physiological pH:*

b. *There are _____ carbon atoms in α-ketoglutarate.*

Worked Example #5

1. Step 6 of the citric acid cycle is shown below.

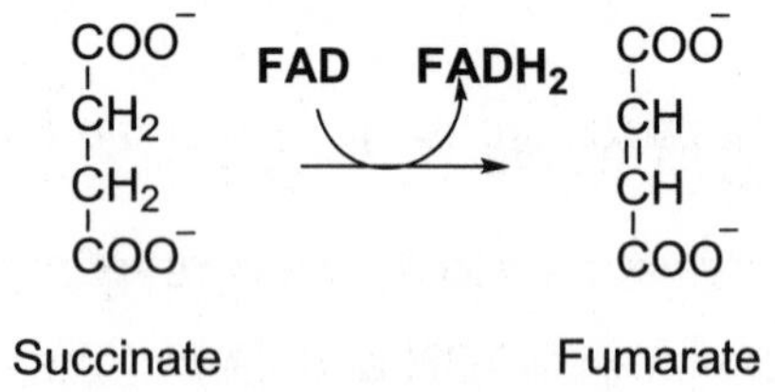

$$\begin{array}{c} COO^- \\ | \\ CH_2 \\ | \\ CH_2 \\ | \\ COO^- \end{array} \xrightarrow[\quad]{\text{FAD} \;\; \text{FADH}_2} \begin{array}{c} COO^- \\ | \\ CH \\ \| \\ CH \\ | \\ COO^- \end{array}$$

Succinate Fumarate

a. What type of reaction occurs in this step?

b. What functional group change occurred to the succinate molecule?

c. Is succinate oxidized or reduced?

d. Is FAD oxidized or reduced?

Solutions

a. *It is an oxidation-reduction reaction.*

b. *An alkane molecule is converted to an alkene.*

c. *Succinate is oxidized.*

d. *FAD is reduced.*

Try It Yourself #5

Step 3 of the citric acid cycle is shown below.

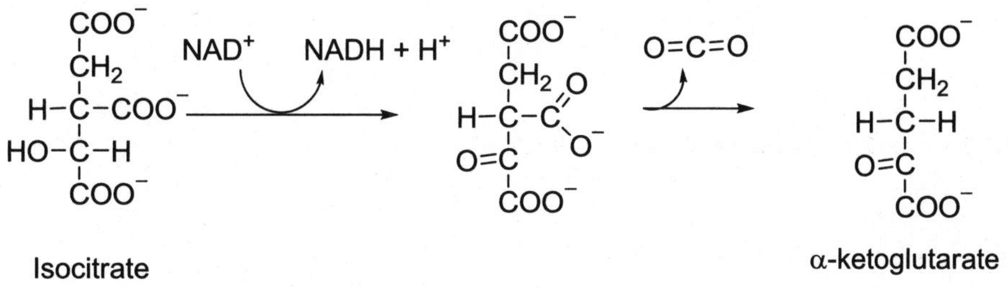

```
   COO⁻                                 COO⁻                                COO⁻
   |          NAD⁺      NADH + H⁺       |                O=C=O              |
   CH₂                                  CH₂  O                              CH₂
   |                                    |   //                              |
 H-C-COO⁻  ------------------------>  H-C-C        ---------------->      H-C-H
   |                                    |   \                               |
HO-C-H                                O=C    O⁻                           O=C
   |                                    |                                   |
   COO⁻                                 COO⁻                                COO⁻

Isocitrate                                                          α-ketoglutarate
```

a. What type of reaction happens in the first reaction of this step?
b. Is isocitrate oxidized or reduced?
c. Is NAD^+ oxidized or reduced?

a. The reaction is a: ______________________________.
b. Isocitrate is: ______________________________.
c. NAD^+ is: ______________________________.

Practice Problems for the Citric Acid Cycle

1. How many molecules of ADP are converted into ATP as the result of one pass through the citric acid cycle, assuming that NADH and $FADH_2$ are used to phosphorylate ADP to ATP?

2. The structure of citrate is shown below.

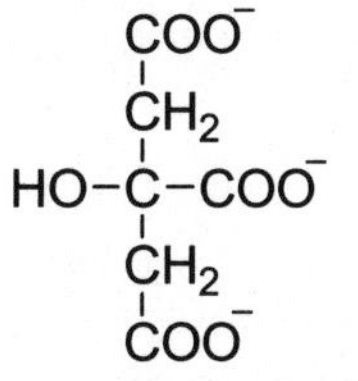

```
    COO⁻
    |
    CH₂
    |
 HO-C-COO⁻
    |
    CH₂
    |
    COO⁻
```

   a. What structure would this compound have at low pH? Write the carboxylic acid functional groups in their unionized form.

b. What other functional group is present in this molecule?

3. Step 4 of the citric acid cycle is shown below.

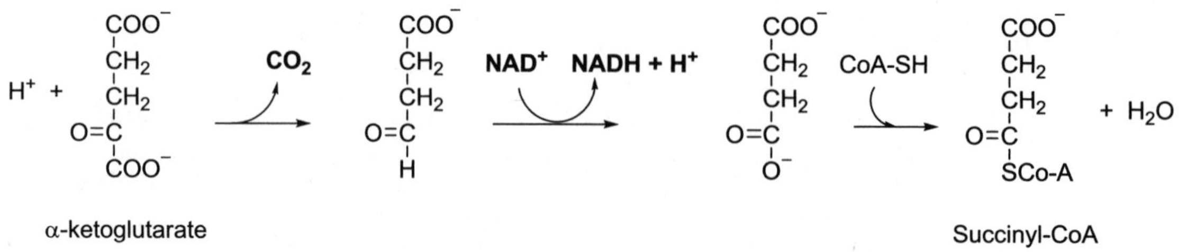

$H^+ + {}^-OOC{-}CH_2{-}CH_2{-}C({=}O){-}COO^- \xrightarrow{CO_2} {}^-OOC{-}CH_2{-}CH_2{-}C({=}O){-}H \xrightarrow{NAD^+ \;\; NADH + H^+} {}^-OOC{-}CH_2{-}CH_2{-}C({=}O){-}O^- \xrightarrow{CoA\text{-}SH} {}^-OOC{-}CH_2{-}CH_2{-}C({=}O){-}SCo\text{-}A + H_2O$

α-ketoglutarate   Succinyl-CoA

a. What type of reaction occurs in the first reaction of this step?

b. How many carbon atoms does α-ketoglutarate contain? How many carbon atoms does succinyl CoA contain?

c. What is the significance of the NADH produced in the second reaction of this step?

15.4 The Electron Transport Chain and Oxidative Phosphorylation

The final phase of catabolism for glucose, fatty acids, and some amino acids occurs when the electrons from NADH and $FADH_2$ generated in the citric acid cycle, glycolysis, and pyruvate oxidation are transferred to the electron transport chain. This transfer of electrons regenerates NAD^+ and FAD, the oxidized forms of these coenzymes, making them available again to shuttle electrons.

The electron transport chain consists of a sequence of electron transfer reactions (oxidation-reduction reactions) within specialized proteins and small carrier molecules, beginning with the oxidation of NADH to NAD^+ or $FADH_2$ to FAD ending with the reduction of oxygen to water. These oxidation-reduction steps collectively supply the energy needed to phosphorylate ADP to ATP.

In a Nutshell: Where It All Happens: The Mitochondria

With the exception of glycolysis, the energy-producing pathways of catabolism occur in the mitochondria. The structure of a mitochondrion is like a large wrinkled bag inside a smaller unwrinkled bag, creating two compartments: one between the two bags and one inside the wrinkled bag. The key parts of a mitochondrion are the outer mitochondrial membrane (the outer bag), the inner mitochondrial membrane (the inner bag), the intermembrane space (the space between the bags), and the matrix (inside the wrinkled bag).

The outer mitochondrial membrane is a phospholipid bilayer containing proteins that make it permeable to molecules, ions, and small proteins. ATP and ADP can readily pass through the outer mitochondrial membrane.

The inner mitochondrial membrane is permeable only to oxygen, carbon dioxide, and water. The proteins involved in the electron transport chain are embedded within or located along the inner mitochondrial membrane. The key enzyme, *ATP synthase*, which catalyzes the phosphorylation of ADP to ATP, lies in the inner mitochondrial membrane, with a significant portion of this large enzyme extending into the matrix.

The inner mitochondrial membrane is a folded membrane that separates the interior of the mitochondrion into two separate aqueous compartments: 1) the intermembrane space located between the outer mitochondrial membrane and the inner mitochondrial membrane; and 2) the matrix, the space enclosed by the inner mitochondrial membrane. The matrix contains all the enzymes of citric acid cycle, as well as dissolved oxygen.

Since ions, including protons (H^+), are unable to diffuse across the inner mitochondrial membrane, the concentration of protons in the matrix is lower (higher pH, less acidic) than the concentration of protons in the intermembrane space (lower pH, more acidic), creating a different concentration of protons on either side of the membrane known as a proton gradient.

Worked Example #6

Which has a lower pH, the intermembrane space or the matrix? Which has a higher concentration of protons?

Solution

The intermembrane space has a higher concentration of protons and a lower pH than the matrix.

Try It Yourself #6

Arrange the following parts of a mitochondrion from the inside to the outside: inner mitochondrial membrane, outer mitochondrial membrane, and the matrix.

The inside of a mitochondrion is: ________________________________.

The middle of a mitochondrion is: ________________________________.

The outside of a mitochondrion is: ________________________________.

In a Nutshell: The Electron-Transport Chain

The coenzymes, NADH and $FADH_2$, carry electrons that are produced during glycolysis, pyruvate oxidation, and the citric acid cycle, and transfer them to the electron transport chain, where they undergo a series of oxidation-reduction reactions. The final step of the electron transport chain is the reduction of molecular oxygen to water: $O_2 + 4\ e^- + 4\ H^+$ 2 H_2O. Oxygen is essential if oxidative phosphorylation is to occur.

The electron transport chain consists of four large multienzyme protein complexes, identified as Complexes I, II, III, IV, along with two mobile electron-carrier molecules—coenzyme Q and cytochrome *c*.

The protein Complexes I–IV are membrane proteins found in the inner mitochondrial membrane. They span part of or the entire inner mitochondrial membrane, extending from the matrix to the intermembrane space. Complex II, which extends into the matrix, receives electrons from $FADH_2$. Electrons from NADH enter the electron transport chain at Complex I. Both Complexes I and II transfer their electrons to the membrane carrier molecule coenzyme Q, thus reducing coenzyme Q. Coenzyme Q, abbreviated CoQ, is a hydrophobic molecule that moves laterally within the inner mitochondrial membrane, transporting electrons from Complex I to Complex III and from Complex II to Complex III. Electrons from Complex III are then transported by the electron carrier molecule cytochrome C, abbreviated cyt C, to Complex IV, where the final reduction of oxygen to water occurs.

As electrons are transferred through the protein complexes of the electron transport chain, energy is released that is used to pump protons from the matrix across the inner mitochondrial membrane and into the intermembrane space, against the proton gradient. The intermembrane space has a higher concentration of protons (higher H^+ concentration, lower pH) than the matrix (lower H^+ concentration, higher pH). Since the natural direction of diffusion is from an area of higher solute concentration to an area of lower solute concentration, energy is required to pump protons in the opposite direction—against the proton gradient.

A proton gradient represents potential energy. The accumulation of protons in the intermembrane space represents both chemical and electrical potential energy since the protons are charged. The charge differential between these two spaces creates an electrical potential. The chemical and potential energy generated in a mitochondrion is known as the proton-motive force, which ultimately provides the energy for phosphorylating ADP to ATP.

The proton pumps in Complexes I, III, and IV are proteins with one face oriented toward the matrix side of the membrane, where they can bind a proton at an acidic side chain of the protein. Energy released in the oxidation-reduction reactions within the complex is used to make a conformational change with the protein that then exposes the proton to the other side of the membrane. The proton is then released into the intermembrane space.

Electrons are transferred between atoms or molecules within a protein complex or electron-carrier molecule in a series of oxidation-reduction reactions. These redox reactions occur at metal-atom centers (copper and iron) and in organic cofactors, such as coenzyme Q. Atoms and molecules have an intrinsic affinity for electrons, a physical property of a substance that determines the direction of electron flow in an oxidation-reduction reaction. Atoms and molecules with a greater electron affinity represent substances with lower potential energy. In the electron-transport chain, electrons are passed to atom centers with increasingly greater electron affinity. Thus, the electron transport chain begins with NADH or $FADH_2$, which have the lowest electron affinity (higher potential energy), and ends with the acceptance of electrons by oxygen, which has the highest electron affinity (lowest potential energy).

Complex I has the lowest electron affinity (highest potential energy) of the four complexes in the electron transport chain. NADH transfers electrons to Complex I. Electrons from $FADH_2$ enter later in the electron transport chain at Complex II, which is lower in potential energy than Complex I. Consequently, no protons are pumped through Complex II. Therefore, $FADH_2$ ultimately supplies less energy than NADH. The amount of energy supplied is often measured in the number of ADP phosphorylated to ATP, which is estimated at 2.5 per NADH and 1.5 per $FADH_2$.

Worked Example #7

Fill in the blanks for the following chain of events, which describes the path electrons take through the electron transport chain when introduced from $FADH_2$:

$FADH_2$ → Complex ___ → Coenzyme ___ → Complex ____ → Cytochrome ___ → Complex ____ → O_2.

$FADH_2$ → Complex II → Coenzyme Q → Complex III → Cytochrome c → Complex IV → O_2.

Try It Yourself #7

Fill in the blanks for the following chain of events, which describes the path electrons take through the electron transport chain when introduced from NADH:

NADH → Complex ___ → Coenzyme ___ → Complex ____ → Cytochrome ___ → Complex ____ → O_2.

Solution:

NADH → Complex ___ → Coenzyme ___ → Complex ____ → Cytochrome ___ → Complex ____ → O_2.

In a Nutshell: Phosphorylation of ADP to ATP

The electron transport chain generates a proton gradient between the matrix and the inner membrane space. The potential energy of this proton motive force is used to phosphorylate ADP to ATP. This process known as oxidative phosphorylation and requires 30.5 kJ/mol.

Within the inner mitochondrial membrane and extending into the matrix is an enzyme complex known as *ATP synthase,* also referred to as Complex V. *ATP synthase* uses the potential energy of the proton-motive force to phosphorylate ADP to ATP—an important coupled biochemical process known as oxidative phosphorylation.

ATP synthase is a multienzyme protein complex that looks like a cylindrical-shaped motor on a stick. The stick is embedded in the inner mitochondrial membrane while the cylinder extends well into the matrix. When the concentration of protons in the intermembrane space reaches a certain level, and ADP and P_i are both bound to *ATP synthase*, an ion channel within the stick opens that allows protons from the intermembrane space to flow freely back into the matrix. Since the proton flow is from a region of higher proton concentration to a region of lower concentration (in the direction of the proton gradient), energy is released that drives the rotation of the motor. Further rotation brings the reactants ADP and P_i together, so a phosphorylation reaction can occur. Then a second turn of the motor releases the product, ATP, from the enzyme.

In a Nutshell: ATP Production from the Complete Oxidation of Glucose

We can calculate the total amount of ATP produced when one molecule of glucose is completely oxidized through the catabolic processes of glycolysis, aerobic oxidation of pyruvate, the citric acid cycle, and oxidative phosphorylation. In glycolysis, 2 ATP and 2 NADH are produced. In pyruvate oxidation, 2 NADH and 2 CO_2 are produced. In the citric acid cycle, 6 NADH, 2 $FADH_2$, 2 ATP, and 4 CO_2 are produced. The net ATP energy produced from one molecule of glucose is 2 ATP (glycolysis) + 2 ATP (citric acid cycle) + 28

ATP ((10 NADH × 2.5) + (2 $FADH_2$ × 1.5) oxidative phosphorylation) = 32 ATP per molecule of glucose.

The ATP produced both directly and during oxidative phosphorylation is available for anabolic pathways, which allow mechanical work, such as muscle contraction, active transport of molecules through membranes to occur.

Worked Example #8

How many ATP are produced from 10 NADH during oxidative phosphorylation?

Solution

There are 2.5 ATP per NADH. Therefore, 10 NADH × 2.5 ATP/NADH = 25 ATP.

Try It Yourself #8

How many ATP are produced from 2 $FADH_2$ during oxidative phosphorylation?

Solution:

There are _______ ATP per $FADH_2$.

Calculation for total number of ATP from $FADH_2$:

Practice Problems for the Electron Transport Chain and Oxidative Phosphorylation

1. Which of the following has the lowest electron affinity? Which of the following has the greatest electron affinity?
   a. O_2
   b. Complex 1
   c. Complex II
   d. Complex III

2. When protons move from the matrix across the inner mitochondrial membrane into the intermembrane space, are they moving with or against the proton gradient?

3. When *ATP synthase* allows protons to move from the intermembrane space back into the matrix, are they moving with or against the proton gradient?

15.5 Fatty Acid Catabolism

In stage 2 of triglyceride catabolism, fatty acids are converted into acetyl CoA in a biochemical process known as β-oxidation. The third stage of catabolism for triglycerides is the same as for glucose: the citric acid cycle and oxidative phosphorylation.

In a Nutshell: Stage 2: β-Oxidation of Fatty Acids

The primary function of fatty acids is to provide energy for cells over the long term. β-oxidation takes place in the mitochondria. Before β-oxidation can take place, a fatty acid molecule must be activated to fatty acyl CoA, a reaction that takes place in the cytosol of muscle and liver cells and requires ATP.

In the cytosol, a fatty acid reacts with a molecule of coenzyme A to form the thioester of the fatty acid, known as fatty acyl CoA. Fatty acyl CoA is similar to acetyl CoA; however, acetyl CoA has a two-carbon acetyl group, while fatty acyl CoA has more carbons in the acyl group. The reaction between coenzyme A and a fatty acid to from fatty acyl CoA is coupled to ATP hydrolysis to produce AMP and two P_i. The activated fatty acid, fatty acyl CoA, is then transported to the mitochondrion where β-oxidation takes place.

β-oxidation is a four-step biochemical pathway that converts fatty acyl CoA into acetyl CoA and a new fatty acyl CoA molecule with two fewer carbons. In this biochemical pathway, one FAD is reduced to $FADH_2$ and one NAD^+ is reduced to NADH.

Two of the four steps of β-oxidation are oxidation-reduction reactions that involve oxidation of a β-carbon, which is where the biochemical pathway gets its name.

In step 1, oxidation occurs and $FADH_2$ is produced. In this step, the carbon-carbon single bond between the α- and the β-carbon atoms of fatty acyl CoA is oxidized into a carbon-carbon double bond. FAD is reduced to $FADH_2$. Step 2 is a hydration reaction. A water molecule in the form of an OH group and an H atom replace the double bond in what is known as a hydration reaction introducing a hydroxyl group at the β-position and a hydrogen atom at the α-position. In step 3, oxidation occurs and NADH is produced. The hydroxyl group introduced in step 2 is oxidized to a ketone and NAD^+ is reduced to NADH. In step 4, a carbon-carbon bond is broken. The single bond between the α-carbon and β-carbon breaks, producing acetyl CoA and a new fatty acyl CoA molecule, which is two carbon atoms shorter than the fatty acyl CoA in step 1. Coenzyme A is required, which forms a thioester with the new fatty acid.

The new fatty acyl CoA molecule then enters another round of β-oxidation, producing another acetyl CoA molecule and a new fatty acyl CoA molecule, two carbon atoms shorter. The cycle then repeats itself until the final step, in which a four-carbon fatty acyl CoA molecule is converted into two acetyl CoA molecules. Thus, a fatty acid containing x carbons is converted into $x/2$ acetyl CoA molecules. The number of passes through the β-oxidation is one fewer than the total number of acetyl CoA molecules produced: $(x/2) - 1$, because in the last pass two acetyl CoA molecules are formed.

Most of the energy obtained from fatty acid catabolism occurs when the acetyl CoA molecules produced in β-oxidation are oxidized in the citric acid cycle: a total of 10 ATP per acetyl CoA. In addition, each pass through β-oxidation produces one NADH and one $FADH_2$ for a total of $(1 \times 2.5) + (1 \times 1.5) = 4$ ATP. To calculate the total amount of energy produced from the complete catabolism of a fatty acid, in terms of ATP, we can sum these values and remember to subtract the two ATP that were initially required to convert the fatty acid and coenzyme A into fatty acyl CoA.

Worked Example #9

Calculate the amount of ATP produced from complete catabolism of myristoleic acid, a fatty acid that contains 14 carbon atoms.

Begin by determining how many acetyl CoA molecules can be produced from the fatty acid by dividing the total number of carbon atoms in the fatty acid by 2. Since myristoleic acid contains 14 carbon atoms, it can form 14/2 = 7 acetyl CoA molecules. Determine the number of passes through the β-oxidation by subtracting 1 from the number of acetyl CoA molecules formed: 7 – 1 = 6. Determine the amount of ATP produced when these seven acetyl CoA molecules go through the citric acid cycle by multiplying by 10. Next, determine the number of ATP produced in each pass through β-oxidation by multiplying by 4. Add these values and then subtract two for the initial activation step:

Myristoleic acid: C-14

7 acetyl CoA molecules through the citric acid cycle	× *10*	=	*70 ATP*
6 β-oxidationsteps	× *4*	=	*24 ATP*
One time activation of myristoleic acid		=	*−2 ATP*
Total			*92 ATP*

Try It Yourself #9

Calculate the amount of ATP produced from complete catabolism of arachidonic acid, a fatty acid that contains 20 carbon atoms.

Arachidonic acid: C-20

Number of acetyl CoA molecules produced: ________________

Number of β-oxidation cycles: ___________________

_____ *acetyl CoA molecules*	× *10*	=	_____*ATP*
_____ *cycles*	× *4*	=	_____ *ATP*
One time activation of arachidonic acid		=	_____ *ATP*
Total			_____*ATP*

Practice Problems for Fatty Acid Catabolism

1. What types of reactions occur in steps 1 and 3 of β-oxidation? Which carbon atoms are involved in these reactions?

2. Calculate the amount of ATP produced from complete catabolism of linolenic acid, a fatty acid that contains 18 carbon atoms.

3. Which step of β-oxidation produces $FADH_2$? Which step produces NADH?

Chapter 15 Quiz

1. a. Why do all the intermediates in glycolysis contain phosphate groups?

   b. How many ATP molecules and NADH molecules are produced in glycolysis?

2. a. Does the reaction that forms lactic acid from pyruvate use or produce NAD^+?

   b. Does the reaction that forms acetyl CoA from pyruvate use or produce NAD^+?

3. What molecules produced in the citric acid cycle provide the energy for the phosphorylation of ADP into ATP?

4. Do the carbon atoms in the carbon dioxide molecules produced in the citric acid cycle come from acetyl CoA or oxaloacetate?

5. In step 5 of the citric acid cycle, succinyl-CoA is converted into succinate. This reaction is coupled to the formation of GTP as shown below.

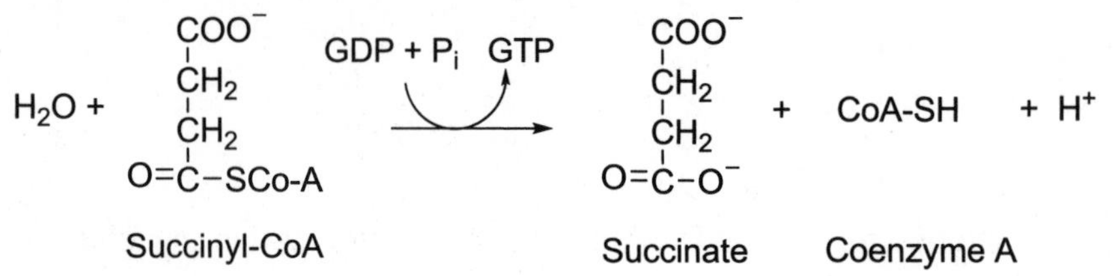

$$H_2O + \underset{\text{Succinyl-CoA}}{\begin{matrix} COO^- \\ | \\ CH_2 \\ | \\ CH_2 \\ | \\ O{=}C{-}SCo\text{-}A \end{matrix}} \xrightarrow{GDP + P_i \;\; GTP} \underset{\text{Succinate}}{\begin{matrix} COO^- \\ | \\ CH_2 \\ | \\ CH_2 \\ | \\ O{=}C{-}O^- \end{matrix}} + \underset{\text{Coenzyme A}}{CoA\text{-}SH} + H^+$$

The formation of succinate from succinyl CoA requires 33.5 kJ/mol, and the formation of GTP from GDP releases 36.8 kJ/mol. Does the coupled reaction absorb or release energy? Show your calculation.

6. Which has a higher pH, the intermembrane space or the matrix? Which has a lower concentration of protons?

7. What is the proton-motive force?

8. How does *ATP synthase* generate energy to phosphorylate ADP into ATP?

9. How much ATP is produced from one molecule of glucose that goes through glycolysis, pyruvate oxidation, the citric acid cycle, and oxidative phosphorylation? Show your calculation.

10. Calculate the amount of ATP produced from complete catabolism of docosahexaenoic acid, DHA, a fatty acid that contains 22 carbon atoms.

Chapter 15

Answers to Additional Exercises

37 The metabolism of triglycerides takes place in muscle and liver cells.

39 B vitamins provide the body with a steady supply of the starting materials needed to produce the coenzymes required for the fat burning biochemical pathways.

41 Catabolic processes produce energy overall.

43 Acetyl CoA is the central molecule of metabolism. It is produced and used in many biochemical pathways.

45 The citric acid cycle, the electron transport chain, and oxidative phosphorylation are the biochemical pathways of the third stage of catabolism.

47 Maltose and dextrins are hydrolyzed into glucose in the small intestine. They are hydrolyzed into glucose.

49 Fructose and galactose are hydrolyzed into glucose in the liver.

51 The hydrolysis of triglycerides produces glycerol and fatty acids.

53 Peptide bonds are broken in the hydrolysis of amino acids.

55 The energy released from catabolism is stored in phosphoanhydride bonds of nucleotide triphosphates.

57 The products are ADP and inorganic phosphate. One phosphoanhydride bond is broken, and 30.5 kJ/mol of energy is released.

59 30.5 kJ/mol is required to phosphorylate ADP to ATP.

61 In coupled reactions, one of the reactants in one of the reactions is a product of the other reaction. The energy transferred in each individual reaction is added together.

63 a. The conversion of phosphenolpyruvate to pyruvate provided the energy for the coupled reaction.

b. The curved arrow represents the coupled reaction of the phosphorylation of ADP to ATP. The straight arrow represents the transformation of the substrate (phosphoenolpyruvate) into product (pyruvate).

Phosphoenolpyruvate → pyruvate

ADP → ATP

c. Reaction

Reaction	
phosphoenolpyruvate → pyruvate	−61.9
ADP → ATP	+30.5
phosphoenolpyruvate —(ADP → ATP)→ pyruvate	−31.4

The coupled reaction releases 31.4 kJ/mol of energy.

65 The biochemical pathways that break down food (carbohydrates, fats, and proteins) produce acetyl CoA.

67 Glycolysis converts glucose into pyruvate. It takes place in the second stage of catabolism.

69 No, it does not require oxygen. It is an anaerobic pathway.

71 Glycolysis occurs in the cytosol, the fluid part of the cell outside the nucleus.

73 Every intermediate in glycolysis contains a phosphate group. The charge on the phosphate groups prevents the intermediates of glycolysis from diffusing out of the cell since polyatomic ions cannot diffuse through the cell membrane.

75 The formation of pyruvate, ATP, and NADH in the second half of glycolysis more than compensate for the energy used in the first half.

77 Two molecules of NADH are produced from one molecule of glucose during glycolysis.

79

$$2\ H{-}\underset{H}{\overset{H}{C}}{-}\overset{O}{\overset{\|}{C}}{-}COO^- \xrightarrow[2\ NAD^+ \rightarrow 2\ NADH + 2\ H^+]{2\ CoA\text{-}SH} 2\ H{-}\underset{H}{\overset{H}{C}}{-}\overset{O}{\overset{\|}{C}}{-}SCoA + 2\ CO_2$$

Pyruvate → Acetyl CoA (carbons from pyruvate: the boxed C–C of acetyl CoA)

81 Carbon dioxide is expelled in aerobic conversion of pyruvate to acetyl CoA. Coenzyme A is the carrier of the two-carbon acetyl group.

83 Anaerobic conditions exist in a marathoner's muscle cells at the end of a marathon.

85 Lactic acid builds up in muscles through lactic acid fermentation. Lactic acid fermentation occurs when oxygen levels are low, as they are during strenuous exercise. Lactate is formed when pyruvate is reduced and NADH is oxidized to NAD^+. NAD^+ is used in glycolysis to produce ATP. Once the energy-demanding

activity is over, lactate is transported to the liver, where it is converted back into pyruvate.

87 The citric acid cycle occurs during the third and last stage of catabolism.

89 The citric acid cycle is an aerobic pathway.

91 For every molecule of acetyl CoA that enters the citric acid cycle, one GTP, three NADH, one $FADH_2$, and two molecules of CO_2 are produced.

93 About 1.5 ADP molecules can be phosphorylated to ATP from the electrons released by one $FADH_2$ molecule during oxidative phosphorylation.

95 a. At physiological pH, the structure of citric acid is:

```
      COO⁻
      |
      CH2
      |
  HO–C–COO⁻
      |
      CH2
      |
      COO⁻
```

b. There are six carbon atoms in citric acid. Two of the carbon atoms come from acetyl CoA, and four of the carbon atoms come from oxaloacetate.

c. It is produced in the first step.

97 Halfway through the citric acid cycle, two carbon atoms have been lost from citrate. Two molecules of CO_2 are expelled to remove these two carbon atoms from citrate. Two molecules of NADH are produced in the first half of the citric acid cycle.

99 The electrons from the NADH and $FADH_2$ produced in the citric acid cycle are used in oxidative phosphorylation to produce ATP.

101 a. Oxaloacetate has a ketone in place of the alcohol in malate. b. Oxaloacetate is one of the reactants in the first step of the citric acid cycle. c. Malate is oxidized. d. NAD^+ is reduced to NADH.

103 The electrons in the electron transport chain come from the NADH and $FADH_2$ produced in the citric acid cycle.

105 The energy-producing pathways of catabolism, except glycolysis, occur in the mitochondria.

107 No, ions cannot pass through the inner mitochondrial membrane.

109 The matrix has a higher pH.

111 The protons are moving against the proton gradient. The energy for this process comes from NADH.

113 Yes, there is a difference in charge between the matrix and the intermembrane space.

115 In the electron transport chain, the affinity for electrons increases as the electrons are transferred though the electron transport chain.

117 The energy needed to pump protons against the proton gradient comes from the energy released in the oxidation-reduction steps within the protein complexes of the electron transport chain.

119 NADH supplies more energy (as measured in the number of ADP that are phosphorylated to ATP). NADH provides energy to phosphorylate 2.5 ADP, while $FADH_2$ provides energy to phosphorylate 1.5 ADP. NADH transfers its electrons earlier in the electron transport chain.

121 No, electrons are not passed between Complexes I and II.

123 $O_2 + 4\ H^+ + 4\ e^- \rightarrow 2\ H_2O$. Oxygen is reduced to water in Complex IV.

125 *ATP synthase* uses the potential energy of the proton motive force to phosphorylate ADP to ATP.

127 Glycolysis: 2 ATP, Citric acid cycle: 2 ATP, Oxidative phosphorylation: 10 NADH (2.5) + 2 $FADH_2$ (1.5) = 28 ATP

2 + 2 + 28 = 32 ATP produced from 1 molecule of glucose

129 The formation of ATP produced directly and during oxidative phosphorylation supply the energy for anabolic processes.

131 Heart and liver cells use the energy provided by fatty acid catabolism.

133 A fatty acid is activated to fatty acyl CoA in the cytosol of muscle and liver cells.

135 Two carbon atoms are removed in the four-step β-oxidation process.

137 Step 4 breaks the carbon-carbon bond. The single bond between the α-carbon and the β-carbon breaks. This reaction produces acetyl CoA and a new fatty acyl CoA molecule that is two carbon atoms shorter than the fatty acyl CoA at the start of β-oxidation.

139 Begin by determining how many acetyl CoA molecules can be produced from the fatty acid by dividing the total number of carbon atoms in the fatty acid by 2. Since linolenic acid contains 18 carbon atoms, it can form 18/2 = 9 acetyl CoA molecules. Determine the number of passes through the β-oxidation by subtracting 1 from the number of acetyl CoA molecules formed: 9 – 1 = 8. Determine the amount of ATP produced when these nine acetyl CoA molecules go through the citric acid cycle by multiplying by 10. Next, determine the number of ATP produced in each pass through β-oxidation by multiplying by 4. Add these values and then subtract two for the initial activation step:

Linolenic acid: C-18			
9 acetyl CoA molecules through the citric acid cycle	× 10	=	90 ATP
8 β-oxidation steps	× 4	=	32 ATP
Activation of 1 linolenic acid molecule		=	−2 ATP
Total			120 ATP

141 Rotenone interferes with the transport of electrons from NADH because rotenone is a competitive inhibitor of Complex I, where the electrons from NADH are delivered.

143 Oligomycin interferes with the proton channel in *ATP synthase* during phosphorylation of ADP to ATP.

145 Cyanide poisoning is lethal because cyanide is an irreversible inhibitor of *cytochrome* a_3. The victim cannot utilize oxygen because electrons are never transferred from *cytochrome* a_3 to oxygen. Consequently, protons are no longer pumped into the intermembrane space, so *ATP synthase* is no longer supplied with the proton motive force that drives the phosphorylation of ADP to ATP. Without ATP, cells lack the energy to drive the processes required for living.

147 Symptoms of cyanide poisoning include headache, lethargy, confusion, nausea, vomiting, and convulsions.